BIOTECHNOLOGIES AND INTERNATIONAL HUMAN RIGHTS

This book follows and complements the previous volume Biotechnology and International Law (Hart 2006) bringing a specific focus on human rights. It is the result of a collaborative effort which brings together the contributions of a select group of experts from academia and from international organisations with the purpose of discussing the extent to which current activities in the field of biotechnology can be regulated by existing human rights principles and standards, and what gaps, if any, need to be identified and filled with new legislative initiatives. Instruments such as the UNESCO Declaration on the Human Genome (1997) and on Bioethics and Human Rights (2005) are having an impact on customary international law. But what is the relevance of these instruments with respect to traditional concepts of state responsibility and the functioning of domestic remedies against misuse of biotechnologies? Are new legislative initiatives needed, and what are the pros and cons of a race toward the adoption of new ad hoc instruments in an area of such rapid technological development? Are there risks of normative and institutional fragmentation as a consequence of the proliferation of different regulatory regimes? Can we identify a core of human rights principles that define the boundaries of legitimate uses of biotechnology, the legal status of human genetic material, as well as the implications of the definition of the human genome as 'common heritage of humanity' for the purpose of patenting of genetic inventions? These and other questions are the focus of a fascinating collection of essays which, together, help to map this emerging field of inquiry.

Studies in International Law: Volume 13

Studies in International Law

Biotechnologies and International Human Rights

Edited by

Francesco Francioni

·HART·
PUBLISHING

OXFORD AND PORTLAND, OREGON
2007

Published in North America (US and Canada) by
Hart Publishing
c/o International Specialized Book Services
920 NE 58th Avenue, Suite 300
Portland, OR 97213-3786
USA
Tel: +1-503-287-3093 or toll-free: (1)-800-944-6190
Fax: +1 503 280 8832
E-mail: orders@isbs.com
Website: www.isbs.com

Hart Publishing, 16C Worcester Place, Oxford, OX1 2JW
Telephone: +44 (0)1865 517530 Fax: +44(0)1865 510710
E-mail: mail@hartpub.co.uk
Website: http://www.hartpub.co.uk

British Library Cataloguing in Publication Data
Data Available

ISBN-10: 1-84113- 703-0 (hardback)
ISBN-13: 978-1-84113- 703-2 (hardback)

Typeset by Compuscript Ltd, Shannon
Printed and bound in Great Britain by
TJ International Ltd, Padstow, Cornwall

Preface

This book follows and complements the previous volume *Biotechnology and International Law*, Francioni and Scovazzi (eds), (Hart, Oxford, 2006) with a specific focus on human rights. It is the result of a collaborative effort undertaken at the European University Institute (EUI) within the framework of the research project 'The Impact of Biotechnology on Human Rights'. This project brought together a selected group of experts from academia and from international organisations with the purpose of discussing the extent to which current activities in the field of biotechnology can be regulated by existing international human rights principles and standards, and what gaps, if any, need to be identified and filled with new legislative initiatives. These problems were first discussed at an EUI workshop in Florence on 25–26 October 2004 on the basis of a questionnaire which outlined two general sets of formal and substantive issues. At the formal level, the questions raised concerned the extent to which current instruments dealing with biotechnology and human rights—such as the UNESCO declarations on the human genome (1997) and on bioethics and human rights (2005)—have become part of customary international law; what is the relevance of these instruments with respect to traditional concepts of state responsibility and the functioning of domestic remedies against misuse of biotechnologies; the extent to which new legislative initiatives are needed; what are the advantages and costs of a race toward the adoption of new *ad hoc* legislation in an area of such rapid technological development, as well as the risks of normative and institutional fragmentation involved in the proliferation of different regulatory regimes? At the substantive level, the workshop addressed the following general issues: what are the core human rights principles that define the boundaries of legitimate use of biotechnology?; what is the legal status of human genetic material and what are the implications of the definition of the human genome as 'common heritage of humanity' for the purpose of patenting of genetic inventions?; what is the meaning of, and how can we implement the emerging right to an equitable sharing of the benefits arising from the commercial use of biogenetic resources?; what is the role of human rights, and in particular of the principle of non-discrimination, in preventing a new 'genetic divide' that would increase the already striking disparities between the industrialised and less developed countries?

Having reflected on these issues, participants in the project were invited to present their preliminary papers at an EUI colloquium held in

Florence in June 2005. This book consists of the revised and edited papers that issued from that colloquium.

As always in the case of a collaborative enterprise, the merit for its completion goes to many people and institutions. I wish to mention here Riccardo Pavoni, from the University of Siena, for his valuable assistance in preparing the background materials and the questionnaire for the workshop; Mario Mendez, PhD candidate at the EUI, for his editorial assistance and linguistic revision; the EUI for providing financial support to this special project; and, above all, the contributors for having accepted my invitation to participate in this challenging project and for their timely response to the many queries during the drafting and editorial process.

Francesco Francioni
European University Institute
Florence

Contents

Table of Cases

European Court of Justice and Court of First Instance

Table of European Legislation

Treaties and Charters

Regulations

Directives

Decisions

Table of National Legislation

NEW ZEALAND

PORTUGAL

SOUTH AFRICA

SWEDEN

THAILAND

UNITED KINGDOM

UNITED STATES

Table of International Instruments

Part I

Overview and Cross-cutting Issues

1

Genetic Resources, Biotechnology and Human Rights: the International Legal Framework

FRANCESCO FRANCIONI*

1. INTRODUCTION

Is it useful to study the interaction between developments in biotechnology[1] and international human rights? After all, isn't science[2] constantly expanding the limits of human freedom and thus compelling us to re-define the substance and scope of such rights? How relevant can it be to look at the present challenges and dilemmas posed by relentless advances in biotech science through the lens of a fixed catalogue of human rights? Is a human rights approach, and indeed a law-based approach, capable of bridging the gap between fundamentally divergent ethical views in this area?[3]

These are not easy questions, and the reflections we have developed in these collected essays do not aim at providing a uniform and definitive

* Professor of International Law and Human Rights at the European University Institute and the Faculty of Law at the University of Siena. Email: francesco.francioni@iue.it.

[1] For the sake of convenience the term 'biotechnology' is used in this chapter in accordance with the definition provided in the Convention on Biological Diversity (see www.biodiv.org/doc/legal/cbd-en.pdf): as any technological application that uses biological systems, living organisms, or derivates from them, in order to make or modify products or processes for specific use.

[2] The term 'science', as used in this chapter, is inclusive of hard science, soft science, technology, engineering and medicine, taking into account the definition provided by UNESCO in 1974 as an enterprise wherein humankind 'acting individually or in small or large groups, makes an organized attempt, by means of the objective study of observed phenomena, to discover and master the chain of causalities; brings together in a co-ordinated form the resultant sub-systems of knowledge by means of systematic reflection and conceptualisation ... and thereby furnishes itself with the opportunity of using, to its own advantage, understanding of the processes and phenomena occurring in nature and in society'. See UNESCO Recommendation on the Status of Scientific Researchers, 18 C/ Res. 40, Nov 1974.

[3] On the limits of law as a regulatory modality in relation to new technologies, see the fundamental contribution of Stanford professor L Lessig, *Code and Other Laws of Cyberspace* (New York, Basic Books, 1999).

set of answers. On the contrary, the approach taken through the lens of human rights is pluralistic and aimed at identifying a broad range of perspectives in which biotechnology regulation can be placed.

But even from the viewpoint of human rights, an evaluation of the impact of biotechnology on international law requires a plurality of epistemological approaches and different levels of inquiry. At a first level, one needs to start with the acknowledgment of the widespread perception that the new genetic science is placing peoples in the difficult position of facing 'something unknown', of not fully grasping the risks and social implications involved in the contemporary process of developing new biotech products and services. In this context, a human rights approach based on transparency, information and participatory rights can contribute to people's empowerment and to raising the awareness of their individual and collective entitlements *vis-à-vis* the blind power of science and industry.

At another level of analysis, looking at biotechnology through the lens of human rights will immediately entail the acknowledgment of the basic freedom of scientific research and the right 'to enjoy the benefits of scientific progress and its applications', to use the words of Article 15 of the International Covenant on Economic Social and Cultural Rights (ICESCR). But at the same time this freedom and this right are not absolute. They must be balanced against certain standards of bioethics respect for which is a condition of the legitimacy of the claim to freedom of scientific research.[4] On this point, one may argue that ethical standards are always responsive to religious and cultural specificity. That is true. But precisely because of that resort to international human rights norms is capable of providing a set of common, objectively defined values, inasmuch as they reflect the universally shared values of respect for life, liberty, human dignity and non-discrimination and, possibly, more specific societal values, such as the right to information and of participation in policy decisions, and the right to share in scientific advancement and its benefits.[5] At the same time, internationally recognised human rights represent the benchmark against which public authorities and international institutions can measure the legitimacy of policy choices or of specific decisions relating to the application of modern science. This aspect is especially relevant in the field of biotechnology. Contrary to the old-fashioned view that human rights depend on states doing nothing, ie, non-interference with individual autonomy, in this field governments have positive obligations to intervene in the sphere of scientific, technological and economic activities

---

[4] On the question of how international human rights may support ethics in scientific research, see F Francioni, 'Valori etici e diritto internazionale' [2004] *Rivista di Studi Politici Internazionali* 567.

[5] See Art 27(1) of the Universal Declaration of Human Rights, n 6 below.

in order to ensure that freedom of research and market freedoms are not abused or distorted in such a way as to cause adverse effects on human rights. This function is consistent with the general provision of Article 28 of the Universal Declaration of Human Rights[6] which, by setting out the right to a 'social and international order where the rights and freedoms set forth in this Declaration can be fully realized', calls upon governments to take positive steps toward the development of a social structure in which human rights can take root and be safeguarded.[7]

At a technical-legal level, a reason for studying the interplay between human rights and biotechnology is that scientific and technological advances have always had the effect of stimulating the development of new law, both in domestic societies and in international law. Thus, it is important to understand what role human rights have in the dynamic evolution of the law. The development of modern biotechnologies has spurred on the elaboration of a considerable number of treaties and soft-law instruments designed to establish standards and oversight procedures in relation to biotechnology related risks. At the global level, the Cartagena Protocol on Biosafety was adopted in response to concerns that modern biotechnology may have adverse impacts on biodiversity.[8] To this end, it provides for stringent risk assessment of 'living modified organisms' and for advance consent by the importing state pursuant to a broad interpretation of the precautionary approach. In the field of agriculture, the 2001 Food and Agriculture Organisation (FAO) Treaty on Plant Genetic Resources has established a framework of international co-operation for the conservation and sustainable use of plant genetic resources for food and agriculture,[9] based on the recognition of the sovereign rights of states over their phyto-genetic resources and on the principle of 'facilitated access' and 'sharing of benefits' arising from the scientific and commercial use of such resources.[10] Concern with biodiversity conservation and with the risk posed to the environment by the deliberate release of genetically modified organisms has had repercussions also on regional international law. Of special relevance in this respect is the EU Directive of 12 March 2001,[11] establishing a common system of authorisation and oversight of the deliberate release of genetically modified organisms into the environment,

---

[6] See UN AG Res. 217A (III), 1948; UN Doc A/810 (1948).

[7] See Art 28.

[8] See 2000 Cartagena Protocol on Biosafety, available at www.biodiv.org//doc/legal/cartagena-protocol-en.pdf.

[9] See 2001 International Treaty on Plant Genetic Resources for Food and Agriculture, available at ftp://ext-ftp.fao.org/ag/cgrfa/it/ITPGRe.pdf.

[10] See Art 10–13.

[11] See Dir 2001/18/EC of the European Parliament and of the Council of 12 Mar 2001 on the deliberate release into the environment of genetically modified organisms and repealing Council directive 90/220/EEC [2001] OJ L/106/1.

as well as of the placement of such organisms or their products on the market. In Latin America, the increasing practice of bio-prospecting has spurred on legislation and administrative practices aimed at regulating access to local biological material and at ensuring a fair distribution of benefits derived from their use and commercialisation. Notable in this regard are Decisions 391 of 1996 and 523 of 2002 adopted by the Andean Community Commission with the objective of laying down conditions for access to the rich repository of genetic resources of the region.[12] These Decisions are informed by the principle of benefit-sharing and the objective of capacity-building in the interest of the Andean countries.[13]

All these instruments are motivated by two distinct but interrelated sets of concerns: protection of the environment, taking into account insufficient knowledge of the long-term effects of genetically modified organisms on natural ecosystems; and the creation of a system of just distribution of benefits arising from the use and commercialisation of genetically engineered material and its products. This second concern is of particular relevance for developing countries: first, because they are the most important repository of biological diversity and, consequently, of potentially useful genetic material; secondly, because they are resisting the development of an international legal regime based on the principle of freedom of access or of common heritage of genetic resources, which they fear—not without good reason—would leave them at the margins of the biotechnology revolution. These concerns are part of the complex dialectic between industrial countries and less-developed ones. However, the problems arising from this complex relationship have not yet been articulated in the language of human rights, but rather in the more elusive language of 'sustainable development' and 'equitable sharing of benefits'. As we shall see in the course of this chapter, collective rights, such as self-determination and peoples' sovereign right over natural resources, as well as individual and community rights may provide a more precise and sound basis for the development of international law in the area of biotechnology regulation.

A closer relationship between the development of biotechnology and human rights can be found in a number of international instruments adopted in the last 15 years in the field of biotechnology applied to human genetic resources. At the global level UNESCO has been at the forefront of an ambitious programme aimed at setting legal and ethical standards applicable to the human genome. The results of this programme are, for the time being, four important soft law instruments: the 1997 Universal

---

[12] The two decisions are available, respectively, at http://216.15.202.3/docs/andeancommunity-decision391-1996-en.pdf and at www.comunidadandina.org/ingles/treaties/dec/D523e.htm.

[13] See F Novak, 'Biotechnology and Regional Integration Systems: Legislation and Practices in the Andean Community Countries' in F Francioni and T Scovazzi (eds), *Biotechnology and International Law* (Oxford, Hart Publishing, 2006), 403 ff.

Declaration on the Human Genome and Human Rights (UDHG),[14] the 1999 Guidelines for the Implementation of such Declaration,[15] and the 2003 International Declaration on Genetic Data and the 2005 Universal Declaration on Bioethics and Human Rights (UDBHR).[16] The UN General Assembly endorsed the UDHG in 1998,[17] and in following years was engaged in the negotiation of a new convention designed to restrict human cloning.[18] At the regional level, the Council of Europe has, since 1997, adopted a variety of legal instruments setting ethical standards in the field of biomedicine and biomedical research, including the Oviedo Convention on Human Rights and Biomedicine,[19] the additional protocol on the prohibition of human cloning,[20] the 2002 Additional Protocol on transplantation of organs and tissues of human origin[21] and the 2005 Additional Protocol on biomedical research.[22]

Against the background of this international legislation, and building on the plurality of legal perspectives outlined above, this chapter will follow a three-step analysis. First, it will try to identify the competing entitlements—property rights, sovereignty, common heritage—that present international law recognises over genetic resources and their use, including their exploitation through biotechnology applications Secondly, it will focus on the general interest that humanity as a whole has in the conservation and management of genetic resources and in the regulation of related biotechnology. Thirdly, it will try to outline a core of international human rights respect for which should be considered a condition *sine qua non* for the legitimate exercise of the freedom of science and business in the development and application of modern biotechnology.[23]

---

[14] See Universal Declaration on the Human Genome and Human Rights, 11 Nov 1997, available at the UNESCO website, at www.unesco.org.

[15] See C/Resolution 23 of 16 Nov 1999, endorsing the guidelines for the implementation of the Universal Declaration on the Human Genome and Human Rights.

[16] The International Declaration on Human Genetic Data was adopted on 16 Oct 2003, and the Universal Declaration on Bioethics and Human Rights on 19 Oct 2005. Both are available in the UNESCO website, at www.unesco.org.

[17] See A/RES/53/152 of 9 Dec 1998.

[18] See the draft text addressed to the UN Secretary General by the Government of Costa Rica on 2 Apr 2003, UN doc. A/58/73 of 17 Apr 2003.

[19] See Convention for the Protection of Human Rights and Dignity of the Human Being with regard to the Application of Biology and Medicine: Convention on Human Rights and Biomedicine, Oviedo, 4 Apr 1997, CETS n 164.

[20] See Additional Protocol to the Convention for the Protection of Human Rights and Dignity of the Human Being with regard to the Application of Biology and Medicine, on the Prohibition of Cloning Human Beings, Paris, 12 Jan 1998, CETS n 168.

[21] See Additional Protocol to the Convention on Human Rights and Biomedicine concerning Transplantation of Organs and Tissues of Human Origin, Strasbourg, 24 Jan 2002, CETS n 186.

[22] See Additional Protocol to the Convention on Human Rights and Biomedicine concerning Biomedical Research, Strasbourg, 25 Jan 2005, CETS n 195.

[23] Of course, in this analysis one cannot ignore the fact that modern genetic science produces a new form of technological power very different from that exercised by the State and to which traditionally human rights abuses are related. On this aspect see CG Weeramantry

2. SOVEREIGN RIGHTS AND BIO-GENETIC RESOURCES

Central to the discussion on how modern biotechnology affects international human rights is the identification of who has rights over the genetic resources that form the raw material from which biotechnology develops new products and new processes. A fundamental distinction in this respect is necessary between plant and animal genetic resources on the one hand and the human genome on the other. While the latter is covered by conventional human rights law and by specific soft law instruments—to be examined later—the former, insofar as they belong to the natural environment, may be brought under the general rule of international law according to which the physical space of the world is allocated to national spheres of jurisdiction coinciding with the territory of a given state. Counterparts of this rule are the regime of the high seas,[24] where no sovereignty is recognised and freedom of access is guaranteed to all states, and the special regime of the common heritage of humankind that has emerged with regard to the international seabed area.[25] If we leave these exceptions aside for the time being (they will be dealt with in section 3), the question we must address is the following: is the principle of sovereignty, and in particular the post-colonial principle of 'permanent sovereignty' over natural resources, applicable to plant and animal genetic material that constitutes the object of biotechnology investigation and commercial application? This question is preliminary to any further discussion of the right of access to genetic material because there is a fundamental distinction between natural resources understood as minerals or as biological resources the utilisation of which entails depletion and consumption in the course of economic activities, and bio-genetic resources whose *genotype*, rather than *phenotype*, is targeted for sampling and biotechnological application with negligible impact on the environment. This distinction, although well-founded in science, has not fitted comfortably into existing categories of international law. At the beginning, in the early 1980s, recognition of the enormous potential of modern biotechnologies for agriculture led the FAO to proclaim that plant genetic resources are an exception to the principle of permanent sovereignty, insofar as they constitute, by their very nature, part of the common heritage of humankind. The International Undertaking on Plant Genetic Resources adopted in 1983[26] recognised that plant germoplasm is a public good of economic and social value to be 'explored, preserved, evaluated and made available for plant breeding and scientific purposes',[27]

(ed), *Human Rights and Scientific and Technological Development* (Tokyo, United Nations University Press, 1990).

[24] See 1982 United Nations Convention on the Law of the Sea (UNCLS), 1833 UNTS 397, part VII.
[25] Ibid, part XI.
[26] See FAO Res. 8/83, Rome, 1983, available at ftp://ext-ftp.fao.org/cgrfa/Res/C8-83E.pdf.
[27] See Art 1.

consistently with *'the universally accepted principle that plant genetic resources are a heritage of mankind and consequently should be available without restriction'*.[28] In spite of this unambiguous recognition of plant genetic resources as part of humanity's collective genetic estate, subsequent developments in international law have fallen short of implementing the principle of the common heritage of humankind with respect to this type of resource. In sharp contrast to developments in the law of the sea—which led to the implementation of the principle of the common heritage of humankind with respect to the mineral resources of the deep seabed—the FAO gradually departed from its initial position and progressively turned toward a cautious recognition of 'sovereign rights' as a legal model to regulate the exploration and development of genetic resources.[29] This legal *revirement* was undoubtedly influenced by the objective difficulty of developing, within the structure of the FAO, effective institutional mechanisms capable of managing the principle of the common heritage of humankind;[30] but it was also related to the major change of policy perspective introduced by the negotiation and subsequent adoption of the Convention on Biological Diversity.[31] This convention, while proclaiming in its preamble that biodiversity constitutes 'a common concern of humankind', explicitly recognised in Article 3 that 'States have …the sovereign right to exploit their own resources'. This provision was reinforced by Article 15, which recognises that access to genetic resources is subject to 'the sovereign rights of States over their natural resources' and that 'the authority to determine access to genetic resources rests with the national government and is subject to national legislation'. Since the entry into force of the biodiversity convention in 1993, the paradigm of 'sovereign rights' over biological resources, including genetic resources, has influenced negotiations within the FAO for the adoption of a multilateral framework of facilitated access and benefit sharing as regards genetic resources important for agriculture. The so-called Seed Treaty adopted by the FAO Conference in 2001[32] has clearly followed a sovereignty-based approach towards access to, and exchange and exploitation of, genetic material. Thus, it has departed from the initial common heritage approach embraced in the 1980s.

Read against the background of this evolving practice, the question we raised at the outset—ie, whose rights are involved in the governance of

---

[28] Ibid (emphasis added).

[29] See the amendments of the Undertaking by subsequent 'agreed interpretations' in 1989 (Res. 4/89, available at ftp://ext-ftp.fao.org/cgrfa/Res/C4-89E.pdf; Res. 5/89, available at ftp://ext-ftp.fao.org/cgrfa/Res/C5-89E.pdf) and 1991 (Res. 3/91, available at ftp://ext-ftp.fao.org/cgrfa/Res/C3-91E.pdf). See M Footer, 'Agricultural Biotechnology, Food Security and Human Rights' in Francioni and Scovazzi (eds), 255 ff n 13 above, and F Francioni, 'International Law for Biotechnology', 3ff in ibid.

[30] See Footer, n 29 above.

[31] See n 1 above.

[32] See Resolution 3/01 of the FAO Conference, available at http://pgrc3.agr.gc.ca/itgrfa/conference_e.html; see n 9 above.

biotechnologies—prompts a preliminary answer: at least with regard to plant genetic resources and, by analogy, animal genetic resources found within state territory, national governments and non-state actors involved in the development of relevant international law have not accepted the application of the principle of the common heritage of humanity. Instead, they have preferred to follow the established sovereign rights approach, which guarantees their role as gate keepers in this new possible field of economic development. This practice must be taken into account in assessing the role that human rights play in the regulation of genetic resources and biotechnology. In the field of biogenetic resources for agriculture, international law still recognises the central role of the state as source of authority and of regulation of access and economic utilisation of resource-related activities. Naturally, states are free to transfer their authority, or if we prefer their 'sovereign rights', to international organisations, as in the case of the EU. But this means only that the identification of human rights involved in biotechnology governance will need to take place in the context of powers (and regulatory competence) transferred to an international or supra-national organisation. In the case of the EU this task is facilitated by the existence of a specific 'charter of rights', now incorporated into the Constitutional Treaty.[33]

3. COMMUNITY INTERESTS AND RIGHTS

3.1 The Human Genome

In sharp contrast to the re-assertion of sovereign rights over bio-genetic resources relevant to food and agriculture, human genetic resources—the subject of investigation and application in medicine and pharmacology—have increasingly been perceived as part of the common heritage of humanity. As such, they are not deemed to fit the category of 'natural resources', so as to fall within the 'sovereign rights' of the territorial state. As is known, developments in this field are due mostly to the ground-breaking research done in the last decade to complete the so-called mapping of the human genome. The results achieved thus far open possibilities of application of gene technology to the life sciences, with the promise of improving the health, longevity and welfare of many human beings. At the same time, the prospect of biotechnological applications to human genetic material has raised justifiable fears that human beings may be reduced to 'means' as a function of biological experimentation and possibly of commercial utilisation of the knowledge derived from the former.[34]

---

[33] See further E Righini in this volume.
[34] See F Lenzerini, 'Biotechnology, Human Dignity and the Human Genome' in Francioni and Scovazzi (eds), n 13 above, and H Boussard in this volume.

Against this problematic background, international practice has, in less than 10 years, evolved toward the robust affirmation of human rights standards that rest on the extension of the principle of the common heritage of humankind from the domain of resources to the new concept of the human genome. Thanks to the vigorous effort of UNESCO, whose mandate in the field of science and culture is linked to the guarantee of 'the democratic principle of the dignity, equality and mutual respect of men',[35] the UDHG was adopted in 1997.[36] Article 1 of the Declaration states: [t]he human genome underlies the fundamental unity of all members of the human family, as well as the recognition of their inherent dignity and diversity. In a symbolic sense, it is the heritage of humanity'. The use of the qualifying phrase 'in a symbolic sense' has been understood as weakening the legal strength of this Article.[37] However, a more convincing explanation is that the adjective 'symbolic' is rather intended to stress that the human genome is not to be treated in a patrimonial sense, like the mineral resources of the sea bed, and that it is not subject to forms of individual or collective appropriation.[38] Its value for humanity is thus not so much in its potential to yield economic benefits, as is the case for the tangible natural resources to which the same concept had previously been applied,[39] but rather in its reflexive capacity to establish an ethical obligation, owed to humanity as a whole, to preserve and safeguard the continuity of the human species when faced with the unfathomable applications of biotechnologies to human genetic engineering. This interpretation is buttressed by the general context of the Declaration, which conclusively confirms an intention to proclaim the human genome as the common heritage of humanity. The Preamble to the Declaration rejects any manipulation of the human genome for social and political purposes in a manner that is incompatible with the inherent human dignity of all 'members of the human family'.[40] Article 4 provides that the human genome in its natural state shall not give rise to financial gains. This makes the human genome an asset *extra commercium*, not subject to appropriation and patenting in its natural form.

[35] See the UNESCO Constitution, available at http://www.unesco.org, Preamble and Art 1.

[36] See Universal Declaration on the Human Genome and Human Rights, 11 Nov 1997, available in the UNESCO website, at www.unesco.org.

[37] R Pavoni, 'Biodiversity and Biotechnology? Consolidation and Strains in the Emerging International Legal Regimes' in Francioni and Scovazzi (eds.), n 13 above.

[38] See L Sturges, 'Who Should Hold Property Rights to the Human Genome? An Application of the Common Heritage of Humankind' (1997) 13 *American University International Law Review* 219, at 249; Lenzerini, n 34 above.

[39] See Part XI of the 1982 Law of the Sea Convention (UNCLOS), n 25 above; see also the 1979 Agreement Governing the Activities of States on the Moon and Other Celestial Bodies, UN GA Res. 34/68 (1979), available at www.lunarregistry.com/treaties/treaty_1979.shtml.

[40] See Preamble, para 4, which states 'the recognition of the genetic diversity of humanity must not give rise to any interpretation of a social or political nature which could call into question "the inherent dignity and the equal and inalienable rights of all members of the human family", in accordance with the Preamble to the Universal Declaration of Human Rights'.

Article 10 requires scientific genomic research in biology and in medicine to respect human dignity and the fundamental rights of individuals and peoples. Further, the Declaration requires a commitment to international co-operation in the assessment of risks and benefits deriving from genomic research and in the promotion of developing countries' capacity to carry out such research and to benefit from its technological applications.

Obviously, the UDHG is not a binding treaty. Its text can at best be understood to reflect emerging principles of international law which, though expressed in the soft law form of the Declaration, are designed to model the evolution of customary law and eventually to harden into more detailed and exacting standards. In any event, it is difficult to deny that the Declaration has already affected the *opinio iuris* of the international community. Its text emanates from the UNESCO General Conference, a body of universal character, where states can express their opinion and cast their vote. Its adoption by acclamation was preceded by extensive consultations and technical preparatory work, with the participation of civil society and the epistemic community with all its scientific, legal, ethical components. No objections or reservations were put on the record at the time of its adoption. After its adoption, the UN General Assembly endorsed its text by Resolution of 9 December 1998.[41] Further, the Universal Declaration has not remained an isolated act. In 1999 UNESCO adopted a resolution laying down implementing measures designed to facilitate the interpretation and application of the Declaration in domestic law.[42] In October 2003 the General Conference adopted the International Declaration on Human Genetic Data,[43] a document that confirms the status of the human genome as the common heritage of humanity.

All the documents discussed above have received broad support from the international community. In 2005 they provided the necessary background against which UNESCO adopted the UDBHR. Most importantly, they are providing principles and criteria which regional organisations and domestic legal systems are drawing upon in drafting legislation and codes of ethics for the exploration and use of the human genome consistent with its nature as a public good.

3.2 Bio-genetic Resources in Common Spaces

Can the principle of common heritage be applied to genetic resources other than the human genome? Can it provide a normative model, in certain circumstances, for the regulation of plant and animal genetic resources also? These questions arise because, although most genetic resources are located

---

[41] See n 17 above.
[42] See n 15 above.
[43] See n 16 above.

in areas subject to national jurisdiction, biotechnology research and industry are increasingly attracted by the genetic material found in organisms that have developed in spaces beyond national jurisdiction—such as the deep sea and Antarctica.[44] There is no state sovereignty in these areas, or at least no generally recognised sovereignty; therefore, there cannot be any uncontested 'sovereign right' within the meaning of section 2 of this chapter. A lack of such a right does not, however, entail that the applicable regime must necessarily be that of common heritage. An alternative model could be that of freedom, as is applicable to the high seas and comparable spaces beyond national jurisdiction. Two arguments might support the application of the principle of freedom in these areas. The first is the close analogy of genetic prospecting and development with fishing, which is one of the classic freedoms of the high seas.[45] The second argument is that bio-prospecting is a manifestation of scientific research, which is also subject to the regime of freedom under customary international law and under UNCLOS.[46] However, these arguments are not conclusive. In our view, exploration and collection of genetic material in areas beyond national jurisdiction cannot be assimilated to fishing. Fishing consists of harvesting biological resources for human consumption, or agricultural or commercial use, and has no relation to the identification and possible technological development of the intangible genetic patrimony contained in living organisms. Freedom to fish entails freedom for the fisherman to appropriate, process and sell the catch, on the assumption that the resources in question are renewable and that, accordingly, anyone can have access to them as long as the equal freedom of others is respected. In the case of genetic resources it is the genetic information contained in the targeted living organism that is at stake. Access to such information does not necessarily entail automatic appropriation of the knowledge that will form the basis of biotechnological application. On the contrary, such knowledge should be considered part of a global common because of its nature as open knowledge available to everyone, and because of its location in common areas where no one can assert property rights or 'sovereign rights' within the meaning of section 2 of this chapter. By the same token, application by analogy of the regime of freedom governing marine scientific research also does not lead to the conclusion that such freedom entails the right to appropriate the bio-genetic resources of common spaces. On the contrary, rules relating to marine scientific research

---

[44] See T Scovazzi, 'Bioprospecting on the Deep Seabed: A Legal Gap Requiring to Be Filled' in Francioni and Scovazzi (eds), n 13 above, as well as P Vigni, 'Bioprospecting in Antarctica: The Economic Value of a Natural Reserve' in ibid; and AI Guyomard, 'Bioprospecting in Antarctica: A New Challenge for the Antarctic Treaty System' in ibid.

[45] T Scovazzi, *La pesca nell'evoluzione del Diritto del Mare* (Milan, Giuffrè, 1979).

[46] R Pisillo Mazzeschi, 'La ricerca scientifica ed il nuovo diritto internazionale del mare' in T Treves (ed), *La ricerca scientifica nell'evoluzione del diritto del mare* (Milan, Giuffrè, 1978).

are activity-related, in the sense of establishing rights and obligations applicable to the conduct of science operations at sea. But in no way can such rules, customary or contained in Part XIII of UNCLOS, be used to establish ownership or sovereign rights over resources. This is made clear by Article 241 of UNCLOS, which provides that:

> Marine scientific research activities shall not constitute the legal basis for any claim to any part of the marine environment or its resources.

Seen against this background, the issue of which regulatory model should govern access to and exploitation of genetic resources located in common spaces cannot be laid to rest with either the 'sovereign rights' model or the free-for-all regime. The correct solution, therefore, must be found in a public common regime, based on the recognition that genetic material found in such spaces constitutes the common estate of humanity, for the conservation and exploitation of which international mechanisms are needed, ensuring co-operation and institutional oversight. No such specific mechanism exists today. However, if we were to follow a simple criterion of competence *ratione loci*, it would be logical to identify the competent institution for marine genetic resources as the International Sea Bed Authority. The mandate of this body is, it must be conceded, limited to the management of mineral resources in the international seabed area. However, nothing would prevent the states party to UNCLOS from formally extending jurisdiction to this new type of resource, unforeseen at the time of the UNCLOS negotiations. In the alternative, an 'evolutive' interpretation of Part XI of the Convention could be adopted, taking into account the criterion of 'proximity' of the most important genetic resources to the hydrothermal vents in the deep seabed area.[47] Or this issue might be considered in the context of current initiatives for UN system reform with a view to establishing a new International Environmental Organisation,[48] the mandate of which would also include standard-setting and the monitoring of prospecting and exploitation activities aimed at genetic resources in the seas beyond national jurisdiction.

However, entrusting the implementation of common heritage to an existing international institution, or to one to be constituted *ad hoc*, is not the only solution. The principle of common heritage in its substantive aspect is, like any norm of international law, perfectly capable of being applied in a decentralised manner by states. Even in the absence of *ad hoc* institutions every state is under an obligation to respect and fulfil the

---

[47] For an assessment of the potential in the genetic traits of organisms living in extreme environmental conditions, with scarcity of light and temperature variations, see Scovazzi, n 44 above.

[48] See F Francioni, 'The Role of the EU in Promoting Reform of the UN in the Field of Human Rights and Environmental Protection' in the European Union, *The European Union and the United Nations* (Paris, Institute for Security Studies, 2005), 31 ff.

principle of the common heritage by ensuring that subjects within its jurisdiction do not act contrary to its object and purpose. This would be the case if a state authorised or negligently failed to prevent bio-technological activities in common spaces that had the effect of causing severe and irreversible damage to the unique biodiversity of that space. Similarly, a state would fail the common heritage if it authorised exclusive appropriation of genetic resources without requiring equitable sharing of pertinent scientific knowledge and without ensuring that a fair portion of economic benefits accruing from their exploitation be devoted to the conservation and sustainable development of such common resources.[49] Similar criteria apply to the genetic resources of Antarctica. Here, the 40 plus years of uninterrupted co-operation within the framework of the Antarctic Treaty (1959)[50] would give the Antarctic Treaty Consultative Meeting (ATCM) undisputed authority to regulate bio-prospecting in the Antarctic Treaty Area, in the interest of humankind and in conformity with the principle of free exchange of scientific information.[51]

The criteria above do not entail that the international seabed or Antarctica may not be treated like great laboratories open also to private scientific operators interested in the biotechnological development of the respective resources. On the contrary, it simply entails that (a) access to these resources occurs within a regulatory framework capable of preserving the interest of humankind in the conservation and sustainable development of these areas; (b) the technological advances and financial return produced by bio-prospecting be equitably shared, under the authority of relevant international institutions or multilateral regimes such as the International Sea Bed Authority or the ATCM; and (c) in the absence of multilateral mechanisms, individual states regulate bio-prospecting and exploitation of the genetic resources of common spaces with full respect for their character as part of the common good of humanity, so as to avoid recognition of ownership and appropriation simply on the basis of earlier finding and discovery.[52]

4. INDIVIDUAL AND COLLECTIVE HUMAN RIGHTS

Having clarified the manner in which international law allocates rights of control over bio-genetic resources, we can now proceed to the examination of

---

[49] See B Conforti, 'Notes on the Unilateral Exploitation of the Deep Seabed' (1980) 4 *Italian Yearbook of International Law* 3.

[50] See http://sedac.ciesin.org/entri/texts/acrc/at.txt.html.

[51] See F Francioni, 'Antarctica and the Common Heritage of Mankind', in Francioni and Scovazzi (eds), *International Law for Antarctica* (Milan Giuttré, 1987), 101 ff.

[52] For a precedent applying these criteria in the domestic law context, see *Edmonds Institute et al v Bruce Babbit* (US District Court for the District of Columbia, 24 Mar 1999, available at www.edmonds-institute.org/ yellowstone98561.pdf), concerning Yellowstone National Park.

the way in which biotechnology applied to such resources affects internationally recognised human rights. The scope of this part of the chapter is limited to a general overview of human rights guaranteed under general international law. Other contributions in this volume will address human rights issues arising from the regulation of biotechnology in specific treaty regimes and in international organisations.[53] In this perspective, one must recognise at the outset that the content and scope of the category of international human rights under customary international law remain somewhat elusive. Faced with a vast array of treaty and soft law instruments on the protection of an infinite variety of human rights, states continue to debate what constitutes the universally shared core of human rights that they must respect and protect as a matter of customary international law. Globalisation, with its powerful integrative force at the economic, social and cultural level, has the effect of raising levels of rights awareness in the most diverse legal systems, thus fostering recognition of basic human rights as the mainstay of an open and democratic society. At the same time, for the recurrent law of 'unintended consequences', the historical process of globalisation is also fuelling a centrifugal trend toward the search for specific identity, often found in opposition to cosmopolitan values in the traditions and moral beliefs of the nation, of minorities or groups. This phenomenon is particularly evident in the area of 'cultural rights', where claims to the enactment and respect of a specific world view and practice may be pitted against internationally recognised human rights, and even the rights of individuals within the group. Such antinomy between the universal and the particular complicates, but does not exclude, the identification of a core of generally recognised human rights rooted in the inherent value of human dignity and shared humanity. The International Court of Justice,[54] the practice of international criminal tribunals[55] and state practice, including that of national courts,[56] recognise the existence of a body of customary international law on human rights binding upon states independently of their consent to specific treaties. This body of law has been constantly expanding since the adoption of the UN Charter and of the Universal Declaration of Human Rights. It today includes the

---

[53] See S Millns, E Righini, A Yusuf, K Abbott, K Mechlem and T Rainey, E-U Petersmann and D Galligan in this volume.

[54] See *Barcelona Traction, Light and Power Company, Limited (Belgium v Spain)*, [1970] ICJ Rep 32; *Military and Paramilitary Activities in and Against Nicaragua (Nicaragua v United States of America)* [1986] ICJ Rep 14; Advisory Opinion on the *Legality of the Threat or Use of Nuclear Weapons* [1996] ICJ Rep 226; Advisory Opinion on the *Legal Consequences of the Construction of a Wall in the Occupied Palestinian Territory*, 2004, all available at www.icj-cij.org/idecisions.htm.

[55] See A Cassese, *International Law* (Oxford, Oxford University Press 2003), 393 ff.

[56] For a comparative overview of national courts' treatment of international human rights obligations, see B Conforti and F Francioni (eds), *Enforcing International Human Rights in Domestic Courts* (The Hague/Boston/London, Martinus Nijhoff, 1997).

prohibition of the most egregious violations of human dignity, such as genocide, slavery, torture and racial discrimination, as well as of violent suppression of the right to self-determination of peoples, of prolonged and widespread deprivation of personal liberty and of so-called 'gross violations of human rights'.[57] The emergence of these rights has contributed to the transformation and modernisation of international law from a legal order governing diplomatic relations between states to a more mature legal order applicable also to non-governmental actors. The implications of this transformative process are far-reaching. First, states can be held internationally accountable also toward individuals at the level of primary human rights obligations, even if secondary rules on responsibility and remedies may still be lacking or limited to regional human rights regimes, notably the European Convention and the Inter-American System. Secondly, human rights obligations are not reciprocal, like most classic customary international law obligations, but are integral, owed to the international community as a whole; thus, they operate *erga omnes*.[58] Thirdly, the assumption that every state has an interest in the respect for basic human rights as a matter of international public policy has contributed to the 'constitutionalisation' of a core of fundamental human rights norms in terms of *ius cogens*, or peremptory norms, endowed with inherent normative strength so that no single state, alone or in conjunction with others, may dispose of them at will. In this perspective, *jus cogens* represents the most powerful legal tool to support the concept of 'international community', as a collective entity that transcends the sovereignty of states and encompasses them *uti universi*. Fourthly, the idea that fundamental human rights constitute a common concern of the international community has led to the development and enforcement of the principle of international criminal liability of individuals who commit serious violations of human rights falling within the category of international crimes.[59]

The considerations above are especially relevant in the context of a discussion on the role of human rights in the international regulation of biotechnology. First of all, they indicate that, even though the status of and access to genetic resources may still be subject to 'sovereign rights', the legitimacy of their biotechnological applications must be gauged

---

[57] *Restatement (Third) of the Foreign Relations Law of the United States* (St Paul, Minn, West, 1987).

[58] See *Barcelona Traction* and *Advisory Opinion on the Wall*, n 54 above; B Simma, 'International Human Rights and International Law' in Academy of European Law (ed), *Collected Courses of the Academy of European Law*, Vol. IV (Leiden and Boston, Mass, Martinus Nijhoff, 1995).

[59] This happened, in particular, with the institution of the International Criminal Tribunal for the Former Yugoslavia (ICTY) in 1993 (the text of the statute is available at www.un.org/icty/legaldoc/index.htm), of the International criminal tribunal of Rwanda (ICTR) in 1994 (the text of the statute is available at www.ictr.org/ENGLISH/basicdocs/statute/2004.pdf), and of the International Criminal Court (ICC) in 1998 (the text of the statute is available at www.un.org/law/icc/statute/romefra.htm).

in accordance with human rights standards, in respect of which the international community as a whole has a legal interest. Secondly, the very notion of international human rights entails that relative standards are binding not only upon states, whose sovereignty is thereby limited, but, at least in an indirect manner, also on private actors, especially powerful new scientific and economic entities—science concerns and business corporations—which can command the technological power necessary to develop and market genetically engineered products. Finally, the introduction of human rights discourse into biotechnology regulation will necessarily entail a deconstruction of the unity and indivisibility of the sovereign state to identify whose individual or collective human rights are actually affected by biotechnology applications.

In the following sections I will look through the lens of this complex normative development to try to discern which human rights are most directly affected by biotechnologies. The focus will be on the following set of rights: (1) human dignity; (2) non-discrimination, (3) self-determination, (4) rights pertaining to the human body, such as life, integrity, health, (5) economic and social rights, including intellectual property rights and sustainable development. This is by no means an exhaustive catalogue of human rights potentially affected by bio-engineering techniques. But it represents the preliminary legal framework within which a more detailed analysis of the human rights involved in biotechnology applications can be developed. This will be the task of the specific contributions in this volume which are devoted to particular regulatory regimes and to specific categories of human rights.[60]

5. HUMAN DIGNITY

The broadest human right concept invoked in the context of biotechnology is human dignity. This is a fundamental concept in international human rights law. The 1948 Universal Declaration refers to it in the Preamble as 'the foundation of freedom, justice and peace in the world'[61] and incorporates it in Article 1, which states that 'human beings are born free and equal in dignity and rights'. Subsequent human rights instruments have systematically referred to human dignity as the foundation and

---

[60] See Parts II, III, IV, V and VI of this volume.

[61] See the first recital, which reads as follows: '[w]hereas recognition of the inherent dignity and of the equal and inalienable rights of all members of the human family is the foundation of freedom, justice and peace in the world'. Human dignity is also referred to in the fifth recital, which states: '[w]hereas the peoples of the United Nations have in the Charter reaffirmed their faith in fundamental human rights, in the dignity and worth of the human person and in the equal rights of men and women and have determined to promote social progress and better standards of life in larger freedom'.

wellspring of specific human rights.[62] In Europe, the value of human dignity constitutes the cornerstone of the 2000 Charter of Fundamental Rights (Article 1), now incorporated into the Treaty adopting a Constitution for Europe (Article II–61).[63] In the well known case of *Netherlands v European Parliament and Council*, Advocate General Jacobs stated that human dignity is 'perhaps the most fundamental right of all, and is now expressed in Article 1 of the Charter'.[64] In the field of biotechnology the concept of human dignity works as a threshold standard against which to test the different applications of genetic engineering techniques. In this role it performs a dual function: (1) on the one hand, it may provide the ethical and legal justification for the development and application of new biotechnologies; and (2), on the other, it is the guiding principle in setting boundaries to the permissibility of the variety of policy options offered by biotechnologies in fields such as bio-medicine and agriculture.

(1) As an ethical justification for the development and application of new biotechnologies, human dignity can play an important role in supporting the legitimacy of cutting edge scientific research in the field of medicine and genetic therapy for hereditary or otherwise incurable diseases, and generally in promoting participation in scientific progress consistent with Article 27(1) of the Universal Declaration of Human Rights.[65] The potential benefits of such progress, especially for people who suffer, or may be born suffering, from severe diseases and disabilities of a genetic nature constitutes a powerful ethical and human rights argument to counter-balance the cultural or religious objections of those who are opposed to playing with a matter of life or the design of nature.[66] Similarly, in the field of agriculture, respect for and protection of human dignity can be an important factor in adopting a policy favourable to the introduction of genetically modified crops or the distribution of genetically modified food when this represents the most effective way

---

[62] See, eg, the Preamble to the Supplementary Convention on the Abolition of Slavery, the Slave Trade, and Institutions and Practices Similar to Slavery (1956), 266, UNTS, *Vol.* 3; the Preamble to the International Convention on the Elimination of All Forms of Racial Discrimination (1966), *Vol.* 660 UNTS, 195; the Preamble to the two 1966 UN covenants on human rights (see International Covenant on Economic, Social and Cultural Rights (ICESCR), *Vol.* 993, UNTS, 3; International Covenant on Civil and Political Rights (ICCPR), *Vol.* 999, UNTS, 171); the Preamble to the 1973 International Convention on the Suppression and Punishment of the Crime of Apartheid, available at www.unhchr.ch/html/menu3/b/11.htm; the Preamble to the 1984 Convention against Torture and Other Cruel, Inhuman or Degrading Treatment or Punishment (1985) 24 ILM 535.

[63] Further on human dignity and common values in Europe, see S Millns in this volume.

[64] Opinion of AG Jacobs of 14 June 2001 in Case C–377/98 *Netherlands v European Parliament and Council* [2001] ECR I–7079, para 197.

[65] See n 6 above.

[66] C Campiglio, 'Human Genetics, Reproductive Technology and Fundamental Rights' (2005) 14 *Italian Yearbook of International Law* 83.

to deal with situations of severe poverty, famine or malnutrition that endanger the dignity, subsistence and the very life of people.[67]

(2) As a constraint, human dignity has already begun to perform a specific role in relation to the manifold applications of biotechnologies, notably in the field of biology and medicine. The UNESCO UDHG, with related instruments, the Council of Europe Biomedicine Convention and the Charter of Fundamental Rights included in the Constitution for Europe are all based upon the primacy of human dignity over the interests of scientific research and technological innovation. In particular, respect for human dignity entails that biotechnological applications in the field of medicine shall: (1) respect the uniqueness and diversity of human beings and, accordingly, avoid a reduction of individuals to their genetic characteristics;[68] (2) respect the free and informed consent of interested persons, in accordance with the modalities established by law; (3) avoid eugenic practices, especially those aimed at the selection of human beings; (4) be based upon the principle that the human genome and parts of the human body may not be disposed of for monetary gain; and (5) shall conform to the basic prohibition of reproductive human cloning.[69] Of course, recourse to such a fluid and open-ended concept as human dignity leaves undecided what is 'human' and whether technological application on the stem cells of human embryos is permissible in view of therapeutic benefits, as discussed above. This remains a contentious area, where national legislation and, more importantly, fundamental ethical standards in different societies continue to diverge. In particular, there is no consensus on the question of when human life begins,[70] whether human embryos are protected under the principle of human dignity, whose consent is relevant—that of the parents?, the spouse?, the future human being?, the beneficiary?—and, ultimately, on how to balance protection of the nascent life of the embryo with other legitimate objectives, such as protection of the health of others, the self-determination of the mother, the rights of the spouse or the utilisation of the embryo cells for scientific and therapeutic purposes. Given the highly subjective concept of 'human dignity' and differing ethical perceptions of the stage of life formation at which the term 'human' and the empowering notion of 'dignity' may apply, it is impossible, at least in the short term, for a human rights approach to develop solely on the basis of a universally shared notion of human dignity. This, however, does not mean that the concept is useless.

[67] T Mechlem and K Raney in this volume.

[68] See Art 2 of the Universal Declaration on the Human Genome and Human Rights, n 14 above. On this point, see N Lenoir, 'La Declaration Universelle sur le genome humain et les droits de l'homme de l'UNESCO' in Conseil d'Etat, *Rapport public du Conseil d'Etat* (1998).

[69] See Art II–63 of the 2004 Treaty establishing a Constitution for Europe [2003] OJC 169/01.

[70] See *Vo v France*, App no 53924/00, [2004] ECHR 326.

In my view, it provides an important legal tool for establishing a dialogue between different and sometimes radically opposed ethical camps. It permits a better understanding of the interests and reasons involved in the moral claims of others, whether to absolute respect for the sacredness of life or to the need to make use of the opportunities science offers to prevent or remedy severe genetic diseases capable of impairing or destroying the dignity of the bearers.

6. NON-DISCRIMINATION

One of the positive consequences, in moral and social terms, of genetic science, and in particular of the Human Genome Diversity Project (HGDP), is the production of scientific evidence that there is no biological basis for the concept of 'race' and that persons belonging to the same racial-ethnic group may indeed have a more diverse genetic patrimony than people who may be profiled as belonging to different racial groups. This disclosure of the 'universality' of the human genome is, no doubt, a significant contribution to the consolidation of the ethical basis of the principle of non-discrimination. This has been acknowledged by the UDHG, which states that the human genome 'underlies the fundamental unity of all members of the human family'[71], this principle also provides the rational justification for the inclusion of a non-discrimination norm in virtually all human rights treaties.[72]

At the same time, genetic science and technology, especially in the field of medicine, are raising new possibilities of discrimination. From a general point of view, the most threatening type of discrimination can come from a new conceptualisation of 'normality' based, rather than on the natural definition as a state of physical and mental wellbeing, on a genetic connotation, which includes the hidden predisposition to some health impairment or, conversely, the search for a certain quality of life. In this context, it is clear that the more genetic tests and therapies are made available, the greater the gap will grow between the fortunate who have access to such tests and therapies and those who do not. This new 'discrimination' would run along the fault line that separates the rich world from the less developed world.[73]

At a more practical level, the principle of non-discrimination may play an important role in genetic patenting. A recent case brought before the

---

[71] Art 1.

[72] See, eg, Art 14 of the ECHR with additional Potocol 12, Art 1 para 1 of the American Convention, Arts 2 , 14 and 26 of the ICCPR.

[73] For an in-depth analysis of these implications, see the yet unpublished PhD thesis by A Rouvroy, *Human Genetics and Justice: Sustaining Uncertainty* (Florence, European University Institute, 2005).

European Patent Office offers an example of race utilisation in patent specification. Myriad Genetic claimed a patent relating to a gene probe 'for diagnosing a predisposition to breast cancer in Ashkenazi Jewish women'. The relevant gene mutation related to ovarian and breast cancer and was found to be prevalent in Ashekenazi Jewish population in the order of 1 per cent as compared to 0.1 per cent of the general population. The European Society of Human Genetics strongly opposed diagnostic targeting of a racial group in a gene patent application. In particular, it argued that genetically discriminating considerations are contrary to *ordre public* and public morality. The European Patent Office decided to uphold the patent in amended form, stating that it 'relates to use of a particular nucleid acid carrying a mutation of the BRCA 2-gene, which is associated with a predisposition to breast cancer for in vitro diagnostic of such predisposition in Ashkenazi Jewish women'.[74]

But the area where the risk of discrimination on a genetic basis is the highest and most disturbing is that of insurance and employment. Here the questions arise: (1) whether insurers and employers may be allowed to require genetic tests as a condition of insurance or employment; (2) whether insurers or employers may require disclosure of prior genetic tests by the applicant; and (3) whether insurers or employers may give weight for business purposes to genetic information voluntarily provided by applicants. *Prima facie*, the answer to these questions appears to be negative in the light of the norms contained in universal and regional instruments on bioethics. For example, Article 11 of the Council of Europe Convention on Biomedicine stipulates that 'any form of discrimination against a person on grounds of his or her genetic heritage is prohibited'. More specifically, Article 12 prohibits predictive tests except for health or scientific research reasons. The same principles are upheld in the UDHG (Article 6), in the International Declaration on Human Genetic Data (Articles 7 and 14) and in the ECOSOC Resolution on Genetic Privacy and Non-Discrimination, of 21 July 2004, which '[u]rges States to ensure that no one shall be subjected to discrimination based on genetic information; [a]lso urges States to protect the privacy of those subject to genetic testing and to ensure that genetic testing and the subsequent processing, use and storage of human genetic data is done with the prior, free, informed and express consent of the individual or authorization obtained in the manner prescribed by law consistent with international law, including international human rights'.[75] As we can see, the international standards on non-discrimination are clear. Thus genotypic differentiations resulting

---

[74] EPO press release, 'Patent on Breast Cancer Gene-2 maintained in amended form after public hearing', 29 June 2005.

[75] The text of the resolution is available at: www.un.org/docs/ecosoc/documents/2004/resolutions/eres2004-9.pdf.

in discriminatory treatment in the field of employment and insurance are not permissible. Naturally, to translate these standards into enforceable prohibitions in domestic law requires precise regulation and a considerable degree of public intervention in insurance and employment markets where private lobbies may show considerable resistance. However, it is fair to say that so far, even in those countries in which heath care is funded by private insurance, there is no indication that genetic science may be leading to systematic discrimination and to the creation of a 'genetic underclass'[76] of unemployable and uninsurable people.

7. SELF-DETERMINATION

Self-determination, originally conceived as the right of peoples to accede to self-government, has become an important component of international human rights. The two UN Covenants, on Civil and Political Rights[77] and on Economic, Social and Cultural Rights,[78] are both premised on recognition, in identical terms, of the right to self-determination in their respective Articles 1. Similar recognition can be found in the 1982 African Charter of Human and Peoples' Rights.[79] At the core of this right is the entitlement of all peoples to 'freely determine their political status and freely pursue their economic, social and cultural development'.[80] But how can this rather indeterminate right be relevant to the governance of biotechnology in the post-colonial world?

First, as indicated in section 2 of this chapter, self-determination complements and reinforces the sovereign right of all peoples to 'freely [to] dispose of their natural wealth and resources',[81] including biogenetic resources within their territorial jurisdiction. As a collective right of the 'peoples', self-determination also entails the right freely to pursue economic, social and cultural development.[82]

Secondly, in its external dimension, this right also entails that states, especially developing states, are entitled to pursue economic policies aimed at protecting their populations against the damaging impacts and unwanted risks of biotechnology applications. This is all the more true given that the spread of biotechnology and of its products, especially in the field of agriculture, depends on the business practices of a relatively

---

[76] This expression is used by Rouvroy, n 72 above, at 139.
[77] See n 62 above.
[78] Ibid.
[79] See Art 20.
[80] See Art 1(1) of the two UN Covenants on human rights, n 62 above.
[81] See Art 1(2) of the ICCPR, n 62 above.
[82] For a recent reaffirmation of this right, see the ICJ *Advisory Opinion on the Wall*, n 54 above.

small number of corporations, all based in the advanced industrial world and increasingly characterised by a high degree of vertical integration.[83] These corporations have an important role as vectors of scientific progress and economic development. Their inventions and know-how can enhance agricultural productivity, provide more nutritious food[84] or new pharmaceutical products, and generally improve the welfare of people. But, at the same time, one cannot ignore that these companies all belong to the private sector and are commercially driven toward the development of biotech products and services capable of ensuring satisfactory financial returns for their conspicuous investments.[85] Besides, they operate in a markedly asymmetrical relationship with developing countries. They rely on structurally superior knowledge of the technological processes and products they market, and consequently on superior knowledge of risks. They enjoy the bargaining advantage of having at their disposal large finance capital for investment, for which less-developed countries desperately compete. And, most importantly, they claim that, at least in a strict legal sense, they are not 'subjects' of international law, so as to be able 'legally' to elude international human rights standards binding upon states.[86] This may lead to abuses and unfair market practices in their relations with host countries in the planning and conduct of foreign operations. While this is a general problem arising in relation to the activities of all trans-national corporations, the impact on a sphere of interest protected by the principle of self-determination can be more substantial in the case of biotech companies. New and

---

[83] Following a process of mergers and consolidation there are now five large companies dominating the biotech market in the area of food and agriculture. See T Mechlen and K Raney in this volume.

[84] As in the case of so-called 'golden rice', a biotech rice enriched with vitamin A, capable of providing a low cost alternative to a more diversified but often unaffordable diet in many poor countries of the world where rice represents a main staple.

[85] The top 10 multinational biotech corporations invest US $ 3 billion per year for agricultural biotechnology research and development only. In contrast the total FAO budget for research in crop improvement within the Consultative Group on International Agricultural Research amounts to US $ 300 million. See K Mechlem and T Rainey in this volume, and FAO, *The State of Food and Agriculture 2003–04—Agricultural Biotechnology: Meeting the Needs of the Poor?* (Rome, FAO, 2004), 32 ff.

[86] We cannot undertake a discussion of the question whether business corporations, as international 'actors', may *de facto* be subject to international human rights or environmental standards here. The concept of 'actorship' in international law is still undefined, and is often used in a less than rigorous manner to mean international law as 'global' law, transcending the traditional distinction between the domestic and international legal orders. It is worth mentioning, however, that a step towards the recognition of corporations as economic entities capable of being accountable under international human rights standards has been made by the UN Sub-Commission on the Protection and Promotion of Human Rights with the adoption in 2003 of a set of 'norms on the responsibility of transnational corporations and other business enterprises with regard to human rights'. See UN Doc. E/CN.4/Sub.2/2003/12/Rev.2 of 26 Aug 2003.

untested biotechnology experimentation on plants or animals may take place in a foreign country without the prior informed consent of local authorities and people, taking advantage of a lack of legislative regulation or infrastructure, or inadequate administrative control.[87] Aggressive marketing strategies aimed at introducing new biotech products, such as genetically modified seeds that farmers are not allowed to reuse, may cause dependence on foreign supply and consequent indebtedness, while at the same time disrupting long established and socially sound patterns of farming techniques.

All states, especially less-developed states, are entitled to invoke the right of self-determination of (their) peoples to protect societal values and sustainable economic structures from the adverse impact of unethical or unfair business practices on the part of international biotech corporations. This, of course, may raise problems with obligations of free trade and market access within the WTO, especially now when so many developing countries are, or are becoming, members of the Organisation. However, this problem cannot be addressed by advocating an inflexible application of free trade principles. On the contrary, as other contributions to this volume will discuss,[88] it requires a human rights approach to trade, based on a broad construction of every state's freedom and responsibility to set an appropriate level of protection for its fundamental societal values, of which the principle of self-determination constitutes the essential core.

A third way in which self-determination can play a role in developing a human rights approach to the legality of modern biotechnology is in relation to the special protection of distinct minorities, groups or peoples whose genetic characteristics or special environmental resources are targeted by bio-science research and industry in view of the development of new products and commercial applications. In this 'internal' dimension, the principle of self-determination guarantees a certain degree of autonomy to the peoples concerned, within the constitutional structures of existing states. This entails the obligation, from the point of view of collective human rights, on every state to take into account the interests of such distinct groups, and especially of indigenous peoples, in maintaining and managing their distinct culture and special and sometimes unique relationship with their land and its biological resources. This particular dimension of the right to self-determination entails a limitation on the sovereign rights of the territorial state, in the sense both that (a) biotechnological projects involving indigenous peoples, or other distinct groups, should be based on the effective participation of these

---

[87] See T McGarity, 'International Regulation of Deliberate Release Biotechnologies' in F Francioni and T Scovazzi (eds), *International Responsibility for Environmental Harm* (London, Graham and Trotman, 1991), 319 ff.
[88] E-U Petersmann in this volume.

peoples in decisions that affect them and their environment;[89] and that (b) eventual economic benefits accruing from indigenous peoples' genetic patrimony, from the biological resources of their environment and from the traditional knowledge that has permitted conservation and development should be equitably shared with such peoples.[90]

8. THE HUMAN BODY

These last remarks introduce us to the most sensitive aspect of biotechnology applications: the bio-prospecting and engineering of parts of the human body in the expectation of finding useful genetic material for diagnostics and therapy for certain inherited diseases. Research has been booming in this field for a number of years; and experience already shows that, while advances in gene therapy may hold the promise of improving the life and health of people, a number of potential adverse impacts on human rights may result. A particularly telling example is that of the experimental use of cell lines—with a living cell proliferating, under appropriate laboratory conditions, into multiple cells that will form a durable cell line—to be studied and manipulated for possible medical applications. An early experiment on cell lines at the end of the 1980s involved the taking of human tissues from a small and fairly remote tribe of indigenous people from Papua New Guinea in order to study their 'unique' characteristics and their possible application in the early detection and eventual cure of adult leukaemia and other degenerative disorders.[91] This and similar initiatives, which were part of the well-known HGDP,[92] were undertaken in the exercise of unfettered freedom of scientific research, in the absence of genuine capacity on the part of

---

[89] For an important application of this principle, see Human Rights Committee, General Comment No 23 of 6 Apr 1994 on Art 27 of the ICCPR (available at <http://www.ohchr. ch/english/bodies/hrc/comments.htm>), and the decision of the Inter-American Court of Human Rights of 31 Aug 2001 in *Awas Tingni Mayagna (Sumo) Indigenous Community v Nicaragua* (Communication No. 167/1984, 26 Mar 1990, available at heiwww.unige.ch/ humanrts/undocs/session45/167-1984.htm) recognising that the collective right of indigenous people to their ancestral land and resources prevailed over the Government's sovereign power to dispose of them by way of licensing their exploitation to foreign investors. Further on this case and on the general question of the biogenetic resources of indigenous peoples, see F Lenzerini in this volume.

[90] See Art 8(j) of the Convention on Biological Diversity, n 1 above, which, however, uses the word 'encourage' with regard to the sharing of benefits deriving from the utilisation of traditional knowledge and practices relating to biological resources.

[91] For a full account of this case see J Rifkin, *The Biotech Century: Harnessing the Gene and Remaking the World* (New York, Penguin, 1998), 50 ff.

[92] The HGDP is a consortium of scientists from North America and Europe aimed at collecting live tissue from hundreds of different human groups throughout the world in order to map the human genome. See 'The Human Genome Diversity Project', *GenEthics News*, issue 10, available at www.hgalert.org/topics/personalInfo/hgdp.htm.

the tribe to provide prior and informed consent, and on the questionable assumption that the people whose bodies provided the valuable tissue samples were mere 'objects', rather than persons endowed with inherent human rights. No wonder, then, that such precedents have spurred a widespread movement among indigenous populations, especially in Latin America and Asia, in opposition to the HGDP, suspected of opening the door to abuse of human genetic material for commercial and even military purposes.[93] To avoid such *a priori* opposition to genetic research a more cautious approach, taking into account respect for human dignity, a right to personal integrity, and the individual and collective right to maintain control over genetic heritage and to decide whether to make (their) DNA available for scientific experimentation, is necessary. This approach has been followed by UNESCO since the adoption of the 1997 UDHG,[94] the first truly universal instrument[95] to set ethical standards on human genetic research and practice. The Declaration carefully balances freedom of scientific research against the need to safeguard human rights and the general interest of humanity against possible abuses. Besides proclaiming the human genome 'the heritage of humanity',[96] the declaration establishes, in Article 5, that research, treatment or diagnosis affecting a person's genome must be undertaken only on the basis of 'the prior, free and informed consent of the person concerned'. More important, the same Article provides that when 'a person does not have the capacity to consent, research affecting his or her genome may be carried out for his or her direct health benefit, subject to the authorization and the protective conditions prescribed by law'.[97] This formulation leaves an ample margin of appreciation for national law-makers to decide when and under what specific conditions research and technological applications affecting someone's genome are permissible. But, as in the case of human dignity, reference to the paramount importance of a direct health benefit to the individual permits bridging the gap between different ethical views, leading to possible convergence in a shared ethical conception of the human person as an end in herself (and not a means to achieve technical or economic goals).

---

[93] Further on this problem see F Lenzerini in this volume.

[94] See n 14 above.

[95] The Convention on Human Rights and Biomedicine, n 19 above, is a regional instrument that remains of limited geographic scope and application.

[96] See Art 1.

[97] This provision appears to be particularly important in meeting the concerns of indigenous people. See the chilling statement by Victoria Tauli-Corpuz, representative of the Cordillera People's Alliance, Philippines: '[a]fter being subjected to ethnocide and genocide for 500 years, which is why we are endangered, the alternative is for our DNA to be stored and collected . . . Why don't they address the causes of our being endangered, instead of spending $ 20 million for five years to collect and store us in cold laboratories?' (cited by F Lenzerini in this volume).

9. ECONOMIC RIGHTS AND BENEFITS

Economic rights and benefit-sharing in relation to modern biotechnologies are 'transversal' issues arising from all biotechnology applications in fields as diverse as human genetics, plant genetic resources, pharmacy, agriculture and industry. Given the growing importance of biotech business in these different fields, it is no wonder that the ethical question of who is to benefit from the commercial application of such science has been cast increasingly in human rights terms. The High Commissioner for Human Rights' Expert Group on Human Rights and Biotechnology has focused especially on the problematic relationship between genetic patenting and equitable sharing of the economic benefits accruing from commercial exploitation of the patented material. In its 2002 report, the Group goes as far as to consider 'the linked issues of the ability to patent genetic material and the sharing of benefits deriving from commercial exploitation of that material to be the most important issue in the area of human rights and biotechnology at this time'.[98] From a human rights perspective this issue requires that we determine what the conditions and limits of property rights over genetic material are, on the one hand, and what the legal basis for recognising the economic rights of individuals or groups from whose body or natural environment the material has been extracted may be, on the other. This preliminary determination is by-passed by those commentators who are satisfied with the reference to a generic principle of 'equitable benefit sharing', sometimes even considered a veritable rule of customary international law.[99] Although reference to equity in this field is welcome and can indeed be useful, inasmuch as it opens the way toward pragmatic accommodation of differing competing interests, it can be of only limited use in a human rights approach to the problem. The latter approach posits a use of equity *infra legem* and not in a legal vacuum. Thus, it requires the prior identification of the legal entitlements that are at stake under international and human rights law and permits the equitable balancing of conflicting legal interests by appropriate techniques of interpretation and implementation of international norms. In this perspective, 'equitable benefit-sharing' is the problem to be addressed, rather than the normative tool providing a key to any solution. Benefit-sharing cannot be 'de-contextualised' from the individual and collective rights that form its basis. As I have tried to indicate in the first part of this chapter, the identification of relevant titles and rights—peoples', humanity's, community's,

---

[98] See High Commissioner's Expert Group on Human Rights and Biotechnology, 'Conclusions', Geneva, 24–25 Jan 2002, available at www.unhchr.ch/biotech/conclusions. htm, para 19.

[99] See R Pavoni, *Biodiversità e biotecnologie nel diritto internazionale e comunitario* (Milan, Giuffrè, 2004) esp chaps III and IV.

individuals'—is a prerequisite for the determination of legal conditions of access to genetic resources and to the sharing of economic benefits among relevant stakeholders. In this context, 'equity' is a variable element the function of which is to infuse considerations of justice and fairness into the balancing of competing rights. Variation depends on the type, location and origin of the relevant genetic stock.

In relation to plant and animal genetic resources found in the territory of states, the function of equity is quite clear: since present international law[100] recognises the territorial state's sovereign rights over such resources, equity has the function of striking a fair balance between, on the one hand, the claim of the investor to protect biotechnological inventions, including property rights arising therefrom, and, on the other, the sovereign right of the source state to obtain equitable remuneration for the exploitation of its biodiversity, including remuneration for local communities' traditional knowledge, which permitted or facilitated the identification and utilisation of the relevant genetic material in the first place.

By contrast, in the context of biotechnological development of genetic resources originating in common spaces beyond national jurisdiction— such as the international sea and seabed, and Antarctica—the role of equity is totally different. Here, equity is called on to accommodate the claim to exclusive property rights of the biotech investor and the general interest of humanity in the identification, conservation and sustainable development of such resources pursuant to common heritage or common concern principles. This entails that the grant of patents over biotechnological applications to such common resources, besides respecting the usual conditions of patentability—novelty, inventive step, capability of industrial application—must be compatible with the global common nature of these resources and the public interest of humanity in maintaining knowledge and control over their development. In this context, the practical requirements to achieve such compatibility ought to include: (1) the duty of the patent applicant to disclose the provenance of the genetic material;[101] (2) the possibility of invalidation of the patent in the event of intentional misrepresentation of the origin of genetic resources; (3) effective use of the patent to support scientific progress, rather than simply produce genetic enclosure with the effect of blocking the development of knowledge and innovation (as in the case of dormant

---

[100] See, in particular, the Biodiversity Convention, n 1 above, and the FAO Treaty on Plant Genetic Resources, n 9 above.

[101] A mandatory requirement of disclosure of source and origin of genetic resources in the TRIPS Agreement and WIPO treaties is advocated *de lege ferenda* by F Abbott in this volume.

patents);[102] (4) peaceful use of the genetic resources; and (5) the equitable sharing of benefits in the form of international pooling of knowledge and, if practicable, by payment of reasonable royalty fees to recognised international institutions competent in the management and conservation of the relevant common resource.[103]

Finally, in relation to human genetic resources, the concept of equitable sharing of economic benefit must take into account the proclaimed nature of the human genome as 'common heritage of humanity'[104] —with its corollary that the human genome in its natural state shall not give rise to financial gains[105]—and the competing claims of researchers, bio-banks and other biotech investors to proprietary rights in the genetic material and in given biotech inventions. The state of play today reveals that overwhelming consideration is given to proprietary and scientific interests of those who carry out research and commercially develop biotechnological inventions, over the general interest in safeguarding open genetic knowledge and the individual and collective rights of donors of genetic samples. Domestic legislation[106] and case law[107] confirm this trend. This is clearly the result of a widespread assumption that it is in the interest of scientific progress not to inhibit bio-technological experimentation by considerations of proprietary or privacy rights of the individuals or groups who have provided genetic material.

But is this assumption correct? Besides the radical critique directed at gene patenting, based on the argument that DNA does not fulfil the requirements of patentability because it occurs in nature, the emerging judicial practice in this area creates some doubt about this assumption. Rather than advance the public interest in the progress of knowledge and the enhancement of health, gene patenting may easily become a tool of

---

[102] A more radical view holds that genes are not patentable because, by allowing such property rights, we would permit private constructive control of the genetic code, since 'the gene is the static chemical compound and the dynamic template executed through the genetic code': E Kane, 'Splitting the Gene: DNA Patents and the Genetic Code' (2004) 71 *Tennessee Law Review* 707.

[103] Obviously, this may be the most difficult issue to address, given the sensitive nature of creating new institutions competent to administer funds in the common interest. However, one should keep in mind that institutions or fora already exist that could perform the function of trustees of the common genetic heritage of humankind: they are: (1) the International Sea Bed Authority, which, under Art 157 of the LOS Convention, n 25 above, shall have such powers 'as are implicit and necessary for the exercise of those powers and functions with respect to activities in the Area'; (2) the Antarctic Treaty Consultative Meeting, with regard to the genetic resources of Antarctica (see Vigni, n 44 above); and (3) the Biodiversity Convention, n 1 above.

[104] See Universal Declaration on the Human Genome and Human Rights, n 6 above.

[105] Ibid, Art 4.

[106] For reference to domestic legislation, see R Brownsword in this volume.

[107] See the famous case of *Moore v Regents of the University of California*, 793 P2d 479, as well as *Greenberg v Miami Children's Hospital Research Institute, Inc.* 208 F Supp. 2d 918, cited by R Brownsword in this volume.

enclosure of knowledge and an obstacle to the legitimate pursuit of health care by patients. In the recent case of *Greenberg v Miami Children's Hospital*, several families affected by a rare genetic disorder (Canavan's disease) had provided research institutions with their children's tissue samples for research purpose and in the legitimate expectation that genetic tests could be developed in order to diagnose and treat the disorder. The Hospital identified the gene mutation which caused the disease, patented the gene, and started charging fees on tests for the syndrome. This led to the same families that had provided the genetic material necessary to identify the origin of the disease being charged fees for having their members tested. Is this correct? I doubt whether it is. And more clearly this outcome would not be consistent with a general principle of justice and equity if those who provided the biological samples enabling the genetic cause of the disease to be discovered were left without access to the results of the tests by reason of inability to pay the fees. The response of the families affected in this case is quite interesting and reveals a sort of Pavlovian reflex in terms of propertisation of the legal thought on the matter. Rather than arguing on the basis of a claim to open knowledge and fundamental right of access to health care, the families chose to base their complaint on the alleged breach of their proprietary data and misappropriation of ownership rights over their biological samples. These claims failed and the United States court decided that the defendant hospital was under no obligation to disclose the financial interests involved in the prospect of a commercial exploitation of the results of the genetic trials.[108]

10. CONCLUSIONS

The foregoing analysis shows that at least a preliminary answer can be given to the question we raised at the beginning of this paper, whether it is useful to look at the challenges posed by genetic science in the perspective of human rights. The answer is clearly yes, it is useful and necessary. The current asymmetry of knowledge and power between scientific and technological actors, on the one hand, and the traditional institutions of government and of civil society, on the other, cannot be redressed by a concurrent race to the privatisation and propertisation of genes, the human body, plants, new discoveries and everything else. A more rational approach is that based on the universally shared value of international human rights. In this chapter, we have identified the role that in this area can be played by human dignity, non-discrimination, self-determination of peoples and groups, the integrity of the human body and the equitable

---

[108] See, ibid, for a precedent see *Moore v Regents of the University of California*, n 107 above.

balancing of property rights and the general interest in the advancement and diffusion of knowledge. In this respect an important role can be played by Article 15 of the International Covenant on Economic Social and Cultural Rights, which proclaims the rights 'of everyone to enjoy the benefits of scientific progress and its applications'. This provision calls for the maximisation of open knowledge and of the benefits of its application, rather than supporting the tendency toward extreme forms of ownership in intellectual property, on the one hand, and in the sources of genetic material, on the other. But this is a long term project. In the short term it may not be so easy. This is an epoch that celebrates the myth of property. And, as has been lucidly put, '[a] time is marked not so much by ideas that are argued about, but by the ideas that are taken for granted ... the idea of property is just such a thought, or better, just such a non-thought; when the importance and value of property is taken for granted; when it is impossible, or at least for us, very hard, to get anyone to entertain a view where property is not central; when to question the universality and inevitability of complete propertization is to mark yourself as an outsider. As an alien.'[109] In the field of biotechnology the human rights discourse is a way to question this central thought. It is a way to avoid becoming an alien.

---

[109] L Lessig, 'The Architecture of Innovation' (2002) 51 *Duke Law Journal* 1783, at 1783–4.

2

State Responsibility for Violations of Basic Principles of Bioethics

PIERRE-MARIE DUPUY[*]

G IVEN THE FACT that biotechnologies should basically be ruled by
ethics rather than by law, certain authors are inclined to say that
this new field of human activity, precisely because it raises many
new moral problems and potential social conflicts between different and
competing objectives, does not stand to gain much from international
law.[1] As a consequence, international state responsibility, as recently codified by the International Law Commission of the UN, would have little or
nothing to do with the violation of basic principles of bioethics.[2]

Such an assumption seems to be far too restrictive. As demonstrated by
Professor F Francioni,[3] an impressive number of legal instruments, both
'soft' and 'hard' (ie, creating either non-binding or binding rules), aimed
at subordinating the use of biotechnologies to the respect of some major
principles, have been adopted during the last decade. Ethical in nature,
most of these principles have been incorporated into the rule of law by
means of legal instruments, either laws within national legal systems or
conventions and/or declarations or recommendations at the international
level.

Reacting to the old school of 'natural law', different strains of the
positivist school of law, including that of Hans Kelsen's 'pure theory
of law', have promoted the idea of distinguishing law, on the one side,
from ethics and morality, on the other. Nevertheless, one of the major
features of modern international law, as I have argued elsewhere, lies

---

[*] Professor of International Law at the European University Institute and at the
Universitéde Paris II. Email: Pierre-marie.dupuy@iue.it.

[1] On the limits of law as a regulatory modality in relation to new technologies, see the
seminal contribution of Stanford professor L Lessig, *Code and Other Laws of Cyberspace* (New
York, Basic Books, 1999).

[2] On this discussion, as on all the matters here at stake, see R Pavoni, *Biodiversità e biotecnologie nel diritto internazionale e comunitario* (Milan, Giuffrè, 2004).

[3] In this volume.

in the fact that there is a fundamental trend towards the 'substantial' unification of the international legal order around a material *corpus juris* of certain fundamental rules, the purpose of which is precisely to integrate a number of basic social values, reputed to be shared by 'the international community as a whole',[4] into the international rule of law. Respect for human dignity is one such principle.[5] Even if at the same time, adverse trends, both legal and political, still considerably slow down this integrative dynamic towards the moralisation of the law, the contemporary co-operation in this field tends progressively to build a future international code of bioethics sustained by international law.

The growing number of rules grounded in both law and ethics should have the following consequence: if states breach theses rules or even let them be breached by people acting under their jurisdiction, this should trigger their legal responsibility at least at the international level. As will be shown below, this might occur in the future in a growing number of cases (1). Nevertheless, for a number of reasons, some technical, others socio-political in nature, invoking and enforcing the international legal responsibility of states for breaching some of the major principles of bioethics will most probably often remain a difficult task (2). As a result, other more flexible forms of social responsibility which are easier to use should be considered (3).

1. AN INCREASED BODY OF INTERNATIONAL LEGISLATION LOGICALLY CREATES INCREASED POSSIBILITIES FOR RAISING THE LEGAL RESPONSIBILITY OF STATES IN BREACH OF THEIR INTERNATIONAL OBLIGATIONS

International law governing biotechnologies lies at the intersection between human rights law and the international law of the protection of the environment.[6] I shall not review the picture of international legislation my colleague Francesco Francioni has already drawn. Let us simply recall, from the human rights perspective, the importance of the 1948 Universal Declaration of Human rights, the 1997 UNESCO Universal Declaration on the Human Genome and Human Rights (UDHG),[7] the 1999 Guidelines for the implementation of that Declaration,[8] and the 2003 International

---

[4] See P-M Dupuy, 'L'unité de l'ordre juridique international, Cours général de droit international public' (2002) 297 *Recueil des Cours de l'Académie de Droit International de La Haye* 000.

[5] On the question of how international human rights may support ethics in scientific research, see F Francioni, 'Valori etici e diritto internazionale' [2004] *Rivista di Studi Politici Internazionali* 47.

[6] See Pavoni, n 2 above.

[7] Adopted 11 Nov 1997, available at www.unesco.org.

[8] See C/Resolution 23 of 16 Nov 1999, endorsing the guidelines for the implementation of the UDHG.

Declaration on Genetic Data.[9] The UN General Assembly endorsed the UDHG in 1998,[10] and in the past three years it has been engaged in the negotiation of a new convention designed to restrict human cloning. At the regional level, the Council of Europe has since 1997 adopted a series of instruments setting out ethical standards in the field of biomedicine and biomedical research. In that respect, the Oviedo convention on human rights and biomedicine[11] and the additional protocol on the prohibition of human cloning,[12] the 2002 Additional Protocol on transplantation of organs and tissues of human origin[13] and the 2005 Additional Protocol on biomedical research[14] demonstrate the normative activism of this institution, which clearly establishes and develops a series of principles in light of its general system for the protection of human rights, as arising from the 1950 European Convention for the Protection of Human Rights (ECHR), further interpreted and actively implemented by Member states thanks to the case law of the European Court of Human Rights.

On the environmental front, instruments like the Cartagena Protocol on Biosafety[15] and, in the field of agriculture, the 2001 FAO Treaty on Plant Genetic Resources[16] demonstrate the importance of international legislation in regulating the impact on the environment of biotechnologies. The 2001 FAO Treaty, in particular, has established a framework for international co-operation and sustainable use of plant genetic resources for food and agriculture. At the regional level, the EU Directive of 12 March 2001[17] set up a common system of authorisation and oversight of the deliberate release into the environment of genetically modified organisms.

All this expanding normativity has as its consequence the creation of new international obligations for states, whether directly, in the case of binding instruments already in force, or indirectly, when it comes to the process of new customary rules of international law, as they emerge from the accumulation of soft law instruments and/or norms established at the national level but converging in setting the same principles. See, for instance, the prohibition on obtaining financial gains out of any treatment

---

[9] See International Declaration on Human Genetic Data, 16 Oct 2003.

[10] See A/RES/53/152 of 9 Dec 1998.

[11] Convention for the Protection of Human Rights and Dignity of the Human Being with Regard to the Application of Biology and Medicine: Convention on Human Rights and Biomedicine, Oviedo, 4 Apr 1997, *CETS* no 164.

[12] Additional Protocol to the Convention for the Protection of Human Rights and Dignity of the Human Being with regard to the Application of Biology and Medicine, on the Prohibition of Cloning Human Beings, Paris, 12 Jan 1998, *CETS* no 168.

[13] Additional Protocol to the Convention on Human Rights and Biomedicine concerning Transplantation of Organs and Tissues of Human Origin, Strasbourg, 24 Jan 2002, *CETS* no 195.

[14] Additional Protocol to the Convention on Human Rights and Biomedicine, concerning Biomedical Research, Strasbourg 25 Jan 2005.

[15] Available at www.biodiv.org//doc/legal/cartagena-protocol-en.pdf.

[16] Available at ftp://ext-ftp.fao.org/ag/cgrfa/it/TPGRe.pdf.

[17] Directive 2001/18/EC [2001] OJ L/106/00.

of the human genome, a principle which, among others, is also to be found in Article 4 of the UDHG.

All these rules are to be considered as 'primary rules' of international law as they require states 'to do' or 'not to do'. Parallel to this increased number of new obligations, some already existing principles, well established either in human rights law or in environmental law, like the 'right to life' or the 'precautionary principle' may be applied in the context of the regulation of biotechnology. An example of that situation will be given below.

As a consequence, from a theoretical as well as from a technical point of view, there is no reason why the breach of such legal obligations would not give rise to the international responsibility of the state committing these violations. This is particularly the case when the norm in question is established in a binding instrument endowed with its own system of 'secondary rules of adjudication', to use the vocabulary of Herbert Hart.[18] In such cases, for instance any EU Directive or a human rights principle such as that set out in Article 2 of the ECHR, according to which every person has the right to life, there is an institutional and judicial framework defining the procedural and substantial conditions under which the responsibility of the state concerned may be triggered. In the first case (EU law), both the European Commission and the European Court of Justice provide, under certain conditions, potential and qualified claimants with the right to raise the responsibility of the failing state. This is also the case for European human rights law as established in the ECHR, over which the European Court of Human Rights in Strasbourg has jurisdiction.

In the ECHR context, an illustration was given as recently as 2004 by the case of *Vo v France* before the European Court of Human Rights sitting as a Grand Chamber.[19] The case originated in an application against the French Republic lodged with the Court under Article 34 of the ECHR. It dealt with the death of a foetus *in utero* which had not been classified as unintentional homicide by the highest competent jurisdiction in France, the *Cour de cassation*. The Court ruled that it had jurisdiction and delivered its judgment on whether France was responsible for not having sanctioned a doctor whose negligence had led to the death *in utero* of Mrs Vo's foetus, more than 20 weeks after its conception. Although this case deals more with a situation of medical liability than a true case of biotechnology, it is worth mentioning inasmuch as the crucial point at stake was precisely the legal status of the embryo and whether it is to be considered as a legal person within the meaning of '*toute personne*' ('everyone' in the English text) as stated in Article 2 of the ECHR.[20]

---

[18] In his book *The Concept of Law*, Oxford University Press, re-edited several times.

[19] *Vo v France*, App n° 53924/00, [2004] ECHR 326.

[20] See N Lenoir, 'La Déclaration Universelle sur le génome humain et les droits de l'homme de l'UNESCO', *Rapport public du Conseil d'Etat*, 1998.

With the growing number of conventions incorporating a self-contained system of adjudication, in particular in the field of the protection of the environment, it is to be foreseen that claims of state responsibility, whether brought before established courts or only before quasi-judicial organs such as the WTO Appellate Body, are likely to grow in number in the near future.

Against the background of the classical distinction between treaty law and customary international law, one could even argue, as far as the latter is concerned, that the key principle of respect for 'human dignity', at the core of any rule aimed at governing the use of biotechnologies,[21] already belongs to the category of *jus cogens*, ie, peremptory norms from which no state can derogate. The legal regime established by the International Law Commission's Draft Articles on state responsibility, endorsed by the UN General Assembly in December 2001, makes it possible, in principle, according to Article 48, for 'any State other than the injured State' 'to invoke the responsibility of another State' if 'the obligation breached is owed to the international community as a whole'.[22] This provision completes another in Article 42 of the same Draft, which provides that 'a State is entitled as an injured State to invoke the responsibility of another State if the obligation breached is owed to the international community as a whole and the breach of the obligation specially affects that State'.[23] These two provisions (Articles 48 and 42), by pointing to the legal regime of state responsibility for breach of obligations 'due to the international community as a whole', open interesting avenues for the future development of the international law of state responsibility in the context of the promotion of basic principles for regulating the use of biotechnologies.

To the extent that it is widely acknowledged that certain activities in this field, such as human cloning for reproduction purposes, are in evident conflict with respect for 'human dignity', then if they are undertaken in a given country this could trigger a reaction by any state 'directly affected', for instance through damage caused to one of its nationals (Article 42 of the ILC Draft) or even the invocation of its international responsibility by 'any other State' (Article 48 of the same Draft) acting individually or

---

[21] See Preamble, para 4, which states '-the recognition of the genetic diversity of humanity must not give rise to any interpretation of a social or political nature which could call into question 'the inherent dignity and . . . the equal and inalienable rights of all members of the human family', in accordance with the Preamble to the Universal Declaration of Human Rights'.

[22] See J Crawford, *The International Law Commission's Articles on State Responsibility, Introduction, Text and Commentaries* (Cambridge, Cambridge University Press, 2002), 276 ff.

[23] For a general comment of the ILC Draft on State Responsibility, see' Symposium: Assessing the Work of the International Law Commission on State Responsibility' (2002) 13 *European Journal of International Law*, in particular P-M Dupuy, 'General Stocktaking of the Connections between the Multilateral Dimension of Obligations and Codification of the Law of Responsibility' (2002) 13 *European Journal of International Law* 1053.

collectively. These states could argue that the country which carried out such wrongfull practices or let them develop by lack of due diligence would be responsible for the breach of a prohibition which is—at least in principle—established and recognised at a worldwide level.

Nevertheless, there are a number of reasons why one should certainly not overestimate the possibilities of such claims succeeding.

2. DIFFICULTIES PERSIST IN TRIGGERING INTERNATIONAL RESPONSIBILITY OF STATES IN THE FIELD OF BIOTECHNOLOGIES

There are at least three reasons why the use of the international responsibility of the state, as a social means of enforcing major principles of bioethics embodied in legal norms, will most probably remain limited in the future. The first reason deals with the very nature of some of the instruments which incorporate such rules. The second lies in the general, if not even vague, quality of the formulation of the basic principles aimed at regulating the use of biotechnologies and biological research. The third reason, even more important and problematical, is provided by the fact that the ethical, political and social background against which these obligations are established may lead to very different interpretations of their very content in today's multicultural world.

The first reason may be referred to as 'heterogeneous normativity', since some of the instruments are already included in 'hard law' instruments. This is the case, for instance, with the 2001 FAO Treaty on Plant Genetic Resources or the Oviedo convention on human rights and biomedecine and its additional protocols on transplantation of organs and on tissues of human origin and on biomedical research. In such a case, obligations set out in the convention are binding upon the states party to the convention, to the exclusion of third states. Other norms, or even the very same norms but at a larger or even universal level, are still laid down in 'soft law' instruments. This is the case in particular for the UDHG or the 2003 International Declaration on Guidelines for the Implementation of such Declaration.

Taking the content of these two Declarations into consideration, it could even be questioned whether the principles they incorporate are, as such, to be considered as *legal* principles in nature or whether, at this stage of generality and without further elaboration by international as well as national courts, they remain at a pre-legal phase of development, as they basically aim at inspiring bioethical codes of best practice for the deliberations of bioethical committees. Support for this opinion could probably be found in particular in Article 20 of the UDHG which provides that 'states should take appropriate measures to promote the principles set out in the Declaration, through education and relevant means, inter alia through the conduct of research and training in interdisciplinary fields and through the promotion of education in bioethics, at all levels, in particular for those responsible for science policies'.

The contrary opinion, according to which the terms of the 1997 Declaration are basically aimed at promoting true legal principles governing the use of biotechnologies, can also rely on other provisions of the same Declaration. In particular, its Preamble and the way in which parts A and B of the Declaration are drafted tend to indicate that the very purpose of its promoters was legal in nature. The Preamble explicitly refers to a series of international legal instruments starting with the Universal Declaration of Human Rights and the major conventions concluded within the framework of the UN for the protection of human rights. Articles 1 to 9 are drafted in such a way as to be read as statements of legal human rights principles. In reality, it seems that the Declaration contains both kinds of provisions, some aiming at the promotion of core human rights as they apply to the regulation of bioethics, others pointing more generally to the socio-political context in which such principles should be promoted.

Whatever the case may be, from a more general point of view, even if they are potentially able to create new legal rules, the provisions of any 'soft law' instrument need to be followed and completed by converging state practice and renewed expression of *opinio juris* in order to become new binding norms. Indeed, as mentioned in section 1, the process of developing customary law actually makes it possible to incorporate 'softly' enunciated principles into 'hard' customary international law. But this is a long and rather difficult process which makes it in particular unclear *from which period* onwards a determined principle aimed, in our case, at governing biotechnologies has become binding on *all* states at the universal level or even on *some* states at the regional level. Confronted with a socially complex set of issues, the international judge will tend to be careful to state in a clear-cut way, for instance, that the rule according to which genetic data associated with an identifiable person and stored or processed for the purposes of research must be held confidential has already attained the quality of a binding customary principle at the international level. Thus, if a court, whether national or international in character, refuses, as it would very often, if not always, to undertake the difficult task of demonstrating the crystallisation of a principle into a new international customary rule, the 'softness' of the instrument would probably prevail over the potential—but not proven—'hardness' of its content. The judge would then be led to the conclusion that the principles contained in this soft instrument remain optional.

A second element could be termed *excessive generality in formulation*. This obstacle to the use of international state responsibility in the field of bioethics no longer derives from the instrument incorporating the norms, but rather from the norms themselves, ie, the way in which they are drafted. This is particularly well illustrated by the terminology found in the UDHG. As the end result of the work of qualified experts as well as negotiations which have taken place between national delegations, this fundamental Declaration sets out important and even fundamental principles; but, being the product of many compromises, it does so in a general way which leaves much room for different interpretations. This is particularly true for Article 1, which provides that the human genome is 'the heritage of humanity' but is so only 'in a symbolic way'! This statement is no doubt of great importance, as it may be interpreted as prohibiting any appropriation of the human genome by any individual or group of people. But that this quality (common heritage) is so only in a 'symbolic way' seems restrictive indeed when compared, for instance, with the clear and unrestricted attribution of the sea bed to the 'common heritage of Mankind' in the UN Convention on the Law of the Sea adopted in Montego Bay in 1982.

The same could be said of Article 2 of the 1997 Declaration.[24] In particular, at paragraph (a), this text provides that 'everyone has a right to respect for their dignity and for their rights regardless of their genetic characteristics'.

Here again, this seems to be a very important and necessary provision. It remains the case that the concept of 'human dignity' may give rise to a variety of interpretations in relation to different types of conduct said to be either in conformity or in contradiction with it. But the very term 'everyone' leads to difficulties of interpretation, as shown by the judgment of the ECHR in *Vo v France*. Does the term 'everyone' apply to the embryo? And, if so, does it do so 'from its conception', or, absent such precision, from which period is the embryo to be considered as a true legal person endowed with rights and obligations?

In 2004, precisely because of the diversity of answers given by different European legislation, the European Court of Human Rights, while considering the term 'everyone' in the context of the 'right to life' set out in Article 2 of the ECHR, took the view that it was necessary to keep the whole issue within the 'margin of appreciation' that states parties to the Convention enjoy; this, 'notwithstanding an evolutive interpretation of the Convention, a "living instrument" which must be interpreted in the light of present-day conditions'.[25]

Article 2 of the UDHG is in itself a non-binding instrument and this would make things even more difficult: there is—*per essence*—no specific judge in charge of interpreting this instrument or of establishing the liability of any state for its violation (contrary to the case under the ECHR, which is

---

[24] See Lenoir, n 20 above.
[25] *Vo v France*, n 19 above at para 82.

a binding agreement endowed with its own system of adjudication). Any interpretation of this provision would then raise even greater difficulties if it were to be applied by an international judge.

Furthermore, even though established in a treaty in force, major principles governing the use of biotechnologies would *not* necessarily be considered, at the national level, as being by definition 'self executing', ie, directly applicable by the national judge of a Contracting Party to a situation where such a principle has allegedly been violated.

The difficulties mentioned above reflect an even more decisive obstacle to the efficiency of invoking the international responsibility of any state in the field of biotechnologies. This obstacle, which may be called that of 'cultural heterogeneity', is of a sociological[26] nature. Conceptions, perceptions of rules and, even more, ethical visions of the way in which the issues raised by the development of biotechnologies differ from one country to the other according to traditions, prevailing religion, history, and so on. For instance, though they agree on the major principles aimed at governing bioethics, countries as close to each other as the UK, France and Germany have not necessarily adopted the same views on the cloning of stem cells or on whether a human being's right to life begins from the time of its conception. The European Group on Ethics at Community level recommended that the Community authorities 'address these ethical questions taking into account the moral and philosophical differences, reflected by the extreme diversity of legal rules applicable to human embryo research'. And it added:

> It is not only legally difficult to seek harmonization of national laws at Community level, but because of lack of consensus, it would be inappropriate to impose one exclusive moral code.[27]

What is relevant at the European and Community level is even more so at the universal level. Absent a blatant violation of a basic principle of bio-ethics incorporated into international law, such as, for instance, the prohibition on reproductive cloning of human beings, it makes it extremely difficult for any judge to impose a single conception on any state without the clear consent of that state to the same perception of the content and bearing of the concerned rule.

3. TOWARDS NEW FLEXIBLE FORMS OF SOCIAL RESPONSIBILITY?

The absence of a high degree of social consensus at the global, or even regional, level among states on the content and interpretation of basic

---

[26] See eg, J Rifkin, *The Biotech Century: Harnessing the Gene and Remaking the World* (New York, 1998).

[27] Cited by the ECtHR in *Vo v France*, n 19 above, at para 82.

principles to be considered as legally binding is apparently the most serious obstacle to using the international responsibility of states as an efficient tool for ensuring respect for these principles. There are, nevertheless, forms of social retribution (positive or negative) of conduct other than that formally organised within any legal system. The promotion of the corporate social responsibility of private investors, for instance, is an illustration of the way in which active participants in the life of international civil society may play a decisive role in the setting up of new national and international standards aimed at the establishment of a socio-ethical framework, which is increasingly providing private corporations with clear guidelines for the protection of basic human rights and rights of people.

Responsible corporate behaviour is a requirement which has primarily been articulated by non-governmental organisations. They have denounced, and continue to denounce, initiatives taken by private companies which ignore basic human rights. One can consider, for instance, the way in which Nike produced sneakers or footballs, without taking into consideration that they were manufactured by children working more that 10 hours a day in extreme conditions. By organising a boycott on products of this firm, they forced the company in a rather efficient way to change its methods of production. It is mainly under the pressure exerted by such groups that the behaviour of private corporations has been modified and that states have been led to amend their legislation to bring it increasingly in conformity with respect for human dignity.

It is not clear how far this may happen with respect to biotechnologies, but it is almost certain that evolving mentalities are a precondition for using, in a second phase, the most classical tools offered by the law, such as the responsibility of public entities and private actors interested in biotechnologies. Such changes will be greatly encouraged by the diligence of civil society, in every part of the world including the South, in prohibiting any attempt to use such technologies in a way incompatible with a clear vision of what everyone's right to human dignity means. As Professor R Brownsword clearly demonstrates in this volume, the regulation of biotechnologies can only marginally be achieved by law. But the law will remain a necessary tool for consolidating the cosmopolitan commitment to respect for basic principles of bioethics.[28] The process, however, is well under way.

---

[28] See also R Brownsword, 'Regulating Human Genetics: New Dilemmas for a New Millennium' (2004) 12 *Medical Law Review* 14.

Part II

Bioethics and Human Genetics

3

Ethical Pluralism and the Regulation of Modern Biotechnology

ROGER BROWNSWORD*

1. INTRODUCTION

AN INVITATION TO write on the topic of 'Ethics and Law in the Regulation of Modern Biotechnology' might be taken up in many different ways.[1] It might prompt thoughts about regulatory legitimacy and regulatory effectiveness (and the relationship between the one and the other), about the regulability of a fast-moving technology, about the inter-facing of international, regional and local spheres of regulation, and even about the distinctive nature of 'law' as a regulatory modality.[2] However, my focus is much narrower, concerning only the bearing of ethical pluralism on the regulation of modern biotechnology (especially its applications with regard to medical and reproductive research and treatment).

In keeping with the spirit of the chapters in this volume, let me take as my starting point the dual proposition that (i) no matter how modern biotechnology is regulated (whether by legal or non–legal modalities), such regulation must be ethically (morally) defensible[3] and (ii) that it is modern

* Professor of Law at King's College London and Honorary Professor of Law at the University of Sheffield. Email: roger.brownsword@kcl.ac.uk. This is a revised version of a paper given at a workshop on 'The Impact of Biotechnologies on Human Rights' held at the European University Institute, Florence, on 3 June 2005.

[1] For a general survey of the global landscape with regard to regulation and technology (particularly biotechnology), see R Brownsword, 'What the World Needs Now: Techno-regulation, Human Rights and Human Dignity' in R Brownsword (ed), *Global Governance and Human Rights* (Oxford, Hart Publishing, 2004), 203.

[2] Seminally, see L Lessig, *Code and Other Laws of Cyberspace* (New York, Basic Books, 1999). For some remarks about regulatory modalities in connection with biotechnology, see R Brownsword, 'Red Lights and Rogues: Regulating Human Genetics' in H Somsen (ed), *Regulating Biotechnology* (Cheltenham, Edward Elgar, (forthcoming); and, more generally, concerning the choice between transparently prescriptive East Coast 'law' and embedded West Coast 'code', see R Brownsword, 'Code, Control, and Choice: Why East is East and West is West' (2005) 25 *Legal Studies* 1.

[3] There is, it should be noted, a difference between (i) the strong demand that all ethical requirements should be translated into positive legal requirements and (ii) the weaker

human rights thinking (or, at any rate, a rights-led approach that is very close to human rights thinking) that offers the most defensible ethic for this purpose.[4] To be sure, we should not assume that this dual proposition will command acceptance in all regulatory settings—after all, regulators are geared to respond to a range of stakeholder interests[5] and regulatees are liable to be driven by personal or sectional interests (especially economic interests) rather than by broader ethical commitments. Nevertheless, if the dual proposition is accepted, it implies a commitment to the principle that the regulation of biotechnology, whether of local, regional or international application, should be compatible with human rights.[6]

At international level, such an ethico-regulatory principle commands strong support. Characteristically, in its recent work on what is expected to become the Universal Declaration on Bioethics and Human Rights,[7] the drafting group of UNESCO's International Bioethics Committee (the IBC) stresses 'the importance of taking international human rights legislation as the essential framework and starting point for the development of bioethical principles'.[8] Applying such a philosophy to the regulation of biotechnology, human rights set limits to the otherwise unbridled development and application of the technology, constraining even well-intentioned development and application of this kind. As the Director-General of UNESCO, Koïchiro Matsuura, put it when opening an

demand that such positive legal provisions as there are should be ethically defensible. On the latter, see the remarkable judgment given by Lord Hoffmann in *Airedale NHS Trust v Bland* [1993] 1 All ER 821, at 850.

[4] The most defensible ethic, I believe, is a will theory of rights based, not on humanity, but on agency. This is the thesis seminally argued for in A Gewirth, *Reason and Morality* (Chicago, ILL, University of Chicago Press, 1978) and elaborated in *Community of Rights* (Chicago, ILL, University of Chicago Press, 1996).

[5] See, eg, A Mundy, *Dispensing with the Truth* (New York, St Martin's Press, 2001) which describes the marketing of the Fen-Phen diet drug and the failure of the FDA to exercise adequate regulatory oversight when alerted to the risks presented by the drug. Essentially, the story is of two cultures within the FDA: one focused on safety and user protection, the other treating the pharmaceutical companies as clients whose products are to be brought to market.

[6] In this chapter, I am focusing on human genetics. However, modern plant biotechnology would be covered by the same regulatory injunction. From a human rights perspective, there is no justification, for eg, for putting at risk the welfare of vulnerable Third World humans in order to promote the economic wellbeing of First World humans or to facilitate the making of informed choices with regard to the food consumed in Europe: see Nuffield Council on Bioethics, *The Use of Genetically Modified Crops in Developing Countries* (London, Nuffield Council on Bioethics Jan 2004).

[7] The UNESCO Preliminary Draft Declaration on Universal Norms on Bioethics (Paris, 9 Feb 2005), SHS/EST/CIB-EXTR/05/CONF.202/2 was taken forward as a Universal Draft Declaration on Bioethics and Human Rights (Paris, 24 June 2005), SHS/EST/05/CONF.204/3 REV.

[8] UNESCO Explanatory Memorandum on the Elaboration of the Preliminary Draft Declaration on Universal Norms on Bioethics (Paris, 21 Feb 2005), SHS/EST/CIB-CIGB/05/CONF.202/4.

international Round Table of Ministers of Science in October 2001, there is (within the group) a:

> firm commitment to international solidarity in scientific progress, and to safe-guarding human rights and human dignity from the misuse of science and technology, particularly in the life sciences[9]

This has now been said and repeated so many times that I take it as axiomatic that, for the IBC as for many other international agencies, bioscience and biotechnology should be regulated in such a way that human rights and human dignity are fully respected.[10]

Given such a cosmopolitan commitment to human rights, and assuming that this commitment cascades down to all regulatory levels, everything seems relatively straightforward: the injunction to regulators is, quite simply, to put in place regulatory frameworks that enable humans to benefit from the development and application of biotechnology while, at the same time, ensuring that human rights are fully respected. Of course, the declaration of such a regulatory objective does not also determine which regulatory strategy should be adopted. Smart regulators, charged with

---

[9] UNESCO, *Bioethics: International Implications* (Paris, UNESCO, 2003), 2.

[10] The work of the IBC in relation to what is now the Universal Draft Declaration on Bioethics and Human Rights (n 7 above) underlines this axiom. In Oct 2003, the General Conference of UNESCO, having resolved that preparatory work on a 'Declaration on Universal Norms on Bioethics' should continue, agreed that it was 'opportune and desirable to set universal standards in the field of bioethics *with due regard for human dignity and human rights* and freedoms, in the spirit of cultural pluralism inherent in bioethics' (32C/Res. 24) (emphasis added). The first draft of the Declaration, published in June 2004, was peppered with references to human rights and human dignity; and one of the general principles (headed 'Human Dignity, Human Rights and Justice') read: '[a]ny decision or practice in the field of bioethics [at all levels] shall be made in the respect [sic] of the human dignity [sic] and in accordance with the universal principles of justice, human rights and fundamental freedoms'. In a second draft, prepared for discussion at the Eleventh Session of the IBC in Paris in Aug 2004, the emphasis on respect for human rights and human dignity persisted. So, eg, the declared aims included 'to ensure the respect for human dignity and the protection of human rights and fundamental freedoms' and 'to recognize the great benefit derived from scientific and technological developments whilst ensuring that such development occurs within the framework of ethical principles that respect human dignity and protect human rights and fundamental freedoms, and to prevent practices contrary to human dignity'. Art 3(iii) of the Preliminary Draft Declaration expresses this aim in similar terms as follows: 'to recognize the importance of freedom of scientific research and the benefits derived from scientific and technological developments, whilst ensuring that such developments occur within the framework of ethical principles that respect human dignity and protect human rights and fundamental freedoms'. See A Yusuf in this volume. In the June 2005 draft Declaration, the Preamble rehearses the importance of checking the applications of (inter alia) biotechnology for their compliance with respect for human rights and human dignity; Art 2 includes in its statement of aims promotion of respect for human dignity and protection of human rights (Aim (iii)) as well as recognition of 'the importance of freedom of scientific research and the benefits derived from scientific and technological developments, while stressing the need that such research and developments . . . respect human dignity, human rights and fundamental freedoms' (Aim (iv)); and Art 3(a) provides that '[h]uman dignity, human rights and fundamental freedoms are to be fully respected'. For discussion of the shift of frame from bioethics to human rights, see H Boussard in this volume.

human rights objectives, will mix and match their regulatory interventions in whatever way seems most appropriate.[11] If non-legal modalities (say, some form of self-regulation or amplified and orchestrated social pressure) seem more effective than traditional legal prescriptions, then the former will be (rightly) preferred (provided that they are not incompatible with human rights). Where traditional legal forms are employed, there are again smart choices to be made. For instance, regulators may adopt a legal model that ring-fences human rights in a general way—such as a patent regime that provides for an exclusion against patentability where the exploitation of an invention would be inconsistent with respect for human rights; or the ring-fencing may be more specific—for example, reference may be made to particular human rights (say, to the right of privacy in the context of the circulation of genetic information) or to a particular activity that is judged to violate a background human right (say, to a requirement of genetic testing that is made by insurers or employers). Such issues of regulatory design are important.[12] However, as I have already indicated, they are not the issues that I propose to address in the present chapter.

Rather, my focus in this chapter takes us back to the ethical context in which we find it being asserted that human rights should set the regulatory agenda. Even allowing for the impact of globalisation, it would be a nonsense to claim that we have, as it were, reached the era of the end of ethics, that human rights is now the only available ethical approach. Far from it, in practice, we find that the ethic of human rights is just one in a plurality of competing views. Indeed, even in the familiar commitments of the IBC—to the advancement of modern biotechnology for the benefit of all humans, while at the same time fully respecting human rights and human dignity—we have three competing ethical approaches in play.

In the first part of the chapter (section 2), I will sketch the three principal constituents in this plurality before going on, in the second part (section 3), to indicate some of the ways in which the plurality obstructs the articulation and application of regulatory regimes that are fully in line with defending the integrity of human rights.[13]

2. SINGULARITY OR PLURALITY?

In the ideal-typical context (presupposed by the dual proposition at the start of this chapter), the legitimacy of regulation has to answer to just one

---

[11] See N Gunningham and P Grabosky, *Smart Regulation* (Oxford, Clarendon Press, 1998).

[12] See, eg, S Leader, 'Collateralism' in R Brownsword (ed), *Human Rights* (Oxford, Hart Publishing, 2004), 53; and DJ Galligan in this volume.

[13] Compare the excellent discussion in A Plomer, *The Law and Ethics of Medical Research—International Bioethics and Human Rights* (London, Cavendish, 2005), esp chaps 1 and 2.

ethical constituency, that of human rights. Such is singularity. However, this is not the context that actually obtains for the regulation of modern biotechnology. Indeed, one of the distinctive characteristics of this particular regulatory space is that a plurality of ethical views is pressed upon regulators. Having said this, it is as well to repeat that the ethical plurality is itself situated within a much larger plurality of competing stakeholder interests—it is a plurality within a plurality. In other words, to assert a human rights perspective, we first have to argue for an *ethical* approach (as against various non-ethical approaches), and then we have to argue for human rights against its rivals *within* the ethical plurality.

The ethical plurality, the 'bioethical triangle'[14] as I will term it, is made up of three ethical constituencies: utilitarians who advocate the pursuit of human welfare (human health, wealth and happiness); human rights theorists; and a constituency that I have previously termed 'the dignitarian alliance'.[15] Each of these constituencies has its own distinctive ethical perspective. Occasionally, the perspectives converge to invite regulators to act on a consensus (as is the case at present with human reproductive cloning); sometimes, a synthesis of utilitarian and human rights thinking will emerge;[16] but, too often for regulatory comfort, these perspectives generate competing and conflicting views (as is the case, for instance, with therapeutic cloning and human embryonic stem cell research).[17]

Neither utilitarian nor human rights thinking requires any extended introduction. Utilitarians count utility (encompassing individual pleasure and preference satisfaction, and the like, together with convenience and economy), the sum of utilities being aggregated in the credit column; and they count disutility (encompassing individual pain, suffering and

[14] See R Brownsword, 'Three Bioethical Approaches: A Triangle to be Squared', paper presented at international conference on the patentability of biotechnology organised by the Sasakawa Peace Foundation, Tokyo, Sept 2004 (available at www.ipgenethics.org/conference/transcript/session3.doc).

[15] See R Brownsword, 'Bioethics Today, Bioethics Tomorrow: Stem Cell Research and the 'Dignitarian Alliance' (2003) 17 *Notre Dame Journal of Law, Ethics and Public Policy* 15.

[16] See R Brownsword, 'Regulating Human Genetics: New Dilemmas for a New Millennium' (2004) 12 *Medical Law Review* 14.

[17] All 191 members of the United Nations support a prohibition on human reproductive cloning. However, after four years of deliberation, during which time efforts have been made to achieve a consensus covering the regulation of all uses of cloning technology (reproductive and therapeutic) in humans, the nations remain divided. On 18 Feb 2005, the Legal Committee voted 71 in favour, 35 against, with 43 abstentions, to recommend to the General Assembly that members should be called on 'to prohibit all forms of human cloning inasmuch as they are incompatible with human dignity and the protection of human life' (UN press release GA/L/3271, available at www.un.org/News/Press/docs/2005/gal3271.doc.htm). On 8 Mar 2005, the General Assembly accepted this recommendation, 84 members voting in favour of the (non-binding) UN Declaration on Human Cloning, with 34 against and 37 abstentions (UN press release GA/10333, available at www.un.org/News/Press/docs/2005/ga10333.doc.htm). For discussion, see R Brownsword, 'Stem Cells and Cloning: Where the Regulatory Consensus Fails' (2005) 39 *New England Law Review* 535.

distress, and the frustration of preferences and the like, together with cost, inconvenience and the general expenditure of resources), such disutilities being totalled in the debit column. For utilitarians, the maximisation of utility and the minimisation of disutility are all that counts. Once the entries have been made, the columns added and the full range of options compared, it does not need a high-powered computer to confirm the logic of utilitarian computation.

By now, we are all familiar with the tension between utilitarian promotion of the general good (where little or no attention is paid to the *distribution* of utility or disutility) and the constraints imposed if individual rights are to be taken seriously.[18] Where human rights has made its mark, it is axiomatic that best practice demands careful attention to free and informed consent, that the capacity for autonomous decision-making should be respected, that privacy and confidentiality should be protected, and so on. However, if bioethics was once a two-way contest between utilitarians and human rights theorists, this is no longer the case.

Where a technology impacts on the human body, as is particularly the case with the human genetics applications of biotechnology, this is widely seen as raising concerns about human dignity. Even in the case of information and communication technology which, issues of (Internet) content regulation apart, largely raises questions within and between utilitarian and human rights thinking, we find concerns about human dignity being expressed once a bio-application is proposed—witness, for example, the recent discussion by the European Group on Ethics in Science and New Technologies with regard to ICT implants in the human body.[19] At all events, what we now have is a third ethical constituency, a determined alliance of dignitarian views, making up the plurality.[20] This perspective condemns any practice, process or product—human reproductive cloning, therapeutic cloning and stem cell research using human embryos being prime examples[21]—which it judges to compromise human dignity. Such condemnation (by reference to human dignity)

---

[18] For some vivid examples of researchers putting their projects before their research participants, see Plomer, n 13 above, chap 2 (US human radiation experiments), chap 3 (UK Porton Down experiments), and chap 6 (research trials in developing countries).

[19] Opinion No 20 of the European Group on Ethics in Science and New Technologies to the European Commission, *Ethical Aspects of ICT Implants in the Human Body* (adopted 16 Mar 2005).

[20] Compare R Brownsword, 'Biotechnology and Rights: Where are We Coming from and Where are We Going?' in M Klang and A Murray (eds), *Human Rights in the Digital Age* (London, Cavendish Glasshouse, 2005), 219; and S Millns in this volume.

[21] See, eg, JMZ. Makdisi, 'The Slide from Human Embryonic Stem Cell Research to Reproductive Cloning: Ethical Decision-making and the Ban on Federal Funding' (2003) 34 *Rutgers Law Journal* 463.

operates as a 'conversation stopper';[22] but the dignitarians are not troubled—to say that something violates human dignity is the ultimate condemnation. The emergence of the new dignitarian view creates a genuinely triangular contest, the dignitarians disagreeing as much with the utilitarians as they do with the human rights constituency—with the former because they do not think that consequences, even entirely 'beneficial' consequences (that is, 'beneficial' relative to a utilitarian standard), are determinative; and with the latter because they do not think that informed consent cures the compromising of human dignity.

Somewhat confusingly, human dignity is a key idea in two corners of this three-way contest.[23] Utilitarians can squeeze some disutility out of concerns about human dignity, but the idea that we should disallow a practice because human dignity is compromised is not a prominent feature of utilitarian thinking. With the human rights constituency, however, matters could scarcely be more different. Human rights theorists believe that the entire apparatus of human rights is premised on the principle of respect for human dignity. This premise is written into the historic human rights instruments of the mid-twentieth century. According to the Preamble to and Article 1 of the Universal Declaration of Human Rights, the fundamental premise is that we each have inalienable and intrinsic human dignity. This is why we have human rights. The jump from human dignity to human rights actually bears closer examination, but it is the move that is standardly made: if the question is why we have human rights, the stock answer is: '[b]ecause we have human dignity'.[24]

In the dignitarian corner, too, respect for human dignity is fundamental, but not as the underpinning of human rights and individual autonomy. Drawing on a mixture of Kantian, Catholic and communitarian credos, this constituency registers its discomfort with various aspects of new technology by contending that human dignity is compromised. If we value a rights-driven conception of autonomy, this is bad news; if we take a utilitarian view, it is also an annoyance; but if we fear, say, genetic discrimination or if we sense a certain insouciance about the way in which human embryos are committed for research, the bold red lines drawn by the new dignitarianism may have some appeal.

---

[22] See D Birnbacher, 'Do Modern Reproductive Technologies Violate Human Dignity?' in E Hildt and D Mieth (eds), *In Vitro Fertilisation in the 1990s* (Aldershot, Ashgate, 1998), 325. Compare Nuffield Council on Bioethics, *Genetically Modified Crops: the Ethical and Social Issues* (London, Nuffield Council on Bioethics, 1999), 96 (those who contend that genetic modification is intrinsically wrong or unnatural present views that 'have something of an "unarguable" quality, inasmuch as no amount of information, explanation or rationalisation would move a person with such views from their position'.).

[23] See D Beyleveld and R Brownsword, *Human Dignity in Bioethics and Biolaw* (Oxford, Oxford University Press, 2001); D Feldman, 'Human Dignity as a Legal Value' [1999] *Public Law* 682 and 61–76; and Plomer, n 13 above, chap 4.

[24] See Brownsword, n 15 above.

3. THE SIGNIFICANCE OF THE PLURALITY

What are the implications of the ethical plurality for the aspiration that the regulation of biotechnology should be compatible with respect for human rights? In what follows we can consider three kinds of case in which respect for human rights is not focal and, in consequence, the ethic loses ground in the contest for regulatory endorsement. In the first type of case, human rights simply fails to register or be sufficiently supported and properly recognised in the pre-legislative debates (the problem of non-recognition); in the second, human rights is recognised but there is some slippage in the drafting (the drafting deficit); and, in the third, the slippage is in the interpretation of the regulation as drafted (the interpretive deficit). Relative to a human rights perspective, each of these cases represents a particular example of regulatory failure.

Before discussing these three problems/deficits (concerning pre-legislative (non-)recognition, drafting and interpretation), it is as well to speak briefly to the idea of regulatory failure. Regulatory failure, like regulatory success, is a matter of degree; but, whether we are plotting success or failure, we will start measuring it in terms of regulatory effectiveness. We can start with a measurement model that is largely content-neutral (at any rate, it is not loaded with a particular substantive regulatory content, ethical or otherwise, let alone specifically a human rights content); but, before we can characterise our three problems/deficits as cases of regulatory failure, we have to introduce a specifically human rights content into our measurement model.

If we frame a model of regulatory process in neutral terms, we can identify four key stages in the regulatory cycle at which failure or success can be tested. These stages are:

— (stage one) the identification of a recognised or authoritative regulator (particularly where new technologies emerge, or where there are cross-border disputes, it might not always be clear who has authority);[25]
— (stage two) the issuing of 'guidance'[26] by a recognised regulator;
— (stage three) the response of regulatees to the guidance issued (whether the guidance is in the nature of a requirement or prohibition, or a permissive facilitation); that is, whether or not regulatees act on, or comply with, the guidance; and

---

[25] Compare, eg, the early experience with domain name disputes and a range of other cross-border Internet issues (eg, *LICRA v Yahoo! Inc.*, where it was claimed that internet auction sites, hosted by a US-based ISP, violated French law by offering Nazi memorabilia for sale: see DW Vick, 'Regulating Hatred' in Klang and Murray (eds), n 20 above, 41).

[26] Regulatory guidance may be of more than one orientation, protective as well as facilitative. Drawing on the logic of a rights-based ethic, the three essential guiding directions are prohibition, permission and requirement.

— (stage four) the response made by regulatory agencies if and
when regulatees do not act on or comply with the guidance; that
is, whether or not remedial steps are taken (whether by way of
enforcement or by making adjustments to the guidance).

So stated, these key stages leave a great deal to be unpacked. In particular, it is implicit in the second stage that the guidance issued (whatever
its particular content) is at least clear and intelligible, that it coheres with
other guidance that has been issued, and that it is properly communicated
to regulatees;[27] and, at the third stage, it is implicit that effective regulation presupposes an accurate monitoring of compliance. Once we add in
to this content-neutral model a specific human rights ethic for regulatory
projection, we can identify the following critical points.

Assuming that we have a recognised regulator in place, the first
opportunity for regulatory failure is at and around stage two, the
determination of the regulatory content and the issuing of the regulatory
guidance. One problem is that the regulator may fail to respond to the
human rights case; it may simply fall on deaf ears. Another is that, even
if the regulator intends to issue guidance that is designed to promote the
human rights view, the drafting may be unsatisfactory. This is not just
a matter of being clear, of giving an unequivocal signal; what matters
is that the right signal (the human rights signal) is clearly given. If the
guidance is defective in its drafting, this stores up problems for those
who seek to act on the guidance—whether they are regulatees who wish
to comply, regulatory agencies which wish to apply, or interpreters who
wish to give effect to the guidance. This is particularly problematic
for those who are committed to the ethic of human rights, a drafting
deficit being unhelpful insofar as it allows in the voices of opposition. It
follows that where there is a drafting deficit, judgments about regulatory
effectiveness at stages three and four will turn on how well the problems
set in train by the defective drafting are handled. Is it possible to limit
the damage or even rectify the drafting? Finally, even if the drafting
is properly attuned to human rights considerations, there is still the
possibility that it may be misinterpreted or misapplied. Accordingly, it
is in this context that we can address the potentially distorting effect of
the ethical plurality.

3.1 The Problem of Non-recognition

In a plurality, even in an ethical plurality, human rights will not always
prevail. Where human rights has to compete with both utilitarian and

---

[27] Compare the seminal analysis in L Fuller, *The Morality of Law* (New Haven, Conn, Yale
University Press, 1963); and see, too, K Yeung, *Securing Compliance: A Principled Approach*
(Oxford, Hart Publishing, 2004), at 33–4.

dignitarian views, there will be different winners and losers in different places at different times. Where a community is constituted by commerce, utilitarian views are liable to prevail; and, where a community has a strong cultural identity, the dignitarian view may be dominant. On occasion, therefore, regulation may be drafted, quite intentionally, in a way that declines to recognise a particular human rights argument. This does not entail that all is lost from a human rights perspective; but it does mean that regulatory failure is already implicated in the regulatory baseline. Let me put some flesh on this with some examples.

One place where utilitarian ethics puts serious pressure on human rights is in setting the regulatory framework for the operation of the criminal justice system (and, by extension, the terms of exceptional provisions designed to respond to threats to national security). In liberal democracies, committed to the principles of due process but also concerned to manage crime in an efficient and effective manner, there is a constant tension between the background ethics of utilitarianism and human rights. Where biotechnology has applications in the criminal justice system, and particularly where its applications are believed to be unusually powerful, there will be a temptation to run with utilitarian thinking. For example, the development of DNA profiles and databases is seen as an important forensic advance, enabling the police to match suspects to crimes and improving the chances of the prosecution proving its case beyond reasonable doubt. From a human rights perspective, such moves are to be welcomed so long as they serve to protect rights-holders who would otherwise be the victims of crime and to identify those who have engaged in serious violations of rights; but implementation of such measures must also respect the rights of those who are presumed to be innocent—hence the trenchant criticism of the UK regulatory framework which authorises the retention of 10-marker DNA profiles even though the person has not actually been prosecuted for, let alone convicted of, an offence.[28] Similarly, if access to the full underlying samples (which are also retained under UK law) is not restricted or regulated in a way that satisfies human rights considerations (of privacy and confidentiality), there is additional cause for concern.[29] At all events, as biotechnology (sometimes in conjunction with ICT) continues to produce ever more effective ways of biometrically identifying, tracking and locating citizens, we cannot expect that proponents of human rights will have it all their own way in turning back the tide of utilitarian thinking.

---

[28] H Kennedy, *Just Law* (London, Chatto and Windus, 2004).
[29] For critical comments concerning the oversight of the National DNA Database in the UK, see HC 96–1, Science and Technology Committee, *Forensic Science on Trial* (Seventh Report of Session 2004–05) (London, TSO, 2005), chap 4.

When we switch from the context of criminal justice to that of health care, we find a rather different pattern. If the utilitarian ethic tends to represent the default position in the former, it is the ethic of human rights that tends to be the default in the latter. Such has been the influence of bioethics that it is relatively unusual to find regulators displacing human rights in order to allow medical research or treatment to be pursued without the free and informed consent of patients and participants, or without due regard to their rights of privacy and confidentiality. However, one notable (and, not surprisingly, much criticised) exception is section 60(1) of the Health and Social Care Act, 2001, according to which:

> The Secretary of State may by regulations make such provision for and in connection with requiring or regulating the processing of prescribed patient information for medical purposes as he considers necessary or expedient: (a) in the interests of improving patient care, or (b) in the public interest.

The import of this controversial provision is spelt out in paragraph 291 of the explanatory notes accompanying the legislation:

> This section enables the Secretary of State to make regulations for and in connection with requiring or regulating the processing of patient information in prescribed circumstances. This will make it possible for patients to receive more information about their clinical care and for confidential patient information to be lawfully processed *without informed consent* [my emphasis] to support prescribed activities such as cancer registries. The Government places importance on the consistent use of informed consent as the basis for handling confidential patient information. The regulation-making power in this section is therefore intended to provide for exceptional situations where essential services cannot, having regard to the present NHS systems and available technology, operate on that basis.

It is one thing for information to be circulated, in practice, without proper attention to respect for privacy and confidentiality, or for informed consent to be handled in an entirely perfunctory fashion; but it is unusual for such displacement to be expressly authorised—but, in a plurality, this can and does happen.

Another example concerns the rights of those who donate biological samples for large-scale public health projects such as that of the UK Biobank.[30] After the much-publicised criticism of the Icelandic De-Code project, regulators are likely to insist that participation must be on the basis of free and informed consent.[31] However, this does not, in itself, reach the vexed question whether those who so participate have property

---

[30] See www.ukbiobank.ac.uk; and, for comment, see JV McHale, 'Regulating Genetic Databases: Some Legal and Ethical Issues' (2004) 12 *Medical Law Review* 70.

[31] See, eg, L Schulz, 'Genetic Data Banks' (2001) 9 *Jahrbuch für Recht und Ethik* 43; but compare V Árnason, 'Coding and Consent: Moral Challenges of the Database Project in Iceland' (2004) 18 *Bioethics* 27.

rights of any description in their biological samples either *ex ante* (ie, prior to participation) or *ex post* (ie, after enrolment and once the materials have been banked).[32] The *ex ante* issue is the crucial one; and, from any ethical standpoint, there are some poor arguments in circulation on this particular question. For instance, those who advocate in favour of an *ex ante* property right in one's own removed body parts and samples cannot derive such a right from the settled and undoubted right to bodily integrity; for it is clearly a non sequitur to hold that, because A has a claim-right to his bodily integrity, it follows that A has a *property* right in relation to his removed body parts.[33] Even if B does wrong by lopping off A's arm (against A's will), this does not entail that A has proprietary rights over the removed arm. Conversely, those who advocate against an *ex ante* property right cannot make out the case by relying on the settled and undoubted understanding that donation involves an *ex post* abandonment of property. Clearly, A, a donor, has no property rights *ex post* (otherwise this would not be a case of donation); but this says nothing about whether A has property rights in removed body parts *prior* to donation—indeed, the argument might be turned on its head, on the basis that, strictly speaking, A could not be a 'donor' (merely a possessor who transfers possession) without an *ex ante* property title.

For present purposes, it suffices to say that there is at least a plausible human rights argument in favour of recognising an *ex ante* property right and that, if such a right were recognised, there might be three significant practical effects. First, participants might be less willing to donate their biological samples; the limits of altruism would be put to the test. Secondly, participants might be willing to donate their biological samples provided that a satisfactory benefit-sharing agreement was in place. Thirdly, participants might be willing to donate their biological samples provided that they retained a measure of control over the way that the biobank was developed as a research resource, including who would have access to the data and on what terms. From a utilitarian standpoint, each of these outcomes has a cost; and, from a dignitarian perspective, it is unacceptable that donors should be seeking to profit by exploiting the value of their biological materials; commodification (or commercialisation) of the human body is one of the paradigmatic examples of the compromising of human dignity.

In the light of these remarks, it will come as no surprise to find that, under the influence of a combination of utilitarian and dignitarian

---

[32] For excellent philosophical discussion, see SR Munzer, *A Theory of Property* (Cambridge, Cambridge University Press, 1990), esp. at 37–58.

[33] As pointed out in JW Harris, 'Who Owns My Body?' (1996) 16 *Oxford Journal of Legal Studies* 55.

thinking,[34] the prevailing view is that no such *ex ante* property rights should be recognised on the part of donors. Most famously, in *Moore v Regents of the University of California*,[35] the lobby against recognition of such rights was getting its way before the case was settled; and, more recently, in *Greenberg v Miami Children's Hospital Research Institute, Inc*,[36] the view (from *Moore*) that there is no property in one's own body parts has been followed.[37] Extending this reasoning to the *ex post* position, under the governance rules for the UK Biobank,[38] one of the points for the information of participants is as follows:

> The fact that the UK Biobank will be the legal owner of the database and the sample collection, and that participants have no property rights in the samples.[39]

This point is underlined by emphasising that participants will not be offered any significant financial or material inducement to participate.[40] And, as if to reassure the research community, it is reiterated that participants 'will not have property rights in the samples'.[41]

While donors are to have no proprietary entitlements, it is accepted that it is perfectly appropriate for those who carry out research on such materials to assert proprietary rights (just as the Biobank does) of both a physical and intellectual nature. Putting this in utilitarian terms, the issue is not so much whether property rights should be recognised; rather, the issue is whether recognising *ex ante* property rights (which might then force concessions in relation to *ex post* interests) promises to maximise the utility of the research facility, to incentivise the research community, and to encourage commercial investment, and the like. If the utilities favour downstream users being free to proceed without upstream proprietary

---

[34] Eg, Art 21 of the Council of Europe's Convention on Human Rights and Biomedicine provides: '[t]he human body and its parts shall not, as such, give rise to financial gain'—trenchantly criticised by C Delkeskamp-Hayes, 'Respecting, Protecting Persons, Humans and Conceptual Muddles in the Bioethics Convention' (2000) 25 *Journal of Medicine and Philosophy* 147, at 162–4. For discussion and, in particular, analysis of the significance of Art 22 (which requires 'appropriate information and consent procedures' for secondary purposes in relation to removed body parts), see D Beyleveld and R Brownsword, n 23 above, chap 8.

[35] *Moore v Regents of the University of California* 249 Cal Rptr 494 (1988); 271 Cal Rptr 146 (1990), 793 P2d 479 (1990); cert. denied 111 S Ct 1388 (1991).

[36] See 2002 WL 1483266 (ND Ill); 264 F Supp 2 1064 (SD Fla 2003) (02-22244-CIV-MORENO (Miami)). For comment, see G Laurie, *Genetic Privacy* (Cambridge, Cambridge University Press, 2002), at 322–3; and F Bellivier and C Noiville, 'The Commercialisation of Human Biomaterials: What are the Rights of Donors of Biological Material?', paper presented at workshop on 'Property in Human Tissue', University of Tübingen, 21–22 Jan, 2005.

[37] It should be said, however, that insofar as the reasoning relies on the effect of donation (and, thus, focuses on the *ex post* issue), it is unconvincing.

[38] See UK Biobank Ethics and Governance Framework (Version 1.0, 24 Sept 2003).

[39] Ibid, at 9.

[40] Ibid, at 14.

[41] Ibid, at 18.

complications, and if regulators are so persuaded, then whatever the detail of the human rights case the argument will have been lost.[42]

A further example is the claimed right *not* to know (about one's own genetic make-up). This is not a right not to be told *tout court*; it is a right to shield oneself against the receipt of information where receipt is against one's will—a right, in other words, that is akin to a right to refuse to accept unsolicited marketing calls, junk mail, or spam messages, and so on. Gilbert Hottois has denounced such a right, linking it to 'traditional—medieval—scientophobic and anti-progressivist beliefs, and to a nostalgia for the paternalism of oligarchies which keep the monopoly of knowledge and the mission of dispensing appropriate fractions of that knowledge to the people'.[43] However, viewing the claim from a human rights perspective, it seems to me that a plausible case might be made out that such a right is immanent within the privacy or autonomy interest; and I will proceed on the assumption that, from a human rights standpoint, an arguable case for the right not to know may be made out.

In contrast with the claimed property right, much more success has been enjoyed in registering a right not to know. So, for instance, Article 10(2) of the Convention on Human Rights and Biomedicine provides that 'the wishes of individuals not to be so informed [ie, about their health] shall be observed'; Article 5(c) of the UNESCO Universal Declaration on the Human Genome and Human Rights states that '[t]he right of every individual to decide whether or not to be informed of the results of genetic examination and the resulting consequences should be respected'; and this is echoed by Article 10 of the International Declaration on Human Genetic Data which provides that [w]hen human genetic data, human proteomic data or biological samples are collected for medical and scientific research purposes, the information provided at the time of consent should indicate that the person concerned has the right to decide whether or not to be informed of the results'—and, what is more, this Article contemplates the right not to be informed being extended to 'identified relatives who may be affected by the results'. Such support, notwithstanding, it is not difficult to imagine how arguments resisting such a right might be marshalled.

For instance, suppose that a pregnant woman (let us call her Ann) is aware that there is a history of Huntington's disease in her family. She is worried that her baby may have the Huntington's gene. If it does, she would want to terminate the pregnancy. However, Ann does not wish to know whether she herself has the Huntington's gene, that is, she wishes to be insulated against any information concerning her own

---

[42] Compare J Boyle, *Shamans, Software, and Spleens* (Cambridge, Mass, Harvard University Press, 1996).

[43] G Hottois, 'A Philosophical and Critical Analysis of the European Convention of Bioethics' (2000) 25 *Journal of Medicine and Philosophy* 133, at 140.

condition.[44] Ann's medical advisers believe that, in the present state of the art, two types of test can be carried out: a mutation test which will establish conclusively whether or not the foetus has the defective gene; or an exclusion test which is a linkage test targeting genetic markers that are close to the site of the mutation. In relation to the right not to know, the difference between these two tests is extremely significant. If the result of a mutation test is negative, there are no implications for the status of the mother; but, if the result is positive, the mother will have the Huntington's gene. By contrast, whatever the result of the exclusion test, whether positive or negative, there are no implications for the mother— she already knows that she is at risk and, irrespective of the result of the exclusion test, she will know no more and no less than this about her own status. If Ann is to prioritise her interest in not knowing whether she has the Huntington's gene, she will opt for the exclusion test.

Why should anyone object to Ann enjoying the right not to know and, thus, opting for the exclusion test? From a dignitarian perspective, the problem is that the exclusion test will establish only whether the foetus is at risk and, in the case of a false positive, the objection is that Ann might elect to abort a foetus that does not actually have the Huntington's gene. Accordingly, the price that is paid for respecting Ann's right not to know is that we may abort a foetus that is perfectly healthy. There is also the question of how expensive it is to counsel Ann and offer her these choices. If administration of the right not to know is unduly costly, utilitarians will join the dignitarians in resisting regulatory recognition for this claimed extension of the privacy right.[45]

Returning to Ann's case, although she can point to background support in the various international and European instruments already cited, she cannot succeed in a legal claim unless the right has been effectively transcribed into her particular legal regime. If legislators have regulated directly on the point, this will govern any dispute between Ann and her medical advisers. However, in the absence of bespoke regulation, the point may arise in more than one way. For example, if the dealings between Ann and the medical advisers are governed by private law, Ann's grievance (in being denied knowledge of, or access to, an exclusion test) may show up as a novel tort claim;[46] or, if the test is provided as a matter

---

[44] For a Gewirthian analysis, see D Beyleveld, O Quarrell and S Toddington, 'Generic Consistency in the Reproductive Enterprise: Ethical and Legal Implications of Exclusion Testing for Huntington's Disease' (1998) 3 *Medical Law International* 135.

[45] This might be an issue, too, in relation to routine genetic profiling of neonates. Generally, for discussion, see Human Genetics Commission, *Profiling the Newborn: A Prospective Gene Technology?* (London, Human Genetics Commission, March 2005).

[46] Compare R Brownsword, 'An Interest in Human Dignity as the Basis for Genomic Torts' (2003) 42 *Washburn Law Journal* 413.

of public health provision, Ann may seek to get the test that she wants by a process of judicial review.[47] In either case, though, there may be further obstacles standing in the way if judges are nervous about creating broad and vague new claims or exceeding their authority by intervening in decisions about the expenditure of public resources.

Before moving on, it is worth remarking that Ann's case draws attention to two important points. One is that there is a considerable distance between international declaration and concrete domestic provision; the fact that the ethic of human rights is recognised at one level does not entail (particularly given the plurality) that it will be recognised domestically. The other point is that the way that the plurality plays may vary from one domestic regulatory arena to another—for example, in judicial settings, we may find that human rights enjoys more support than in legislative settings, or not, as the case may be. Putting these two points together, it is apparent that a complete account of the ethical and legal dynamics would need to plot the operation of the plurality at all levels from international to domestic, in both soft and hard law dimensions, and taking into account the prevailing culture of executive, legislative and judicial decision-making—a task for a book rather than a single chapter, I think.

3.2 The Drafting Deficit

Ideal-typically, regulation should be expressed in such a way that requires the development or application of biotechnology to conform to human rights standards. As I have said, such expression might be drafted in more than one way; but, in the ideal-typical case, no matter which drafting style is adopted, the focal concern is respect for human rights.

Ethical pluralism apart, there can be routine slippage in the drafting, whether arising from carelessness, the ambiguity or vagueness of language, or the indeterminacy of regulatory intent (particularly where rapidly developing technologies outstrip their regulatory frameworks causing a lack of connection), and so on.[48] Such stock problems are well known. However, with ethical pluralism in play, there are two particular forms of drafting deficit, one involving over-inclusion (especially where human rights is expressed in dignitarian terms), the other under-inclusion (especially where human rights is qualified by utilitarian considerations).

---

[47] Compare *R v Cambridge Health Authority, ex parte B* (1995) 23 BMLR 1 (CA).
[48] Generally, see JN Adams and R Brownsword, *Understanding Law* 3rd edn (London, Sweet and Maxwell, 2003).

3.2.1 Over-inclusion

The problem of over-inclusion tends to arise where, although the ethic of human rights has registered, it is not the only ethical element in the regulatory mix.

There are a number of possible scenarios here. One is that, although human rights is the dominant ethic, there is significant support for an alternative view. In such a setting, regulators might be forced to make compromises—one ploy, for example, is to register minority concerns in the non-operative parts of a legal instrument, in the Preamble or the Recitals, even if such views are not fully enacted as such in the Articles. Where this is the situation, there is a temptation to 'fudge' the drafting in order to conceal the ethical divisions. So, for example, in her perceptive commentary on the background to the Convention on Human Rights and Biomedicine, Aurora Plomer[49] says:

> By opting for general definitions and deferring the specification of key concepts to later protocols, the Bioethics Committee was also undoubtedly aiming to maximise the chances of avoiding outright divisions and reaching a consensus on broad, overarching principles which allowed States which were reluctant to sign up to common European legislation a wide margin of appreciation when implementing the Convention's provisions. But in so doing, the drafters also opened themselves to the charge that the Convention would either be an empty text, devoid of substantive meaning, or a 'conceptual muddle' glossing over sharp ethical divisions.[50]

While drafting for consensus rather than clarity is understandable in the broader scheme of things, it obviously militates against maintaining a clear focus on the ethic of human rights.

Perhaps the outstanding example of over-inclusiveness is in the work of the IBC itself. The problem is that the Committee refuses to adopt an exclusive conception of human dignity, as a result of which its instruments draw on both the human rights conception of human dignity (human dignity as empowerment) and the version of human dignity propounded by the dignitarian alliance (human dignity as constraint).[51] Arguably, while the latter is particularly evident in the Universal Declaration on the Human Genome and Human Rights,[52] the former seems more influential

---

[49] N 13 above See, too, C Delkeskamp-Hayes, 'Respecting, Protecting Persons, Humans and Conceptual Muddles in the Bioethics Convention' (2000) 25 *Journal of Medicine and Philosophy* 147; G Hottois, 'A Philosophical and Critical Analysis of the European Convention of Bioethics' (2000) 25 *Journal of Medicine and Philosophy* 133; and K Schmidt, 'The Concealed and the Revealed: Bioethical Issues in Europe at the End of the Millennium' (2000) 25 *Journal of Medicine and Philosophy* 123.

[50] N 13 above, at 24–5.

[51] See Beyleveld and Brownsword, n 23 above.

[52] Adopted by General Assembly Resolution 152, UN GAOR, 53rd Sess., UN Doc A/53/625/Add.2 (1998).

in the International Declaration on Human Genetic Data.[53] However, what is beyond argument is that these are competing conceptions of human dignity; and that the general rhetoric of respect for 'human rights and human dignity' glosses over a critical ethical faultline.[54]

Very much the same point might be made about the recent United Nations Declaration on Human Cloning that calls on members 'to prohibit all forms of human cloning inasmuch as they are incompatible with human dignity and the protection of human life'.[55] Quite remarkably, this requirement runs together three opportunities for over-inclusion. First, as Bart Wijnberg has observed, the use of the phrase 'inasmuch as' creates a 'constructive ambiguity' in this resolution (allowing for both a narrow ('to the extent that') and a broad ('for the reason that') interpretation of the prohibition on human cloning).[56] Secondly, to make my own point, this ambiguity reaches through to invite reading the prohibition in line with one's favoured conception of human dignity. And, thirdly, as pointed out by several of the members who voted against adopting the Declaration, the invitation to over-inclusiveness is repeated with reference to the protection of human life.[57] To be sure, the dignitarian view (contending for a broad prohibition) is in the ascendancy here. However, insofar as we view the recommendation (and any regulation expressed in such terms) from a human rights perspective, the problem is not that the drafters demand that there should be respect for human dignity and human life (no supporter of human rights could disagree with that) but that the demand is expressed in over-inclusive terms.[58]

---

[53] Adopted by the General Conference on 16 Oct, 2003. According to Art 1, the principal aim of the Declaration is 'to ensure [respect for] human dignity and protection of human rights and fundamental freedoms in the collection, processing, use and storage of human genetic data and of the biological samples from which they are derived . . .'. However, by contrast with the Universal Declaration on the Human Genome and Human Rights (which is peppered with references to human dignity), this is a relatively rare occurrence; and, even when human dignity is mentioned, it is in company (as in Art 1) with human rights and fundamental freedoms. A plausible explanation for this difference of emphasis is that 'the collection, processing, use and storage of human genetic data' raises foreground questions of privacy, confidentiality, linkage, anonymisation and so on, all of which are familiar human rights issues.

[54] Brownsword, n 1 above.

[55] See n 17 above.

[56] B Wijnberg, 'Intergovernmental Activities in Bioethics Worldwide' in Council of Europe, *Meeting the Challenges of Changing Societies* 151 at 155 (paper given at the Eighth European Conference of National Ethics Committees, Dubrovnik, 25–26 April 2005). See, too, the explanatory remarks made by the South African representative, n 17 above, saying that the delegation had abstained because it detected a deliberate ambiguity in the drafting of the text.

[57] See n 17 above, especially the explanatory statements by China, the UK, and Spain. So, eg, the UK representative explained his vote against the Declaration by saying that 'the reference to 'human life' could be interpreted as a call for a total ban on all forms of human cloning': ibid.

[58] Compare Hottois, n 43 above, who (speaking of the Convention on Human Rights and Biomedicine) says (at 135):

3.2.2 Under-inclusion

Regulation is drafted in an under-inclusive way when, notwithstanding a clear human rights commitment, it fails fully to secure the ethic. Commonly, we find the commitment to human rights watered down by the use of qualifiers such as 'wherever reasonably practical' or the like, the effect of which is to allow the fundamental paper commitment to be relaxed in practice.

A more subtle instance of under-inclusiveness can be seen in the European Data Protection Directive.[59] According to Article 1 of the Directive, 'Member States shall protect the fundamental rights and freedoms of natural persons, and in particular their right to privacy with respect to the processing of personal data'. Elaborating on this foundational idea, the Directive outlines how data may be legitimately processed either (a) where the data subject consents to the processing or (b) where one of a number of legitimating conditions is satisfied. Where the data subject consents, any infringement of the data-subject's privacy rights is negated; no wrong is done. Where there is no consent, from a human rights perspective, a wrong is done to the data subject even if, all things considered, it can be justified. There is much that could be said about this; but the short point is that the consent-based justification ought to be privileged in any human rights-led regulatory regime. In other words, it should be explicitly provided that, first, attempts must be made to satisfy the consent condition before turning to any of the alternative conditions. The logic of human rights thinking, so to speak, is that consent has lexical priority.[60] Of course, it would not be too difficult for the Directive to be interpreted in a way that restored the priority of consent; but this assumes a willingness to repair a human rights deficit, and this takes us to a third point at which we can find the plurality inviting regulatory failure.

3.3 The Interpretive Deficit

Ideal-typically, regulation that purports to govern the development, use or application of modern biotechnology should be interpreted in a way

---

'I want to stress that I don't deny the importance and interest of a Convention which, in the universalistic line of human rights, would protect persons against specific risks concerning their fundamental freedoms and rights and their dignity in relation with biomedical practice and research and development. But there are many ways to conceive of these risks and protections, and, for some issues, these conceptions cannot be unified.'

[59] Directive 95/46/EC [1995] OJ L 281/31.

[60] On the 'priority of consent', see D Beyleveld and R Brownsword, *Consent in the Law* (Oxford, Hart Publishing, forthcoming).

that ensures that human rights are respected. However, the ideal-type presupposes that interpreters are dealing with regulation that is drafted in such a way that a straight-line human rights interpretation is facilitated. Where this is not the case, interpreters face a more complex task in articulating a human rights compliant interpretation.

Accordingly, in this part of the discussion, the following three scenarios are taken to be relevant: (i) where the regulation is expressed in terms that make an ethic other than human rights focal (so that we already have a case of regulatory failure); (ii) where the regulation is expressed in (general) terms that do not make an ethic other than human rights focal but neither does the drafting make human rights focal; and (iii) where the regulation is correctly expressed in terms that make human rights focal. In each of these scenarios, of course, we would need to know more about the background culture of adjudication or interpretation.[61] Without such background information, my remarks are necessarily sketchy and schematic.

3.3.1 Where regulation is expressed in terms that make an ethic other than human rights focal

If regulation explicitly embeds a utilitarian or a dignitarian ethic as the guiding approach, a great deal of interpretive work will need to be done to transform the particular regulatory measure into a human rights-compliant provision. For example, when Article 6(2) of the Directive on the Legal Protection of Biotechnological Inventions[62] (a provision that clearly reflects dignitarian thinking) decrees, inter alia, that processes both for cloning human beings and for modifying the germ line genetic identity of human beings shall be unpatentable, it would be a bold interpreter who would read such restrictions as being lifted provided either that no rights were infringed or that any relevant rights-holder had authorised such processes or modification by consenting thereto. Unless the drafting of the regulation is at odds with a strong background culture of human rights thinking, interpreters are liable to run into a catalogue of familiar objections to the effect that they are usurping the legislative function, going beyond their role as interpreters of the law, abusing their independence, and so on. If, however, the background culture is such that it is accepted that drafting of this kind must be 'read down' to render it compliant with human rights values, then interpreters have the support that they need—and, of course, in an ideal-typical

---

[61] Compare, eg, T Cottier in this volume and E-U Petersmann in this volume for valuable insights as to the culture of dispute settlement at the WTO.
[62] Directive 98/44/EC [1998] OJ L 213/13.

world, they would act on it to render an interpretation that is human rights compatible.[63]

3.3.2 Where regulation is expressed in terms that do not make an ethic other than human rights focal but neither does the drafting make human rights focal

Often, regulation will signal that morality (or ethics) matters, but it will not specify the regulative ethic nor will it offer more specific regulatory guidance. Similarly, the drafters may use class descriptors such as 'human being', or 'no one', or 'everyone', but without specifying precisely where the class boundaries lie;[64] or they may use a term such as 'eugenics' but without spelling out precisely what kinds of genetic intervention are prohibited and which interventions, if any, are permitted. For example, it has been questioned whether the prohibition against 'eugenic practices' in Article 3.2 of the EU Charter of Fundamental Rights might be read as covering pre-implantation genetic diagnosis (even where PGD is designed to forestall the implantation of human embryos that have the genetic markers for serious disorders).[65] Where ethics is made relevant in this general unspecified way, it is open to interpreters to read the relevant ethic (or descriptor or term) in a way that is in line with human rights thinking; but, equally, the regulatory blank cheque might be filled out in favour of a rival ethic.

To start with a couple of well-known examples, albeit not involving the regulation of biotechnology, we can recall the local police bans imposed on 'dwarf-throwing' in clubs in France and on the Laserdrome 'killing game' in Germany. In both cases, the initial question was whether the bans were authorised under police powers to maintain *ordre public* (in France) or public order (in Germany). This question led, in both cases, to the core issue of whether human dignity, as a fundamental constitutional value, was violated by the banned activities. In the German case, the *Omega* case, there was also the 'European' question: namely, whether the ban, even if consistent with the value of human dignity as expressed by Article 1 of the Basic Law, was compatible with market freedoms, in particular the freedom to provide services under Article 49EC. Stated shortly, the

---

[63] Under s 3 of the Human Rights Act 1998, there is an obligation to 'read down' domestic legislation in such a way that it becomes Convention compliant. According to Lord Steyn in *Regina v A* [2001] UKHL 25, at paras 44–45, this obligation 'is a strong one' applying 'even if there is no ambiguity in the sense of the language being capable of two different meanings'; and, it should be noted, the limit of interpretive leeway is the mere possibility of the reading rather than its reasonableness.

[64] See, eg, Art 8 of the Preliminary Declaration, n 7 above, which provides that 'no one shall be subjected to discrimination'.

[65] HC 7–1, Science and Technology Committee, *Human Reproductive Technologies and the Law* (Fifth Report of Session 2004–05) (London, TSO, 2005), paras 22–23.

outcome of both cases was that the bans were held to be lawful and in line with respect for human dignity.[66]

There are two ways of reading these decisions, one a straightforward dignitarian interpretation, the other a more complex human rights interpretation. The straightforward reading is that, as held by the dignitarian alliance, each member of the community has a responsibility to act in ways that are consistent with the constitutive value of human dignity. As elaborated by communitarians, the distinctive local take on human dignity (whether relative to the application of biotechnology or the leisure industry) represents its version of civilised society and is the key to its collective (cultural) identity. The alternative reading, which is particularly encouraged by some of the remarks made in the *Omega* case, is to the effect that participation in certain activities (such as dwarf-throwing or killing games) might awaken or strengthen an attitude of disrespect for the rights of others or for others as rights-holders. If so, such activities represent an indirect threat to a community of (human) rights and the value of human dignity that underpins it. For the state to interfere legitimately with the free and informed life-style choices of the participants takes some considerable justification (a point made forcefully, of course, by Manual Wackenheim, the dwarf who challenged the legality of the bans in France). If the consensual activities of the participants directly impinge on the rights of third parties, the case for intervention is clear. However, where the threat is indirect and speculative, prohibition is much more difficult to justify—although it should be said that each community of rights, as a community that constantly reviews whether its practices are in line with the best interpretation of its rights commitments, will need to decide how precautionary it should be in its public policy.[67] But this is to take us away from the point that matters for present purposes: quite simply, this is that, where regulation is framed in terms that are neutral between rival ethics (terms such as 'public order' or '*ordre public*', respect for human dignity, and so on), the ethic of human rights might claim some regulatory support, but it will not be the only ethic that can do so.

Turning to the regulation of biotechnology, reference has already been made to the recent Declaration on cloning at the United Nations. The drafting of the Declaration, I have suggested, is over-inclusive (triply so, in fact). Now, let us suppose, for the sake of illustration, that, whereas

---

[66] For the 'dwarf throwing' case, see *Ville d'Aix-en-Provence* [1996] Dalloz 177 (Conseil d'Etat) req. nos. 143–578; *Cne de Morsang-sur-Orge* [1995] Dalloz 257 (Conseil d'Etat) req. nos. 136–727; and for the Laserdrome case, see (Case C–36/02) *Omega Spielhallen und Automatenaufstellungs-GmbH v Oberbürgermeisterin der Bundesstadt Bonn*, 14 Oct 2004, [2004] OJ C300.

[67] For an extended discussion of this issue, see R Brownsword, 'Happy Families, Consenting Couples, and Children with Dignity: Sex Selection and Saviour Siblings' (2005) 17 *Child and Family Law Quarterly* 435.

human reproductive cloning is incompatible with respect for human rights, therapeutic cloning is compatible. Whatever the intention behind the call, it allows for (indeed, it probably encourages) a dignitarian reading; and, to the extent that a dignitarian rather than a human rights interpretation is given, regulators will support a broad prohibition, including a prohibition on therapeutic cloning. Where this happens, one regulatory failure leads to another, over-inclusive drafting inviting a deviant (non human rights) interpretation.

By contrast, in both the Relaxin Opposition[68] and the Leland Stanford Opposition,[69] the European Patent Office drew the sting from any dignitarian objections to the use of human tissue in research by interpreting the morality clause of the European Patent Convention as though it were a charter (albeit a limited charter) for the protection of human rights.[70] In the former, where the opponents argued that a patent on a human gene sequence, or a copy of a human gene sequence isolated from the body, should be excluded, the researchers had taken tissue from pregnant women; and, in the latter, the objection centred on the use of cells and tissue taken from aborted foetuses and young children.[71] In both cases, the Opposition Division engaged in some detail with the objections, but it clearly assumed that it was an adequate response to point out that the tissue had been donated on the basis of an informed consent. For present purposes, the fact that this misses or misreads the dignitarian point is not the issue. The point is that a general clause is filled out with a particular ethical content, in this case that of human rights—but it might have been otherwise.

3.3.3 Where the regulation is correctly expressed in terms that make human rights focal

Where regulation is drafted in terms that clearly make human rights focal, life is easier for interpreters. For example, in the UNESCO International Declaration on Human Genetic Data, wherever the rights of data-subjects may be overridden, the drafters limit the exception by reference to what is permissible according to the international law of human rights. Where exceptions otherwise appeal simply to compelling reasons, public order, the public interest, national security, and the like, the qualifying reference

---

[68] [1995] EPOR 541.

[69] [2002] EPOR 2.

[70] Compare the argument advanced in D Beyleveld and R Brownsword, *Mice, Morality and Patents* (London, Common Law Institute of Intellectual Property, 1993).

[71] As the Opposition Division concedes in Leland Stanford (at para 50):
'[It] is undeniable that the production of chimeric animals containing human organs grown from human cells isolated from aborted foetuses or deceased persons, whether children or adults, instinctively appears distasteful, if not immoral, to many people at first glance.'

is very important indeed.[72] Even so, in a plurality, we cannot afford to be complacent: other ethics are always in play and liable to be pleaded, directly or indirectly.

A good example of how there can be such a breakdown in regulation is provided by Recital 26 of the Patent Directive, according to which:

> Whereas if an invention is based on biological material of human origin or if it uses such material, where a patent application is filed, the person from whose body the material is taken must have had an opportunity of expressing free and informed consent thereto, in accordance with national law

On the face of it, this is an example of regulation that takes free and informed consent (and, concomitantly, human rights) very seriously. However, there is in the plurality the (utilitarian) view that the consent requirement imposes too great a burden on researchers and, what is more, that requirements of this kind are out of place in a patent system that is geared for granting IP protection to those who make public the secrets of their inventive work. Interpreters seeking to weaken the human rights impact of this provision might do so in several ways.

First, the terms of Recital 26 might be simply ignored on the ground that the operative part of the Directive is limited to the Articles; the Recitals, so it might be contended, have no operative life of their own.

Secondly, Recital 26 may be given a reading that marginalises the significance of consent. For example, the requirement may be read as one of merely having the *opportunity* to consent (or refuse). On this basis, it might be arguable that, provided that there is some kind of opportunity to opt out from the use of one's materials for research (or for commercial exploitation), this satisfies an informed consent standard. Or, if this seems too cavalier, the highly ambiguous phrase 'consent thereto' may be read as simply referring (in the ordinary way) to the need for consensual taking of the material rather than to consent to research let alone consent to patenting.

Thirdly, and notoriously, the idea of free and informed consent is open to a broad range of interpretations. Whereas a utilitarian interpreter might tend to favour conditions of free and informed choice that reflect an ethic of self-reliance, an interpreter guided by human rights considerations might well insist upon conditions that evince a more co-operative or supportive approach.

Fourthly, the question of the scope of a particular consent is also problematic. Again, whereas a utilitarian interpreter might favour so-called broad consents (reaching through to various kinds of research and then to patenting), in the human rights tradition interpreters would restrict the scope of a consent to those matters expressly considered and authorised by the consenting agent.

---

[72] See, eg, Art 8 (consent), 13 (access to data), 14 (privacy and confidentiality), and 16 (secondary purposes).

Fifthly, it might be argued that Recital 26 leaves open the possibility of implying in limitations or exceptions for cases in which there are compelling public interest reasons for proceeding even in the absence of free and informed consent.[73]

As for the utilitarian concern that the patent regime should not get side-tracked by moral dilemmas, the ECJ all but eliminated Recital 26 in its response to the Netherlands' challenge to the Patent Directive.[74] In this case, one of the Netherlands' several claims was that the Directive is morally deficient, failing to ensure that donors of biological samples give a full, free and informed consent (crucially, that donors consent all the way through to possible patenting and commercial exploitation). The ECJ could have given a perfectly plausible response to this element of the Netherlands' challenge by pointing to the way in which Recital 26 of the Directive underlines the importance of informed consent. Far from relying on a robust interpretation of Recital 26, however, the ECJ effectively said that such a matter was not one for the patent regime, thereby seeking to disconnect patent law from the ethics of patenting (even where the ethics at issue involve fundamental human rights values).[75]

We should not assume, therefore, that the expression of regulation in terms that make respect for human rights focal is any guarantee that interpreters will stick faithfully to the human rights script. Even where the baseline regulation is adequate, pluralism can induce regulatory failure at the stage of interpretation.

4. CONCLUSION

In this chapter, I have focused on the dynamics of a particular instance of ethical pluralism, the bioethical triangle. Whatever conclusions we draw from this chapter, however, we should remember that we are focusing on merely a sub-set of pluralism (*within* ethics) in the context of a larger plurality that characterises modern biotechnology as a particular regulatory space.

In fact, if the regulation of modern biotechnology is to be taken forward in a way that is sensitive to human rights considerations, there are three levels of challenge to be overcome. First, in the larger plurality, the battle

---

[73] Compare Art 8 of the International Declaration on Human Genetic Data which states: '[l]imitations on this principle of consent should only be prescribed for compelling reasons by domestic law consistent with the international law of human rights'.

[74] See Case C–377/98 *Netherlands v European Parliament and Council* [2001] ECR I–7079. For commentary, see D Beyleveld and R Brownsword, 'Is Patent Law Part of the EC Legal Order? A Critical Commentary on the Interpretation of Article 6(1) of Directive 98/44/EC in Case C–377/98' [2002] *Intellectual Property Quarterly* 97.

[75] See, further, S Millns in this volume.

for ethics must be won.[76] Secondly, within ethics, the battle for human rights must be won. And, thirdly, procedures for settling issues that might divide a human rights community (for example, as to the priority between privacy and freedom of expression, the status of the human embryo, and so on) must be established.[77]

In the present chapter, I have spoken only to the second of these challenges. Even here, I have done no more than sketch the way in which the conditions of ethical pluralism interfere with a human rights-sensitive regulatory oversight of modern biotechnology. It would take a much more extended analysis to work out precisely how the plurality plays (including how it plays as we move from one regulatory level to another and from one regulatory arena to another).[78] However, precisely because proponents of human rights must contend with rival ethical perspectives, nothing can be taken for granted; at all levels, regulatory failure is a real possibility.

The dual proposition with which I opened the chapter requires us to suspend regulatory belief in two respects: first, by ignoring the larger stakeholding plurality; and, secondly, by ignoring the ethical plurality on which I have focused. Once we are back in the real world, we might have some sympathy with the recently expressed view that, while '[c]harters, declarations and treaties no doubt keep diplomats busy and fulfilled ... [s]uch charters can only produce vague, lowest common-denominator agreements that are of questionable clarity and dubious effectiveness'.[79] At all events, it is clear that, even though the cause of human rights has been greatly advanced in these last 50 years or so, and even though ethics has more than one foothold in the regulation of modern biotechnology,[80] there is still a long way to go before the paper commitments to human rights are fully focused—and even further before those commitments are translated into real remedies.[81]

---

[76] Compare E-U Petersmann in this volume.

[77] Compare T Cottier in this volume, concerning the balancing of rights.

[78] See comments above at p 60.

[79] House of Commons Science and Technology Committee, n 65 above, at para 387.

[80] Arguably, too many different and unco-ordinated footholds, see Wijnberg, n 56 above; and R Pavoni in this volume.

[81] Compare Plomer, n 13 above, chap 6; and A Yusuf in this volume.

4

Consolidating Bio-rights in Europe

SUSAN MILLNS*

THE RELATIONSHIP BETWEEN biotechnological developments, bio-medicine, bioethics and human rights law in Europe has been steadily increasing in its intimacy and intricacies. Over the last two decades, European legal systems (national, as well as those of the European Union (EU) and Council of Europe) have been obliged to consider carefully their responses to biotechnological innovation, as this affects individual human beings and the human species as a whole. This has resulted in the broad recognition of a new species of 'bio-rights'[1] which may bind institutional actors and may be invoked by individuals as they negotiate their way through the whirlwind changes associated with the regulation of biomedicine.

Inevitably, the protection of fundamental human rights at both national and European levels has been a major concern in responding to new technologies and has led to increased attention being paid to the growing number of bio-rights acknowledged in national Constitutions, in the Council of Europe's Convention on Human Rights and Biomedicine and in the EU's Charter of Fundamental Rights. As a consequence, the search for a common European response[2] to the ethical and legal dilemmas to which the use and development of biotechnologies and biomedicine give rise, coupled with the cross-border impact of these technologies, has generated the need for increased legal harmonisation so as to promote

* Professor of Law at the University of Sussex. During 2002–4 she held a Marie Curie Individual Fellowship at the Robert Schumann Centre for Advanced Studies at the EUL. E-mail: s.millns@sussex.ac.uk. This chapter draws on an earlier contribution, 'Bio-Rights, Common Values and Constitutional Strategies' in T Tridimas and P Nebbia (eds), *European Union Law for the Twenty-First Century* (Oxford, Hart Publishing, 2004).

[1] On the emerging concept of 'biolaw', see D Beyleveld and R Brownsword, *Human Dignity in Bioethics and Biolaw* (Oxford, Oxford University Press, 2001); C Neirink (ed), *De la bioéthique au biodroit* (Paris, LGDJ, 1994).

[2] A common European response to biotechnologies and biomedicine does not, of course, imply agreement upon a universal global response. A notable opposition has developed between European and US approaches, with Europeans demonstrating themselves to be rather more cautious with regard to technological development than their US counter-points.

best practices across Europe. The recognition of bio-rights in Europe has in add ition had to tread a fine line between support for innovation and scientific research and the maintenance of high ethical and moral standards—two objectives the relative importance of which is viewed in quite variable terms across the Member States.

Nevertheless, a common European approach towards progress in the field of biotechnology and biomedicine began to take shape during the 1990s. Initiatives by both the Council of Europe and EU sought to reflect national preoccupations with the human rights dimension of bioethics and biotechnological developments. While in the EU early policy initiatives tended to crystallise around the economic nucleus of the single market (with, for example, agreement upon the Biotech Directive[3]), the focus then widened somewhat to consider the interface between biotechnologies and another emerging area of EU law and policy, that is fundamental or human rights. This broadening of perspective seems set to continue, given recent developments in EU constitutionalism, notably the debate on the future of the Union, the elaboration of a Constitution for Europe and the showcasing of human rights as one aspect of this. Likewise, within the Council of Europe, efforts to develop a common European biolaw and policy area have been bound up with the concern to promote fundamental rights in the sphere of biomedicine and bioethics. The context of discussion at the European level is, therefore, very much one of legal pluralism characterised by an interplay of national and European (Council of Europe and EU) constitutional and human rights systems—all of which are now implicated in the regulation of new biotechnologies.

Within the contexts of harmonisation initiatives and multi-level constitutional plurality in Europe, this chapter considers the efforts undertaken to consolidate a common European response to developments in biotechnology and particularly biomedicine. It does so through reference to two widely recognised core values which have become synonymous with European attempts to provide a coherent ethical framework for responding to new technologies, these values being respect for human dignity and the protection of fundamental human rights.

---

[3] Directive 98/44/EC of the European Parliament and Council of 6 July 1998 on the legal protection of biotechnological inventions [1998] OJ L/213/13. The Directive requires Member States, through their patent laws, to protect biotechnological inventions and deter-mines which inventions involving plants, animals or the human body may or may not be patented with a view to ensuring the free movement of patented biotechnological products. See further ER Gold and A Gallochat, 'The European Biotech Directive: Past as Prologue' (2001) 7 *European Law Journal* 331 and Case C–377/98 *Netherlands v European Parliament and Council* [2001] ECR I–7079.

1. THE COMMON VALUE OF RESPECT FOR HUMAN DIGNITY

The Council of Europe, in elaborating the Convention on Human Rights and Biomedicine,[4] brought into being the first legally binding international instrument in the field of biomedicine and bioethics. The Convention, which was signed on 4 April 1997 and entered into force on 1 December 1999, has paved the way for the consolidation of a set of common values which establish the ethical foundation upon which the regulation of biomedicine in Europe is to be conducted. The common values which underpin the Convention have tremendous symbolic importance, in so far as they are designed to make the peoples of Europe feel part of the same social and moral order and to create a sense of belonging to a peaceful, pan-European society. They are a manifestation of the ties that bind us together and an acknowledgement of mutual expectations and commitments with regard to the broader European integration project. It is these values which provide the navigational map with which to formulate European responses to the challenges of scientific and technological progress. Equally, when the pace of scientific change goes beyond the scope of existing legal provisions, the values themselves offer a measure of how European decision-makers and researchers should respond to new developments which take them into previously uncharted waters.

It should come as no surprise that, in the context of biomedicine, one particular value has risen to prominence above all others—that of respect for human dignity.[5] The consensus on the importance of promoting respect for human dignity at the European level reflects the significance of this value which is to be found extensively in the constitutional traditions of those states which are members of the EU and the Council of Europe.[6] Of particular import also for present purposes is the way in which the concept of respect for dignity has been specifically linked to developments in the biomedical sphere for the very reason that in such

---

[4] Convention for the Protection of Human Rights and Dignity of the Human Being with regard to the Application of Biology and Medicine, signed on 4 Apr 1997, available at http://conventions.coe.int/treaty/en/treaties/html/164.htm.

[5] See further the chapters by R Brownsword and H Boussard in this volume.

[6] Eg, many of the constitutions of the 'old' 15 Member States of the EU refer to the value of respect for human dignity either as a foundational aspect or primary obligation of the state (eg the Constitutions of Portugal, Art 1, Sweden, Art 2, Finland, Art 1, Greece, Art 2) or as a core component of the system of protection of fundamental rights (eg the German Basic Law, Art 1). With regard to the new Member States see, eg, the discussion by Catherine Dupré of the importation of human dignity from German into Hungarian constitutional law: C Dupré, *Importing the Law in Post-Communist Transitions: The Hungarian Constitutional Court and the Right to Human Dignity* (Oxford, Hart Publishing, 2003). The European Court of Human Rights has assisted in the creation of a pan-European consensus on the vital importance of human dignity, stating that 'the very essence' of the European Convention on Human Rights is 'respect for human dignity and human freedom' (*SW v United Kingdom* and *CR v United Kingdom* (1995) 21 EHRR 363, paras 44 and 42 respectively).

an area the dignity of the human person and human species may well be in danger of being compromised.

Before discussing the value of human dignity as represented in the Council of Europe's Biomedicine Convention, it is worth noting that an illuminating example of the causal nexus between biomedicine, bioethics and respect for human dignity and fundamental rights is to be found in national constitutional traditions, suggesting a model for the future framing of similar considerations at the European level. For example, in 1994 France introduced three legislative proposals on bioethics following a number of high-profile assisted conception cases.[7] Two of these laws were subsequently submitted to the Constitutional Council in order for their compatibility with the Constitution, particularly its fundamental rights requirements, to be verified.[8] In response, the Constitutional Council adopted a particularly novel solution which offers a blueprint for the subsequent interpretation of European biolaw. The Council found that the ensemble of the texts were constitutional because they were specifically in conformity with the national constitutional principle of safeguarding human dignity. This statement was certainly a revelation, in so far as prior to this case the French Constitution (or 'block of constitutionality' against which legislation is checked[9]) was not known expressly to include any such principle. Nevertheless, despite this lack of textual reference, the Constitutional Council gave a new reading to the opening sentence of the Preamble to the 1946 Constitution (the proclamation by the French people, following their victory over regimes which sought to make servile and *degrade the human person*, that all human being possess inalienable and sacred rights[10]) and from this constructed a new principle of constitutional value of safeguarding human dignity. The discovery of this principle in national law precisely in order to address the biotechnological revolution (where dignity is apparently in so much danger of being compromised) has since been mirrored at the European level.

---

[7] See, in particular, *Parpalaix v Centre d'études et de conservation du sperme (CECOS)*, TGI, Créteil (1re ch. civ.), 1 Aug 1984, *Gazette du Palais*, 18 Sept 1984, 560.

[8] French Constitutional Council Decision no 94–343/344 DC of 27 July 1994 (*Bioethics*) on the constitutionality of proposed legislation on respect for the human body (Law no 94–653 of 29 July 1994) and on the donation and use of elements and products of the human body, medically assisted conception and prenatal diagnosis (Law no 94–654 of 29 July 1994): [1994] JORF 1103.

[9] The *bloc de constitutionnalité* includes the Constitution of 1958, the preamble to the Constitution of 1946, the Declaration of the Rights of Man and the Citizen of 1789 and those fundamental principles recognised by the laws of the Republic: see Decision no 71–44 DC of 16 July 1971, *Freedom of Association*.

[10] '*Au lendemain de la victoire remportée par les peuples libres sur les régimes qui ont tenté d'asservir et de dégrader la personne humaine, le peuple français proclame à nouveau que tout être humain, sans distinction de race, de religion ni de croyance, possède des droits inaliénables et sacrés*'. (emphasis added).

Thus, reminiscent of the French example of the three-way constitutional bond between biomedicine, fundamental rights and human dignity, the Council of Europe made a similar connection in its Convention on Human Rights and Biomedicine. The Convention, which specifically considers biomedicine within a universal human rights framework containing provisions on patients rights, medical research, new reproductive technologies and medically assisted procreation, gene therapy and organ transplantation, explicitly does so from the perspective of promoting the value of human dignity. To this extent, the document has been cited as 'exemplary' in the area, being hailed as the first legal instrument to establish the relationship between fundamental rights and biomedicine.[11] In addition, it is remarkable in its explicit reference to the obligation to protect human dignity, mentioned not only in its full title (the Convention for the Protection of Human Rights and Dignity of the Human Being with Regard to the Application of Biology and Medicine) but on four further occasions in the text, including in particular the first Article which states that the parties 'shall protect the dignity and identity of all human beings'.

The Convention is gradually being supplemented by specific protocols which deal in detail with some of the main themes addressed therein and which act to develop further the principles enshrined within the original text. On 6 November 1997, an additional protocol prohibiting human cloning was adopted which again makes explicit reference to human dignity, setting out in its preamble that 'the instrumentalisation of human beings through the deliberate creation of genetically identical human beings is contrary to human dignity and thus constitutes a misuse of biology and medicine'.[12] Furthermore, an additional protocol on the transplantation of organs and tissues of human origin was opened for signature by member states on 24 January 2002, stressing in its first Article that the dignity and identity of everyone should be protected with regard to the transplantation of human organs and tissues and setting out the criteria for a well-structured system facilitating equitable access by patients, in accordance with clearly defined qualitative and ethical standards.[13]

---

[11] B Maurer, *Le principe de la dignité humaine et la Convention européenne des droits de l'homme* (Paris, La Documentation Française, 1999), 83.

[12] Additional Protocol to the Convention for the Protection of Human Rights and Dignity of the Human Being with regard to the Application of Biology and Medicine, on the Prohibition of Cloning Human Beings, signed on 12 Jan 1998, available at http://conventions.coe.int/treaties/html/168.htm. This Protocol came into force on 1 Mar 2001.

[13] Additional Protocol to the Convention on Human Rights and Biomedicine concerning Transplantation of Organs and Tissues of Human Origin, signed on 24 Jan 2002, available at http://conventions.coe.int/treaties/html/186.htm. This has not yet come into force.

One further additional protocol has been agreed concerning biomedical research on the human being. This was approved by the Council of Europe's Steering Committee on Bioethics (CDBI) in June 2003, adopted by the Committee of Ministers on 30 June 2004 and opened for signature on 25 January 2005. It specifies in detail a harmonised approach to ethical and legal standards in biomedical research, particularly rules on the consent of persons taking part in a research operation and their medical and legal protection. It too sets out in its first Article that the parties to the protocol shall protect the dignity and identity of all human beings with regard to any research involving interventions on human beings in the field of biomedicine. Furthermore, the CDBI is presently preparing an instrument on research involving archived biological materials of human origin which will deal with the use of tissues and personal data employed or archived in 'bio-banks' for research purposes and a protocol on human genetics comprising two components—one relating to genetics in the sphere of health and another dealing with the use of genetics in employment and insurance. Under discussion too is an instrument on the protection of the human embryo and foetus, although the wide variations of opinion in Europe on this issue have made it difficult to identify a common approach.[14]

Overall, the amplification of the references made to human dignity throughout the Biomedicine Convention and the additional protocols is remarkable in its intensity. It indicates an important acknowledgment of the value of human dignity as a fundamental concern in the area of biomedicine at the European level, which must be considered in the exercise of other freedoms such as the carrying out of research, the development of the human knowledge base and the pursuit of technological innovation. That said, the content, meaning and scope of the obligation to respect human dignity, both generally and in the sphere of biomedicine in particular, are deliberately fuzzy. Reflecting back upon the constitutional traditions of the member states, it is evident that expressions of national constitutional values like dignity are capable of producing very different interpretations once they are fleshed out by judicial interpretation. It is inevitable that a similar problem should present itself at the European level, with competing versions of the content of 'common' values such as dignity being put forward. A particularly telling example in this regard is the variation in national constitutional court responses to the question of how far human dignity extends in the context of their review of the constitutionality of legislation on abortion. This example is mentioned as it has clear implications for the legal response to biotechnological

---

[14] On the role of the CDBI see J Michaud, 'Le Comité Directeur de Bioéthique du Conseil de l'Europe' in B Feuillet-Le Mintier (ed), *Normativité et Biomédicine* (Paris, Economica, 2003), 169.

advances, given that the regulation of new reproductive technologies, like abortion, is linked to the status of the foetus and embryo as an object of constitutional protection (as the Council of Europe's strained attempts to agree upon an additional protocol on the protection of the human embryo and foetus indicate).

Thus, an instructive contrast may be identified in the approach of the French Constitutional Council in its Decision no 2001–446 DC of 27 June 2001 and the view of the German Constitutional Court in cases 39, BVerGE 1 (1975) and 88 BverGE 203 (1993) both regarding liberalising changes to national laws on abortion. In the French case it was found that the new abortion law (augmenting the time limit for abortions where the woman is in a state of distress from 10 to 12 weeks) did not disrupt the balance imposed by the Constitution between safeguarding human dignity (interpreted to mean that of the foetus) and respecting the liberty of the pregnant woman.[15] In the German cases, however, the state's obligation to protect any form of human life, born or unborn, which shares fundamental human dignity, whether or not it is conscious of this dignity and capable of defending it, was held to take precedence over the woman's right to self-fulfilment irrespective of any time limit.[16] The fact that human dignity is a vague and malleable concept with no clear outer limits suggests that much discussion will be needed in order to flesh out its contents at the European level. This is not least since Article 1 of the European Biomedicine Convention distinguishes between 'everyone' (the bearers of human rights) and 'all human beings' (those whose dignity and identity are to be protected). In light of the fact that some signatories to the Convention may take the view that embryos and foetuses do not have human rights and are not included in the word 'everyone', but that human dignity and identity are to be respected as soon as life begins, Article 1 may still operate to protect *potential life* from a violation of dignity when it is not eligible for full human rights protection.[17]

What is certain is that the reconciliation of diverging national perspectives in Europe on the appropriate response to biotechnological advances will provide an exciting challenge for European lawyers, researchers and philosophers in the twenty first century as they search for common responses to technological change. Their quest for solutions will clearly put to the test the ethical and legal outer limits of the requirement to give due regard to the value of human dignity in biomedical matters.

---

[15] B Mathieu, 'Une jurisprudence selon Ponce Pilate (constitutionnalité de la loi sur l'interruption volontaire de grossesse et la contraception)' [2001] Dalloz jur 31, 2533–7.

[16] S Walter, 'Thou Shalt Not (But Thou Mayest): Abortion after the German Constitutional Court's 1993 Landmark Decision' (1993) 36 *German Yearbook of International Law* 385.

[17] D Beyleveld and R Brownsword, 'Human Dignity, Human Rights, and Human Genetics' in R Brownsword, WR Cornish and M Llewelyn (eds), *Law and Human Genetics: Regulating a Revolution* (Oxford, Hart Publishing, 1998), 69, at 72.

2. THE COMMON VALUE OF RESPECT FOR FUNDAMENTAL RIGHTS

A second core value which is set out uniformly across European and national constitutional discourses, and which is closely associated with the universal objective of respect for human dignity, is that of respect for fundamental rights.[18] A crucial issue in the consolidation of a common European approach to biotechnological development, therefore, is the elaboration of a response to the potential impact of the new technologies upon fundamental rights, both their protection and their exercise. Hence consideration has now to be given to the content of a new generation of bio-rights, this term being used to denote those fundamental human rights which intersect with the use of new technologies, biomedicine and bioethics. While bio-rights undoubtedly have as their objective the protection of human beings (particularly their dignity) from the dangers of scientific excess, they do also need to be viewed alongside other values, rights and interests such as individual freedom, equality, self-determination and autonomy which can sometimes seem to pull in another direction from respect for the dignity of the human species.

2.1 Bio-rights in the Convention on Human Rights and Biomedicine

It has been noted above that the Council of Europe's Convention on Human Rights and Biomedicine seeks to promote the universal dimension of the rights identified therein.[19] So much is evident in the Convention's preamble, which states that the Convention should be read and interpreted against the background of a number of international human rights documents, including the Universal Declaration of Human Rights, the European Convention on Human Rights, the European Social Charter, the International Covenant on Civil and Political Rights and the International Covenant on Economic, Social and Cultural Rights, the Convention for the Protection of Individuals with regard to Automatic Processing of Personal Data and the Convention on the Rights of the Child.

Yet the relationship between the protection of universal, fundamental rights within the Biomedicine Convention and the commitments to broader ethical principles and values also endorsed therein deserves further exploration. Articulating the nature of this relationship, Chapter 1

---

[18] On the commonality of this value see S Millns, 'Unravelling the Ties that Bind: National Constitutions in the Light of the Values, Principles and Objectives of the New European Constitution' in J Ziller (ed), *The Europeanisation of Constitutional Law in the Light of the Constitutional Treaty for the Union* (Paris, L'Harmattan, 2003), 97.

[19] See also A Plomer, 'In Search of Universals: Rights, Principles and Political Values in Medical Research', Conference paper, American Philosophical Association, 100th Anniversary Conference on Morality in the 21st Century, University of Delaware, 26–28 Oct 2001.

of the Convention goes on to posit that the purpose of the Convention is 'to protect the dignity and identity of all human beings and guarantee everyone, without discrimination, respect for their integrity and other rights and fundamental freedoms with regard to the application of biology and medicine'. Chapter 1 states three additional general provisions or principles which serve to inspire the interpretation of European bio-rights. Article 2 asserts the primacy of the human being (requiring that 'the interests and welfare of the human being shall prevail over the sole interests of society and science). Article 3 provides that contracting states shall take appropriate measures to provide equitable access to health care, taking into account health needs and available resources, and Article 4 states that any intervention in the health field, including research, must be carried out in accordance with relevant professional standards.

In addition to these three general principles, fundamental bio-rights and freedoms are to be respected through the provisions governing the requirement to obtain an individual's free and informed consent to medical interventions and the requirement that the person shall beforehand be given appropriate information as to the purpose and nature of the intervention as well as on its consequences and risks. In this regard interventions (which include scientific research as well as medical treatments) without the person's informed consent are prohibited by Article 5. In the case of an individual who is a minor and lacks the capacity to consent, the intervention may be carried out only for the person's direct benefit and with the authorisation of a legal representative (Article 6.2).

The Convention also contains additional specific rules on research which are aimed at balancing the freedom to engage in research (and social needs) alongside the freedom and autonomy of the individual. Three general requirements on the conduct of research are presented in Article 16. There must be no alternative of comparable effectiveness to research on humans; the risks must not be disproportionate to the potential benefits; and the project must have received prior approval by a multidisciplinary and independent ethics committee.

Where people are unable to consent, Article 17 distinguishes between, on the one hand, research which has the potential to produce real and direct benefit to the individual and and, on the other, research which has the aim of contributing to the ultimate attainment of results capable of conferring benefit on the person concerned or on other persons in the same age category or afflicted with the same disease or disorder or having the same condition (Article 17.2.i). While the terminology used may be different, these two categories of research correspond to the traditional categories of therapeutic and non-therapeutic research.[20] In addition to the general

---

[20] Ibid.

consent requirements imposed by Articles 5 and 6, it is provided in Article 17.1.iv that both types of research are subject to the further evidentiary requirement that the authorisation of the legal representative be given specifically and in writing, and in Article 17.2.ii that non-therapeutic research must carry only minimal risk and burden for the individual concerned.

Undoubtedly the Convention envisages respect for human beings and their fundamental rights and freedoms as comprising both individual respect and respect for the person as a member of the human species— suggesting respect for both *personal* and *human* dignity and a degree of tension between the two. Thus, in considering the uses to which research may be put, the preamble states that 'progress biology and medicine should be used for the benefit of present and future generations', suggesting a utility which serves the human species rather than specific individuals. In this regard too, as mentioned above, the Convention legitimises the use of human subjects for research which will not directly benefit the individual and which may even inflict harm, albeit minimal, on the subject concerned.

On the other hand, in other respects the Convention asserts the moral priority of the rights of the individual over those of society. Individuals cannot be compelled to act for the collective benefit of others. Thus, the rule on informed consent is consistent with liberal regimes where individual interests or fundamental rights take priority over more collective welfarist values. In this respect, informed consent acts as a device to protect individual freedom.

2.2 Bio-rights in Europe beyond the Convention

Given the impetus which the European Convention on Human Rights and Biomedicine creates to elaborate a pan-European approach to biomedicine it is useful finally to position its requirements, especially those regarding individual free and informed consent, alongside more recent and not dissimilar developments in the European Union. In particular, the Charter of Fundamental Rights (now incorporated into Part II of the Constitutional Treaty), like the Biomedicine Convention, makes an explicit connection between the protection of human dignity (the object of its first Title) and developments in the area of biomedicine. Article II–63–2 of the Charter on the right to integrity of the person in the fields of medicine and biology provides four key principles which are to be respected in the name of dignity: free and informed consent; the prohibition of eugenic practices, especially those aimed at the selection of people; the prohibition on commercialisation of the human body; and the prohibition on reproductive cloning of human beings. The remit of these is striking in its overlap with that of the principles enshrined in the Biomedicine Convention and the first additional protocol.

The theme of respecting the value of human dignity and the fundamental rights flowing from it has likewise been reiterated with respect to Directive 98/44 EC on the legal protection of biotechnological inventions which, when reviewed by the ECJ for its legality, was examined specifically for its compatibility with respect for human dignity—a violation of this principle being one of the arguments put forward by the Netherlands in its application for annulment.[21] The alleged violation regarded the patentability of parts of the human body (Article 5(2) of the Directive) which it was suggested undermined fundamental rights and notably human dignity in treating human matter as a means to an end rather than as an end in itself.[22] Interesting in the Court's response on this matter is the stony silence it maintained with regard to the Charter provisions on dignity. This is despite the fact that its decision of 9 October 2001 post-dated the solemn declaration of the Charter in December 2000 by the institutions of the EU and its political approval by the Member States.[23] It is despite the fact, too, that Advocate General Jacobs in his Opinion on the case had made explicit reference to the Charter, finding that '[t]he right to human dignity is perhaps the most fundamental right of all, and is now expressed in Article 1 of the Charter ... It must be accepted that any Community instrument infringing those rights would be unlawful.'[24] Instead, however, in line with its history of the development of the relationship between fundamental rights and EC law, the ECJ referred to the competence of the Court to review the compatibility of acts of the institutions with the general principles of Community law to ensure that the fundamental right to human dignity and integrity was observed. Thus, rather like the rise to prominence of dignity in the French constitutional context, the principle was read into existing EC human rights guarantees—there all along and simply waiting for the biotechnological revolution to bring it to light.

As a result, in applying the principle of respect for human dignity to the Biotech Directive the ECJ found, as had the Advocate General, that there was no violation of the principle. Instead, again as in the French Constitutional Council's *Bioethics* decision,[25] it was positively affirmed that human dignity was respected by the Directive. This finding resulted

---

[21] Case C–377/98, n 3 above.

[22] This reflects the Kantian view that '[h]umanity itself is a dignity: for a human being cannot be used merely as a means by any human being (either by others or even by himself) but must always be used at the same time as an end': I Kant, *The Metaphysics of Morals* (trans., M. Gregor, Cambridge, Cambridge University Press, 1991, first published 1797), 255.

[23] [2000] OJ I364/8.

[24] Opinion of Jacobs AG of 14 June 2001 in Case C–377/98, n 3 above, para 197.

[25] Above n 8.

from a consideration of its Article 5(1), which provides that the human body (at the various stages of its formation and development, and the partial discovery of one of its elements including the sequence of a gene) cannot constitute a patentable invention.[26] Nor are elements of the human body patentable in themselves unless they are capable of being isolated and combined with a technical process for industrial application (Article 5(2)).[27] Also ensuring respect for human dignity, Article 6 of the Directive offers extra security in rendering contrary to *ordre public* or morality—and therefore excluded from patentability—processes for cloning human beings, for modifying the genetic identity of human beings, and the use of human embryos for industrial or commercial purposes.[28]

Following directly on from the dignity dimension of the question, the Netherlands in its legal challenge to the Directive raised a second concern regarding fundamental rights violations, suggesting that there was a violation of the right to human integrity, understood in the context of medicine and biology as including the need for free and informed consent of donors and recipients. Demonstrating the limits of a human rights approach in EC law to such matters, the ECJ in its response held that reliance on this right was misplaced because the Directive concerned only the granting of patents and did not extend to activities before or after they had been awarded.[29] Thus, in a particularly constrained interpretation of the scope of Community law, it was found that the grant of a patent under EC law did not preclude national legal limits on research into, or exploitation of, a patentable product, the ethical rules on which were beyond the scope of the Directive.[30]

Of particular note in the ECJ's discussion of the legality of the Biotech Directive is the close link that is made between the two values discussed in this Chapter—respect for human dignity and respect for fundamental rights. This relationship has already exercised the minds of national constitutional lawyers as well as the drafters of the European Biomedicine Convention. Equally it will interest EU lawyers who seek to make a bridge between Article I–2 of the Constitutional Treaty (setting out the values of the EU to include human dignity) and Part II of the Treaty. Inspiration may be sought once more from national constitutional traditions. In France, for

[26] Case C-377/98, n 3 above, para 71.

[27] Ibid, para 72.

[28] Unsurprisingly, the exclusion of certain patents on ethical grounds proved one of the most controversial questions to be addressed by the Biotech Directive. See further D Beyleveld, R Brownsword and M Llewelyn, 'The Morality Clauses of the Directive on the Legal Protection of Biotechnological Inventions: Conflict, Compromise and the Patent Community' in R Goldberg and J Lonbay, *Pharmaceutical Medicine, Biotechnology and European Law* (Cambridge, Cambridge University Press, 2000) 157.

[29] Case C–377/98, n 3 above, at para 79.

[30] Ibid, para 80.

example, there has been a good deal of discussion amongst constitutional scholars as to whether human dignity can properly be viewed as a fundamental right (as opposed to a value, constitutional principle or objective).[31] A helpful way of conceptualising the problem has been put forward by Bertrand Mathieu who suggests that safeguarding dignity can best be viewed as a sort of meta-value or 'matrix' providing a guiding pathway for the configuration and engendering of other more specific rights and duties.[32] In this way all rights need to be considered in the light of the primary need to respect dignity. In a rather similar vein, David Feldman, looking at the matter from a UK constitutionalist perspective, has argued that dignity is not a right *per se* but a value underpinning all fundamental rights and constitutional principles.[33]

Viewed from an EU perspective, however, the relationship between respect for human dignity and respect for fundamental bio-rights may not be quite as extensive as that envisaged in national law, or indeed in the European Convention on Human Rights and Biomedicine. This is because EU interventions in the area of biotechnologies in particular have, for reasons of the limitations on EU competence, tended to crystallise around internal market and harmonisation considerations. This does not mean, however, that no fundamental rights issues are raised in such instances, as the Biotech Directive case clearly shows. Similarly, increasingly high-profile examples of 'reproductive tourism', where EU citizens seek to utilise their right of free movement to obtain assisted conception services in another Member State which are not available to them in their home state, clearly highlights fundamental rights issues (such as the right to found a family) at stake in such a process, even if these are not yet fully articulated in EU law. This potential clash or consolidation of economic rights and fundamental rights in the area of biotechnology has resonances with previous debates on the construction of abortion as a service under EC law and its potential undermining of national constitutional provisions protecting the unborn child.[34]

The move towards a constitutionalisation of human rights within European law requires now a shift in perspective to reposition the economic rights of EU citizens within the context of a more comprehensive system of rights protection based not only upon the Charter provisions but also the EU's possible accession to the European Convention on Human

---

[31] B Mathieu, 'Pour une reconnaissance de "principes matriciels" en matière de protection constitutionnelle des droits de l'homme' [1995] 27 *Dalloz chron* 211–2.

[32] Ibid. See also B Mathieu, *Génome humain et droits fondamentaux* (Paris, Economica, 2000), chap 2.

[33] D Feldman, 'Human Dignity as a Legal Value—Part I' [1999] *Public Law* 682; D Feldman, 'Human Dignity as a Legal Value—Part II' [1999] *Public Law* 61.

[34] See further DR Phelan, 'Right to Life of the Unborn v. Promotion of Trade in Services: The European Court of Justice and the Normative Shaping of the European Union' (1992) 55 *Modern Law Review* 670.

Rights. It suggests, too, that prevailing constitutional trends demand a consideration of European law measures regarding biotechnological advances within the broader framework of constitutional pluralism and human rights mainstreaming. In this regard there are many other EU Charter rights which intersect with the development of biotechnologies outside the first Title on Dignity and which show some semblance to the Biomedicine Convention (for example, Article II–67's right to respect for private and family life, Article II–68's protection of personal data, Article II–69's right to marry and found a family, Article II–71's freedom of expression, Article II–73's freedom of scientific research and Article II–77's right to property). Also likely to be of particular future relevance, however, notably in the employment sector and the provision of services, is Article II–81. This sets out a prohibition on discrimination on the grounds of 'genetic features' and should be taken to mean that as individuals become more aware of information regarding their genetic make-up, allowing them to look into their own biological futures, employers, insurance companies, schools and governments must be prevented from discriminating against them on the ground of their genetic profile. Combined with Article II–67 this might even suggest the need to develop a right to genetic privacy at the European Union level at least.

In conclusion, therefore, it is evident that the consolidation of a new generation of bio-rights is beginning to take shape at the European level. The debate on the extent of these rights, their content and their interpretation involves many institutional actors, including national constitutional courts, the European Court of Human Rights[35] and the European Court of Justice, as well as individuals and those involved in processes of scientific and technological research. The continuation of a pan-European dialogue on the appropriate response to biotechnological developments is vital for the purposes of harmonisation, but is equally necessary in order to give voice to the multiplicity of cultural and ethical perspectives which inspire the debate on such sensitive matters as biomedicine and bioethics. In the light of such a dialogue, scientific progress in reproductive medicine and research should continue to be made in the context of the promotion of the twin values of respect for human dignity and respect for fundamental rights.

---

[35] The ECtHR has jurisdiction to give advisory opinions on the interpretation of the Biomedicine Convention.

5

UNESCO Standard-setting Activities on Bioethics: Speak Softly and Carry a Big Stick*

ABDULQAWI A YUSUF**

1. INTRODUCTION

IT IS NOT easy to establish normative standards in the area of bioethics. Some even contest the appropriateness of legal instruments to deal with bioethics. According to one of the institutions that has recently responded to a UNESCO questionnaire on a declaration on universal norms on bioethics, '[w]hile it may be appropriate to have legislation and regulations on some bioethical issues, ethics and law must never be conflated. Ethics often calls for higher standards of behaviour than the law requires and in some cases ethics requires disobedience to (unjust) law'.[1] So, how can the law establish standards, principles and rules in the field of bioethics?

Even if it is accepted that legal standards may be established to deal with bioethics, further questions are raised regarding which kind of bioethics should be addressed and what should be the scope of such standards. As was pointed out by one of the respondents to the abovementioned UNESCO questionnaire, '[t]here are different ways to view bioethics and in discussions of bioethics we should be clear which approach we are addressing'.[2] In other words, should normative standards address

---

* 'I have always been fond of the West African proverb "speak softly and carry a big stick; you will go far". It is a homely old adage': US President Theodore Roosevelt, circa 1900.
** Director, Office of International Standards and Legal Affairs, Unesco, Paris; a member of the Institut de Droit International, Geneva, and founder and General Editor, *African Yearbook of International Law*. The views expressed here are not necessarily those of UNESCO.
[1] See International Bioethics Committee, 'Towards a Declaration on Universal Norms on Bioethics—Written contributions', UNESCO doc SHS/EST/04/CIB-EXTR/INF.1 of 25 Apr 2003, hereinafter 'Written Contributions', at 33–4.
[2] Ibid, at 13.

'prescriptive bioethics', or should they address 'interactive bioethics' or 'practical bioethics'? All these aspects are defined in a detailed fashion by the respondent, without referring to other possible divisions in bioethics, such as 'clinical ethics', 'research ethics' and 'public ethics', as pointed out by another respondent.[3]

Even more controversial is the scope—or the coverage—of such normative standards, as was shown by the replies to the UNESCO questionnaire which, it must be emphasised here, was addressed to some of the most prominent institutions and personalities in the area of bioethics from all over the world. Regarding the possible scope of a declaration on universal norms on bioethics, some of the respondents were of the view that such a declaration 'should not be limited to human beings', the argument being that 'human beings are only one of a wide biodiversity, and many applications of life technology touch other living organisms besides human beings' or that 'it should reflect the responsibility of human beings for the well being of all living systems',[4] while others argued that such a declaration 'should only be limited to human beings in order to specifically address the crucial and unique importance of human dignity with which only humans are endowed',[5] or, as another respondent put it, 'because the ethical and moral obligations towards human beings are on a different plane (spiritually and legally) than the ethical considerations one may have toward other living organisms and animals'.[6] So, the question remains: should normative standards on bioethics be limited to humans only as the subjects to whom ethical obligations are owed, or should they encompass also animals, plants and ecosystems or address the ethical implications of the relationship of humans to other living organisms (such as the modification of crops, animals and other non-human life-forms), or should it go even further, as one respondent pointed out, and deal with the ethical implications of the way we humans 'will treat non-organic (e.g. robots), or hybrid (e.g. cyborgs) beings endowed with consciousness or cognitive capacity before they are made'.[7]

Due to the differing conceptions regarding the scope of a normative framework on bioethics, the inherent complexity of the subject-matter which lies at the interface of many disciplines, and the diverse contexts (cultural, social, economic) in which ethical thinking takes root in different parts of the world, the UNESCO General Conference decided from the very beginning of the Organisation's activities in this field to adopt a gradual and prudent approach which would deal with 'one thorn

---

[3] Ibid, at 33.
[4] Ibid, at 43.
[5] Ibid, at 137.
[6] Ibid, at 142.
[7] Ibid, at 194.

bush at a time', to paraphrase a Somali saying. It thus started in 1997 with the adoption of the Universal Declaration on the Human Genome and Human Rights, followed in 2003 by the International Declaration on Human Genetic Data and a Declaration on Bioethics and Human Rights, which was adopted at the 33rd Session of the General Conference in October 2005. It should however be pointed out that, notwithstanding the 'soft law' approach adopted by the General Conference, the normative content of the principles and standards enunciated in the UNESCO 'bioethics declarations' and the values which they express appear to confer on them a legal effect that transcends the role normally ascribed to declarations as normative instruments in the international arena. The above-quoted West African proverb may therefore constitute a fitting characterisation of this standard-setting approach. Equally important is the fact that the declaratory approach has been coupled with follow-up measures incorporated into the normative instruments themselves, whose implementation has been entrusted to institutional mechanisms such as the International Bioethics Committee and national Bioethics Committees. A brief examination of the two existing declarations and their follow-up mechanisms may further shed light on these considerations.

2. THE UNIVERSAL DECLARATION ON THE HUMAN GENOME AND HUMAN RIGHTS

The Universal Declaration on the Human Genome and Human Rights was the first normative instrument on bioethics adopted by UNESCO. It was adopted by acclamation of the General Conference on 11 November 1997,[8] and was endorsed a year later by the General Assembly of the United Nations.[9] Thus, it became the first normative instrument in the field of bioethics that has received collective endorsement from all member states of the United Nations and of UNESCO.[10]

2.1 The Preamble

The preamble refers to a number of international instruments of direct relevance to the subject matter of the Declaration. First and foremost is the

---

[8] See Resolution 29 C/16 in *Records of the General Conference*, 29th session, Paris, 21 Oct–12 Nov 1997, v. 1: Resolutions, at 41–7.

[9] See A/RES/53/152 adopted by the General Assembly on 9 Dec 1998.

[10] The European Convention for the Protection of Human Rights and Dignity of the Human Being with regard to the Application of Biology and Medicine: Convention on Human Rights and Biomedicine was signed on 4 Apr 1997, but it is a regional instrument applicable only among the European states party to it.

Universal Declaration of Human Rights, as well as the two United Nations Covenants whose principles have served as a source of inspiration and as the foundation on which the principles of the Declaration have been constructed. Secondly, the international instruments in force in the field of intellectual property are evoked to indicate that the Declaration does not preclude research findings on the human genome from giving rise to recognition of intellectual property rights, within the limits imposed by observance of the principle set forth in Article 4 of the Declaration. Thirdly, reference is made to the United Nations Convention on biological diversity to emphasise that human genetic diversity should not be considered as an end in itself, nor should it be dissociated from the protection of the inherent dignity and human rights of the individual. Fourthly, it is recognised that research on the human genome and the resulting applications open up vast prospects for progress in improving the health of individuals and of humankind as a whole, while emphasising the need for an ethical framework to guide such work, particularly through respect for human dignity, freedom and human rights as well as the prohibition of all forms of discrimination based on genetic characteristics; thus laying the ground for the operative provisions of the Declaration.

2.2 The Human Genome

The human genome as such is not defined in the Declaration. It is, however, described as encompassing several interconnected notions ranging from the genes of the individual human being, whose diversity is recognised, to the genetic heritage of the human species the unity of which is underscored in order to highlight its reference both to the sum total of the genetic components of humanity and genetic heritage of the individual human being. Thus, the human genome is declared to constitute, in a symbolic sense, 'the heritage of humanity'.[11] The purpose, in my view, is to emphasise that the ethical obligations established in the Declaration are owed both to individual human being as well as to humanity in general, and that it is the duty of the international community as a whole to ensure the preservation of the human species in the face of advances in biotechnology and genetic engineering.

This heritage of humanity—as opposed to the notion of common heritage in international law—is not considered as being capable of some collective form of appropriation or of individual appropriation, in its natural state, for financial gain. It should therefore be manipulated neither for social or political purposes 'which could call into question the inherent dignity and the equal and inalienable rights of all members

---

[11] See Art 1 of the Declaration.

of the human family in accordance with the preamble to the Universal Declaration of Human Rights'[12] nor for financial gain.[13]

With regard to Article 4 which prohibits manipulation for financial gain, it is important to note that it does not preclude the results of genetic research from giving rise to intellectual property rights, in accordance with Article 27(2) of the Universal Declaration of Human Rights. What it precludes is that simple discovery of human genes, or partial sequences of genes, in their natural state, should be directly used to secure financial gains.

2.3 Standards and Basic Principles

The Declaration establishes certain ethical obligations owed to human beings when carrying out research on the human genome or when utilising its resultant applications. One of the fundamental principles enunciated in the declaration, which also permeates most of its provisions, is the principle of respect for human dignity.[14] This is one of the founding principles of bioethics; it is a threshold standard against which all the uses of the human genome have to be tested for compliance with the ethical framework established by the Declaration. It finds its primary expression in Article 2 of the Declaration where it is stated that:

(a) Everyone has a right to respect for their dignity and for their rights regardless of their genetic characteristics [and]
(b) That dignity makes it imperative not to reduce individuals to their genetic characteristics and to respect their uniqueness and diversity.

Thus, a first corollary, which flows from this threshold standard, is that there is a fundamental right to respect for human dignity regardless of the genetic characteristics of the person concerned. Secondly, it is the dignity of human beings that makes it imperative to reject all claims that an individual may be judged solely on the basis of genetic characteristics, usually referred to as genetic reductionism,[15] and to respect the uniqueness and diversity of all individuals. Viewed as such, human dignity may be considered in this context to be conterminous with the attributes of the human personality.

---

[12] See preambular para 4.

[13] See Art 4.

[14] For the role of the concept of human dignity in bioethics, see generally D Beyleveld and R Brownsword, *Human Dignity in Bioethics and Biolaw* (Oxford, Oxford University Press, 2001).

[15] See, eg, N Lenoir, 'La Déclaration universelle sur le génome humain et les droits de l'homme de l'UNESCO', in *Rapport public du Conseil d'Etat*, 1998, at 352.

Article 10 further elaborates on the corollaries of respect for human dignity, particularly with regard to research in the biosciences, by affirming that such research cannot prevail over 'respect for the human rights, fundamental freedoms and human dignity of individuals or, where applicable, of groups of people'. Having thus established the primacy of respect for human dignity and human rights over scientific research on the human genome, Article 11 of the Declaration specifies that scientific and technological practices 'which are contrary to human dignity, such as reproductive cloning of human beings, shall not be permitted'. It also calls upon states and competent international organisations 'to cooperate in identifying such practices and in taking, at national or international level, the measures necessary to ensure that the principles set out in this Declaration are respected'.

The fact that the reproductive cloning of human beings is specifically or explicitly mentioned in the Declaration as an example of a practice contrary to human dignity was perhaps meant to show the existence of a basic consensus at the international level to ban this specific practice. With regard to other practices that may equally be considered as being contrary to the threshold standard of respect for human dignity, Article 24 of the Declaration calls upon the International Bioethics Committee (IBC) of UNESCO to make recommendations to the General Conference of the Organisation regarding the identification of such practices, mentioning as one possible example that of germ-line interventions.

Despite the constraints imposed on scientific research by the above-mentioned standards and principles, the Declaration reaffirms in Article 12 the freedom of scientific research, and recognises that it is necessary for the progress of knowledge and is part of freedom of thought. But, it also provides that research applications, including those in biology, genetics and medicine, concerning the human genome shall seek to offer relief from suffering and improve the health of individuals and humankind as a whole. Inspired by the principles of the UNESCO Recommendation on the Status of Scientific Researchers, the Declaration further calls upon states to foster intellectual and material conditions favourable to the free pursuit of research on the human genome, while at the same time giving necessary consideration to the ethical, legal, social and economic implications of such research.[16]

Other principles enunciated in the Declaration include the prohibition of discrimination on the basis of one's genetic characteristics (Article 6), the principle of free and informed consent (Article 5), the principle of respect for privacy and confidentiality (Article 7), as well as the principles of justice, equity and solidarity (Articles 17, 18 and 19). Solidarity is dealt

---

[16] See Art 14.

with in the Declaration with regard to individuals, families and population groups who are particularly vulnerable to or affected by disease or disabilities of a genetic nature and international co-operation between industrialised and developing countries. Solidarity in the first case implies, inter alia, the promotion by states of research on the identification, prevention and treatment of genetically-based and genetically-influenced diseases, in particular rare as well as endemic diseases affecting large numbers of the world's population.[17] In the framework of international co-operation, solidarity is to be articulated through the international dissemination of scientific knowledge concerning the human genome as well as through specific measures aimed at strengthening the capacity of developing countries to carry out research on human biology and genetics, and at enabling them to assess the risks and benefits of research on the human genome carried out in those countries in order to prevent abuses.[18] Relevant international organisations are called upon to support and promote the initiatives taken by states for the purpose of international co-operation in this field.

3. THE INTERNATIONAL DECLARATION ON GENETIC DATA

The Declaration on Genetic Data was adopted by the General Conference of UNESCO on 16 October 2003. It is a sequel to the Universal Declaration the principles of which it applies to the collection, processing, storage and use of human genetic data. Human genetic data is defined as 'information about heritable characteristics of individuals obtained by analysis of nucleic acids or by other scientific analysis'.[19]

The Declaration has three basic purposes. First, it aims to set out the principles which should guide states in the formulation of their legislation and their policies with respect to the collection, treatment, use and storage of genetic data. Secondly, it is meant to form the basis for guidelines of good practice in those areas for individuals and institutions. It is thus aimed not only at states, but also at individuals and institutions handling genetic data. And thirdly, and perhaps most importantly, its objective is to ensure the respect of human dignity and protection of human rights and fundamental freedoms in the collection, processing, use and storage of human genetic data and of the biological samples from which they are derived, in keeping with requirements of equality, justice and solidarity, while giving due consideration to freedom of thought and expression, including freedom of research.[20]

---

[17] See Art 17.
[18] See Art 18 and 19(i).
[19] See Art 2(i).
[20] See Art 1.

The scope of the Declaration is quite interesting because, while it applies in general to the collection, processing, use and storage of human genetic data, human proteomic data[21] and biological samples,[22] it explicitly excludes, under certain conditions, its applicability to these activities when they are carried out in the investigation, detection and prosecution of criminal offences and in parentage testing.[23] The first condition for the exclusion to be operative is that such activities are subject to domestic law. The second condition, which is even more important, is that, where such domestic law exists, it must be consistent with the international law of human rights. Thus, if there is no domestic law or if the provisions of the domestic law are violative of or contrary to human rights law, then the principles and provisions of the Declaration will apply. This is further elaborated in Article 5 of the Declaration, dealing with the purposes for which human genetic data may be collected and used, where it is stated that this may be done also for purposes of forensic medicine, civil, criminal and other legal proceedings, taking into account the conditions spelled out in Article 1(c). The other legitimate purposes recognised by the Declaration are: diagnosis and healthcare, including screening and predictive testing, and medical and other scientific research, as well as any other purpose consistent with the Universal Declaration on the Human Genome and Human Rights and the international law of human rights.

One of the basic provisions of the Declaration deals with the relationship between a person's identity and his genetic make-up. It states that '[e]ach individual has a characteristic genetic make-up. Nevertheless, a person's identity should not be reduced to genetic characteristics, since it involves complex educational, environmental and personal factors and emotional, social, spiritual and cultural bonds with others and implies a dimension of freedom'.[24]

Another basic provision is the one on non-discrimination and non-stigmatisation aimed at ensuring that human genetic data and human proteomic data are not used for purposes that discriminate against a person or impinge upon the individual's dignity or lead to the stigmatisation of an individual, a family, a group or communities.[25]

---

[21] Human proteomic data is defined in Art 2(ii) of the Declaration as 'information pertaining to an individual's proteins including their expression, modification and interaction'.

[22] Biological samples are defined in Art 2(iv) as 'any sample of biological material (for example blood, skin and bone cells or blood plasma) in which nucleic acids are present and which contains the characteristic genetic make-up of an individual'.

[23] See Art 1(c).

[24] See Art 3.

[25] See Art 7.

The Declaration also stresses the special status of the human genetic data for the reasons indicated therein:

—They can be predictive of genetic dispositions concerning individuals;
—They may have a significant impact on the family, including off spring, extending over generations, and in some instances on the whole group to which the person concerned belongs;
—They may contain information the significance of which is not necessarily known at the time of the collection of the biological samples;
—They may have cultural significance for persons or groups.[26]

Concerning procedures, the Declaration calls for collecting, treating, using and storing data on the basis of transparent and ethically acceptable procedures. It also calls for independent, multidisciplinary and pluralist ethics committees to be promoted and established at national, regional, local and institutional level.

More importantly, the Declaration establishes certain principles and norms to be observed in each of the abovementioned procedures.

At the collection stage,[27] for example, it emphasises a prior, free, informed and express consent 'without inducement by financial or other personal gain' for the people providing the data. Consequent on this principle is the right to withdraw consent and the right to decide whether or not to be informed of research results.

At the processing stage,[28] the principle of confidentiality of human genetic data linked to a person, family or group is enunciated; while at the use stage,[29] the principle of non-use for a purpose that is incompatible with the one for which consent had originally been given is established.

Finally, at the storage stage,[30] the principle of consent is re-affirmed for cross-matching or for destruction of samples.

4. IMPLEMENTATION AND FOLLOW-UP MEASURES OF THE DECLARATIONS

An entire section of the Universal Declaration on the Human Genome and Human Rights entitled 'Implementation of the Declaration' is devoted to the measures to be undertaken by states and by the IBC to promote the dissemination, application and respect for the principles and standards established therein. States are called upon to make every effort

---

[26] See Art 4(a).
[27] See Arts 8–12.
[28] See Arts 13–15.
[29] See Arts 16–19.
[30] See Arts 20–22.

to promote the implementation of the principles, to take appropriate measures—through education, training and information dissemination—for their respect and to foster their recognition and effective application. They are also expected to encourage exchanges and networks among independent ethics committees, as they are established, to foster full collaboration among them.[31]

The IBC of UNESCO is tasked with the dissemination of the principles enunciated in the Declaration and with the examination of the issues raised by their actual application and by the evolution of technologies dealing with the life sciences. It is also mandated to provide advice and recommendations to the UNESCO General Conference concerning the follow-up of the Declaration and the identification of practices that could be contrary to human dignity.[32]

In pursuance of its abovementioned mandate, the IBC drew up Guidelines for the implementation of the Universal Declaration, which were adopted by the General Conference of UNESCO at its 30th session.[33] Although the Guidelines are primarily addressed to states and intergovernmental organisations, they also state the following: '[e]xperience shows that to implement an international instrument, synergy needs to be created between all actors at the different levels. Today, international action is characterized by partnership in which each actor, while retaining his identity and specific nature, complements the role played by others'.[34] It is therefore stressed that the Guidelines are equally intended for, among others, public and private decision-makers, especially in science policy, law-makers, ethics committees and similar bodies, scientists and research workers, and individuals, families and populations with genetic mutations that may lead to illnesses or disabilities.[35]

The Guidelines identify the dissemination of the principles set forth in the Universal Declaration as a priority and a preliminary condition for their effective application. Such dissemination is to be specifically directed at scientific and intellectual circles, people involved in education and training, particularly in universities, and decision-making bodies such as parliaments. They also call for consciousness-raising, education and training as well as exchanges of studies and analyses pertaining to questions of bioethics. Most importantly, the Guidelines call on the

---

[31] See Art 23.
[32] See Art 24.
[33] See Resolution 30 C/ 23 adopted by the General Conference of UNESCO at its 30th Session, *Annex: Guidelines for the Implementation of the Universal Declaration on the Human Genome and Human Rights.*
[34] Ibid, para 4.
[35] Ibid.

UNESCO Secretariat to prepare an evaluation of the worldwide impact of the Universal Declaration five years after its adoption in 1997.[36]

Consequently, an 'Overall Report on, and Evaluation of, the Implementation of the Universal Declaration on the Human Genome and Human Rights' was presented to the UNESCO General Conference at its 32nd Session in 2003.[37] In its conclusions, the Evaluation Report states that the principles of the Universal Declaration are now reflected in 'national legislation and regulations in many countries' and can also be found in the 'terms of reference of bioethics and ethics committees in medical faculties and research institutions alike'.[38] It also affirms that 'it is now clear that the Universal Declaration on the Human Genome and Human Rights has become an authority in bioethics'.[39] Notwithstanding these positive conclusions, the General Conference invited the Director-General of UNESCO to continue to evaluate the impact of the Universal Declaration, thus underlining the importance of continuous follow-up and monitoring for the successful implementation of declaratory instruments.

The International Declaration on Human Genetic Data also contains specific provisions on its implementation. Thus, the Declaration calls upon states to give effect to its principles and standards through appropriate measures at the national level, whether of a legislative, administrative or other character. Such measures should be supported by action in the sphere of education, training and public education so that the widest possible dissemination is given to the provisions of the Declaration. States are equally exhorted to enter into bilateral and multilateral agreements to enable developing countries to build up their capacity to participate in generating and sharing scientific knowledge concerning human genetic data and related know-how. Ethics education and training at all levels, as well as information and knowledge sharing, are singled out as important means for states to promote the principles set out in the declaration.

In the same way as in the Universal Declaration, the implementation of the principles of the Declaration on Human Genetic Data is to be monitored by the UNESCO ICB and Intergovernmental Bioethics Committee (IGCB) on the basis of reports submitted by Member States. In addition, UNESCO as a whole is required to take appropriate follow-up action to the Declaration so as to foster progress of the life sciences and their applications through technologies 'based on respect for

---

[36] Ibid, paras 2.1, 2.2 and 5.
[37] See *Overall report on, and evaluation of, the implementation of the Universal Declaration on the Human Genome and Human Rights*, UNESCO doc 32 C/23 of 26 Aug 2003.
[38] Ibid, para 41.
[39] Ibid, para 42.

human dignity and the exercise and observance of human rights and fundamental freedoms'.[40]

5. CONCLUDING REMARKS

The use of the 'declaration' for standard-setting in the field of bioethics in UNESCO might be ascribed to the need to establish within a short period of time certain principles and minimum standards that are formally accepted by all the nations of the world. Two factors underlie this quest for universality and rapid acceptance.

The first factor concerns the universal nature of the issues to be addressed. Present day scientific practices cross national boundaries. Thus, the protection of the human species from the adverse effects of certain scientific and technological practices or from the misuse of scientific discoveries cannot best be effected on the basis of territoriality, but on the basis of universality. Secondly, rapid scientific advances in biotechnology and genetic engineering require fast and flexible legal responses both internationally and nationally. International conventions may not always be capable of satisfying the realisation of these goals, while declarations may lead to their fulfilment depending on the circumstances of their adoption, their subject matter and the implementation mechanisms established for their follow-up. At the same time, the declarations can serve as the precursors of conventions, laying the groundwork for future codification and translation of the standards and principles into actual rules and regulations. This method of gradual crystallisation of general standards into hard rules has occasionally been used by UNESCO in its standard-setting activities, particularly in the field of culture.[41]

It is however too early to say whether a similar approach might be adopted by UNESCO in the future in the field of bioethics. One more declaration—the Declaration on Bioethics and Human Rights—has recently been adopted and much will depend on the effect it will have, together with the two Declarations discussed in this chapter, on consensus building at the international level and on the process of maturation of the emerging rules and standards in the field of bioethics.

---

[40] See Art 26 of the Declaration.

[41] See, eg, the Recommendation on the Means of Prohibiting and Preventing the Illicit Export, Import and Transfer of Ownership of Cultural Property of 19 Nov 1964, followed in 1970 by the adoption of the Convention on the Means of Prohibiting and Preventing the Illicit Import, Export and Transfer of Ownership of Cultural Property; the Recommendation on the Safeguarding of Traditional Culture and Folklore of 15 Nov 1989, followed by the adoption on 17 Oct 2003, of the Convention for the Safeguarding of the Intangible Cultural Heritage; and the Universal Declaration on Cultural Diversity adopted on 2 Nov 2001, followed by the current negotiations on a 'Convention on the Protection of the Diversity of Cultural Contents and Artistic Expressions' (Resolution 32 C/34 of 17 Oct 2003).

6

The 'Normative Spectrum' of an Ethically-inspired Legal Instrument: The 2005 Universal Declaration on Bioethics and Human Rights

HÉLÈNE BOUSSARD[*]

1. INTRODUCTION

T HE ATTENTION PAID by the international community to the impact of scientific and technological development on human rights, although recent, is not new and can be traced back to the end of the 1960s.[1] Human rights law needs to be constantly refined and redefined so as to keep pace with advances in sciences. With the 'genetic revolution'[2] and human health-related biotechnology[3] in the 1990s, it was the underlying conception of man and humanity in the field which was modified. At the same time we witnessed a shift in the normative

---

[*] PhD researcher of the Department of Law, European University Institute. Email: helene. boussard@eui.eu.

[1] The introduction of the question of the impact of recent scientific and technological development on human rights first received international recognition with the 'Proclamation of Tehran', Final Act of the International Conference on Human Rights, Tehran, 22 Apr–13 May 1968. See S Ogata, 'Introduction: United Nations Approaches to Human Rights and Scientific and Technological Developments' in CG Weeramantry (ed), *Human Rights and Scientific and Technological Development* (Tokyo, United Nations University Press, 1990), 2, available at www.unu.edu/unupress/unupbooks/uu06he/uu06he0c.htm.

[2] The expression 'genetic revolution' refers to discoveries in genetic research and their potential application in medicine such as gene therapy, cells therapy, genetic screening and genetic testing, and use of embryonic stem cells. For use of the notion by scholars, see J Burley (ed), *The Genetic Revolution and Human Rights* (Oxford, Oxford University Press, 1999).

[3] The term 'Biotechnology' is used in this chapter in accordance with the definition provided in the Convention on Biological Diversity (see www.biodiv.org/doc/legal/cbd-en.pdf) as any technological application that uses biological systems, living organisms or derivates from them in order to make or modify products or processes for specific use. Biotechnology is divided into three sectors: industry, agriculture and medicine. This chapter will focus on the last, which is also called 'human health-related biotechnology'.

response to scientific progress from reliance on the general corpus of international human rights law[4] to the elaboration of a more specific and intrusive reaction through the creation of an international law of life sciences based on a human rights approach and influenced by bioethics.[5] The aim was to guide states in their regulation of the world of science, given that the risk of abuse of technology justifies abandoning the system of self-regulation which was predominant in the field. What is innovative is the resort to two bodies of norms,[6] on the one hand bioethics, which governs activities in life sciences and biomedicine[7] and on the other hand international human rights law, which aims to protect the individual.[8] This evolution evidenced a revival in the standard-setting function of intergovernmental organisations at the regional (in particular with the adoption by the Council of Europe of the European Convention on Biomedicine and Human Rights)[9] and international levels. In response to the UN Commission on Human Rights' incitation to the organisations to act under a new agenda item: 'Bioethics and Human Rights',[10] the World Health Organisation (WHO) and the United Nations Educational Scientific and Cultural Organisation (UNESCO) favoured two different approaches.[11] While the former integrated an ethical approach in its

---

[4] The only international human rights instrument specifically related to scientific and technological progress was until recently the Declaration on the Use of Scientific and Technological Progress in the Interest of Peace and for the Benefit of Mankind (10 Nov 1975), which expresses the wish that all states make use of scientific and technological progress for good purpose. Besides, we find the UNESCO Recommendation on the status of scientific researchers, adopted on 20 Nov 1974, which combines the problem of the freedom of researchers and the implication of science and technology for world problems.

[5] L De Castro, 'Towards a Declaration on Universal Norms on Bioethics' Report for the Extraordinary Session of the International Bioethics Committee of UNESCO (IBC), SHS/EST/04/CIB-EXTR/1, Paris, 12 July 2004, 5.

[6] For a clear juxtaposition of the two sets of norms, see the Declaration on Science and Use of Scientific Knowledge and the Science Agenda, World Conference on Science, 1999, Budapest.

[7] The word 'biomedicine' refers to the application of life sciences in medicine.

[8] H Yamane, 'Impact of Scientific and Technological Progress on Human Rights: Normative Response to the International Community' in CG Weeramantry, n 1 above.

[9] Convention for the Protection of Human Rights and Dignity of the Human Being with regard to the Application of Biology and Medicine, Council of Europe, Oviedo, 4 Apr 1997, CETS no 164, (1997) 36 ILM 821.

[10] Commission for Human Rights, *Human Rights and Bioethics*, Res. 93/91, 67th Meeting, 10 Mar 1993. For an example of national response to the call for action by the Commission for Human Rights see American Association for the Advancement of Science, *The promotion of Human Rights in the Life and Health Sciences, Recommendations to the United Nations* (Washington, DC, American Association for the Advancement of Science, 1994).

[11] The two organisations have a statutory mandate to deal with ethics. This raises a problem of competence between them. See WHO Response, Results of the Written Consultation on the Third Outline of the Text of a Declaration on Universal Norms on Bioethics (27 Aug 2004), SHS-2005/WS/15, UNESCO, Paris, 10 Jan 2005.

activities and decentralised work with non-governmental organisations (NGOs),[12] the latter set up an ethics committee[13] to engage in a norm-setting function in bioethics. The International Bioethics Committee (IBC)[14] of UNESCO, which is the sole ethics committee at the international level,[15] is a body of independent and multidisciplinary experts whose tasks is to promote reflection on the ethical and legal issues raised by research in the life sciences and their applications. The IBC not only reports on new scientific practices in life sciences and makes recommendations but also participates in the elaboration of instruments. It was charged with drafting the 1997 Universal Declaration on the Human Genome and Human Rights (UDHGHR) and the 2003 International Declaration on Human Genetic Data (IDHGD).[16] After these two specific instruments, the General Conference of UNESCO invited the Director General to draw up an instrument to elaborate 'universal norms on bioethics'.[17] This chapter will focus on the latter, the Universal Declaration on Bioethics and Human Rights (UDBHR), which was adopted by acclamation on 19 October 2005 by the General Conference of UNESCO, and may be seen as the most ambitious work undertaken at international level to combine ethical and legal norms to create a 'universal law of bioethics'.[18]

The three UNESCO declarations[19] are human rights instruments that have been explicitly developed in line with the Universal Declaration on Human Rights (UDHR) 1948. Although these declarations are legal instruments, since they are resolutions of an international organisation, they are not legally binding *stricto sensu*. Resolutions in fact are not among the international sources of law listed in Article 38 of the Statute

[12] See Department of Ethics, Trade, Human Rights and Health Law, *Ethics at the World Health Organisation*, available at www.who.int/ethics.

[13] As will be further explored (see text following n 52), the expression 'ethics committee' refers to an independent, multidisciplinary and pluralist body of experts whose task is to deal with bioethical issues.

[14] The IBC was established in 1993 for the drafting of the UDHGHR. It was set up as a permanent committee and its statutes were adopted by the Executive Board on 7 May 1998.

[15] At the regional level, the European Commission in Dec 1997, set up an ethics committee, the European Group on Ethics in Science and New Technologies (EGE) to succeed the Group of Advisers on the Ethical Implications of Biotechnology (GAEIB). Moreover, in 1991 the Council of Europe established the Steering Committee on Bioethics (CDBI), which is often considered as an ethics committee, although it is a group of government experts and, as a result, does not fulfil the requirement of independence of an ethics committee.

[16] For a study of the two declarations, see A Yusuf in this volume.

[17] UNESCO Document, *Bioethics programme*: Priorities and Perspectives, General Conference of UNESCO, 31st session, Res. 31C/22, 2003.

[18] N Lenoir, Opening speech, Proceedings of the Round Table of Ministers of Science, UNESCO, Paris, 22–23 Oct 2001, at 20–24.

[19] The adoption of 'declarations' is not provided for by the UNESCO Constitution (16 Nov 1945) and corresponds to a practice which has been developed by the organisation so as to reach an intermediary normative level between 'recommendations', which are only guidelines, and 'international conventions', which are legally binding instruments. The latter are the sole instruments referred to in the Constitution (Art IV.4).

of the International Court of Justice (ICJ). Therefore they impose binding obligations on states not legally, but only politically.[20] These 'soft' obligations will be referred to as 'duties'. The originality of the UNESCO 'bioethics declarations' lies in their drafting process and their aim, which is to translate ethical standards into legal terms in order to protect human rights in scientific research and medical interventions. The new norm-making process which was instigated with these declarations can be seen as a convergence of legal and ethical norm-making processes. Human rights law and bioethics are closely intertwined, and a clear analysis of the normative features of the UNESCO declarations is not an easy task. The aim of this chapter is to clarify the interaction between international human rights law and bioethics through the study of the drafting process, content, form and implementation machinery of the UDBHR. To this end, the two normative systems will be schematically presented, and so will the challenges and possible advantages of their interaction.

It is regrettable that the introduction of bioethics into the agendas of international organisations did not give rise to more studies. Few have been conducted, and the most significant work was completed only in 2002 by the High Commissioner on Human Rights' expert group on biotechnologies, while it discussed 'the complex relationship between ethical and human rights approaches'.[21] Unfortunately no long-term reflection on the subject was carried out, in part due to the lack of necessary budgetary resources.

Bioethics is defined in the Preliminary Draft of the UDBHR for the limited purpose of the instrument as 'the systematic, pluralistic and interdisciplinary study and resolution of the ethical issues raised by medicine and the life and social sciences as applied to human beings and their relationship with the biosphere, including issues relating to the availability and accessibility of scientific and technological developments and their applications' (Article 1).[22] The High Commissioner's expert

---

[20] These declarations belong to the body of standards, commitments, joint statements and declarations of policy or intention created within international organisations and called 'soft law', as opposed to the 'hard law' which makes up international law proper.

[21] High Commissioner's for Human Rights' Expert Group on Human Rights and Biotechnology, 'Human Rights and Biotechnology', Geneva, 24–25 Jan 2002, 2. Although a special *rapporteur*, Ms Iulia-Antonella Motoc, has been appointed by the Sub-commission on the Promotion and Protection of Human Rights to consider the question of bioethics and human rights, Motoc's work focused on specific issues rather than considering the normative approach. See I-A Motoc, `Specific Human Rights Issues: Human Rights and Bioethics´, expanded working paper submitted in accordance with Sub-Commission decision 2002/114, CHR, Sub-Commission on the Promotion and Protection of Human Rights, 55th session, E/CN.4/Sub.2/2003/36, Geneva, 10 July 2003; I–A Motoc, 'Specific Human Rights Issues: Human Rights and the Human Genome', CHR, Sub-Commission on the Promotion and Protection of Human Rights, 56th session, E/CN.4/Sub.2/2004/38, Geneva, 23 July 2004.

[22] Preliminary Draft Declaration on Universal Norms on Bioethics, SHS/EST/05/CONF.204/3, Paris, 4 May 2005. See n 34 below and the text following it for the elaboration of this instrument.

group, alongside the WHO,[23] echoed a widespread idea that bioethics and human rights law pursue the same end. This can be ascertained insofar as they both protect a certain conception of humanity.[24] However the two sets of norms rely on a different value system. On the one hand, international human rights law is monolithic, since it pretends to universalism. It rests on a rights-based approach (RBA) and is individual-centred. 'A [sic] RBA involves the viewing of a particular issue from the perspective of the rights-obligations imposed by international human rights'.[25] The individual is considered in his or her universality: legal human rights are inherent in individuals and are recognised in an entity on the basis of the sole criterion that the such entity is considered as human.[26]

On the other hand, the value system is plural in bioethics in so far as any single culture is fully acknowledged as generating its own normative framework. Bioethical principles are set in the legal, cultural, philosophical and religious bedrock of the various human communities. The common denominator of all bioethical approaches is the consideration of practices rather than actors. In that sense, the bioethics project is a better understanding of the issue, which questions not only the doctor–patient or doctor–doctor relationship, but also issues of public policy on the control and direction of science.[27]

International human rights law and bioethics can both be used as a 'tool for finding solutions for ethical dilemmas in the field of life sciences'.[28] In bioethics, none of the principles provides an overriding justification. In international human rights law, the principle is that none of the rights is absolute, even if there is room for debate on exceptions. In the case of conflict of rights or principles, human rights law and bioethics will seek to balance the various rights/principles in order to maximise respect for all right-holders and interested parties. However, whereas traditional human rights law hardly goes beyond a state–individual relationship in a quite static dialectic between obligations of the former and rights of the latter,[29] bioethics

---

[23] WHO response in Commission on Human Rights, *Human Rights and Bioethics*, 78th Meeting, Res. 2001/71, 25 Apr 2001.

[24] N Lenoir and B Mathieu, *Les normes internationales de la bioéthique* (Paris, PUF, 2004), 9.

[25] High Commissioner for Human Rights' Expert Group on Human Rights and Biotechnology, n 21 above.

[26] Among others, see D Anzilotti, *Cours de droit international*, (Paris, LGDJ, 1929, re-ed 1999), 132; J Rivero, 'Rapport général', in Librairie Générale du Droit et de la Jurisprudence, *Les droits de l'homme, droits collectifs ou droits individuels* (Paris, LGDJ, 1980), 23.

[27] H Kuhse and P Singer, *A Companion to Bioethics* (Oxford, Blackwell 1998), 4.

[28] High Commissioner's Expert Group on Human Rights and Biotechnology, n 21 above.

[29] Such innovations take into account individual–individual relationship by reference to the German '*Drittwirkung* effect' (see for examples European Commission of Human Rights, *Whiteside v UK*, App no 20357/92, 7 Mar 1994 and Human Rights Committee, *Delgado v Colombia*, App no 195/1985, 12 July 1990) and obligations put on non-state entities (see F Francioni and T Scovazzi, *International Responsibility for Environmental Harm* (Boston/London, Kluwer Academic Publishers, 1991); R Provost, *State Responsibility in International*

considers all interests in a spatial (individual, member of a society, member of humanity) and temporal dimension (evolution of the individual on the one hand, past, present and future generations, on the other hand) through a dynamic balance of the rights and duties of different actors. Rooted in 'heuristics of fear',[30] bioethics is characterised by a strong inter- and intra-generational dimension and considers those who can benefit or suffer from science, namely present and future generations.

Given these differences, the introduction of bioethics into international human rights represents a challenge, but also a precious opportunity to bring the claim to universalism closer to reality. In this regard the fact that the word 'universal' was chosen for the 2005 declaration is particularly significant. For the IBC, this word refers 'not merely to the general applicability of the norms but also emphasises the global recognition of bioethical principles'.[31] On the one hand, the issue of recognition recalls the irreducible antagonism between cultural diversity and universalism, which is undoubtedly an issue when we attempt to develop a 'universal code of bioethics'. On the other hand, a 'general applicability', which is understood as the applicability of the declaration to any individual, group, corporation and states, demands a reconsideration of the theory of the subjects of international human rights law so as to extend the categories of the latter to new entities, which are present in bioethics. In the traditional positivist doctrine of international law, the individual human being is never directly a subject of international law, and can only be an object.[32] The basic criterion of positivism is the state.[33] These issues will appear alongside the study of the ethical and legal features of the UDBHR.

Law (Aldershot, Dartmouth, 2002); P Alston, *Non-state Actors and Human Rights* (Oxford, Oxford University Press, 2005).

[30] The word 'bioethics' was invented by Van Rensselaer Potter to refer to the reflection on the risk of auto-destruction generated by certain scientific developments: see V.R. Potter, 'Bioethics: The Science of Survival' (1970) 14 *Perspectives in Biology and Medicine* 127. Hans Jonas theorised bioethics as the expression of this anxiety: see H Jonas, *Le principe respon-sabilité. Une éthique pour la civilisation technologique* (Paris, Le Cerf, 1990). For a presentation of these contributions, see Lenoir and Mathieu, n 24 above, at 11–12.

[31] IBC Drafting Group, *Explanatory memorandum on the elaboration of the preliminary draft declaration on universal norms on bioethics*, SHS/EST/05/CONF.204/4, Paris, 4 May 2005, at 3.

[32] L Oppenheim, *International Law, Vol I, Peace* (9th edn, Harlow, Longman, 1992), 16–22; H Lauterpacht (ed), *International Law and Human Rights* (London, Longman, 1955), para 13.

[33] On the opposite side, the naturalist theory gives prevalence to the individual as a per-manent entity of the international system, while the phenomenon of nation-state is transi-tory and historically conditioned. Some have argued that the notions of 'subject' and 'object' have no credible reality and no functional purpose. Higgins emphasised that it was more helpful and close to perceived reality to return to the view of the international as a particular decision-making process. And within that process (which is a dynamic and not a static one) there are a variety of participants, making claims across state lines, with the object of maxi-mising various values. See R Higgins, *Problems & Process: International Law and How we Use It* (Oxford, Oxford University Press, 1994), 48–55. For an interesting overview of the issue,

During the two-year period devoted to the drafting of the UDBHR (undoubtedly a short timeframe for such an ambitious task),[34] the IBC attempted to establish the conformity of bioethics with international human rights law.[35] However, government experts rejected many of the innovations proposed by the IBC in the Preliminary Draft Declaration ('Preliminary Draft').[36] Although the UDBHR was envisaged to be an ethically inspired legal instrument, it was deemed to be more important to conform to the traditional human rights law approach. Therefore the 'revised' text of the Member States' experts, which constitutes the UDBHR, differs in many respects from the draft of the IBC. The IBC Preliminary draft and the UDBHR will be compared throughout this chapter.

Notwithstanding the differences between these texts, if we consider the content, form, drafting process and means of implementation of the UDBHR, we obtain a 'normative spectrum', which consists of a range of elements from a purely bioethics perspective to a strict human rights law approach. On a procedural level, we witness the dominance of bioethics, insofar as there is a multiplication of the actors involved. On a substantive level, the convergence and reconciliation of ethical and legal principles lead to an evolution, respectively, in the content and in the form of international human rights law. Both of these issues will be elaborated upon in the following sections.

2. DOMINANCE OF BIOETHICS: DIVERSIFICATION OF THE ACTORS INVOLVED IN INTERNATIONAL HUMAN RIGHTS LAW

Given their different material sources, bioethics and international human rights law involve different sets of actors.

International human rights law refers to 'internationally agreed upon set of principles and norms embodied in international legal instruments'.[37] Accordingly, as the main subjects of international law, states are the key decision-makers and duty-bearers in international human rights law.

---

see G Maggio and OJ Lynch, 'Human Rights, Environment and Economic Development: Existing and Emerging Standards in International Law and Global Society', Nov 1997, available at www.omced.org/wri/om_wri.htm.

[34] The Preliminary Draft Declaration on Universal Norms on Bioethics ('Preliminary Draft') was adopted by the IBC on 9 Feb 2005: Preliminary Draft Declaration on Universal Norms on Bioethics, SHS/EST/05/CONF.204/3, Paris, 4 May 2005.

[35] IBC Drafting Group, n 31 above, 2.

[36] The Preliminary Draft was proposed by the IBC to the Intergovernmental Meeting of Experts of the Member States and Associate Members of UNESCO in June 2005. After revising the Preliminary Draft, the Member States' experts adopted the Universal Draft Declaration on Bioethics and Human Rights ('Final Draft'): Universal Draft Declaration on Bioethics and Human Rights, SHS/EST/05/CONF.204/3 REV, Paris, 24 June 2005.

[37] Response of the World Health Organisation, n 11 above, at 6.

The dominance of the role of states is strengthened by the fact that the universality, indivisibility and interrelatedness of human rights laid down in the Vienna Declaration 1993[38] imply a limited number of actors so as to guarantee a harmonised interpretation in the field. This paradigm is challenged by the reality of the multi-level (international, regional, national) sources of human rights law, but the main actors remain states. In contrast, there is no such delimitation in bioethics, and if professionals and their organisations (mainly the World Medical Association (WMA)) have, for a long time, been the only drafters of norms, non-state (scientific researchers, physicians, decision-makers, pharmaceutical corporations) and also potentially state, actors are a potential source of bioethics.[39]

Consulting the different stakeholders gives the UDBHR legitimacy and ensures its efficiency. That is why, departing from the traditional interstate procedure in international human rights law, the drafting of the UDBHR involved a plurality of state and non-state, legal and ethical, actors.

2.1 The 'Institutional Loop' between Legal and Ethical Actors

The following sections intend to focus first, on the actors who were consulted in the drafting of the declaration and, then, on those who are involved in the implementation machinery. This static presentation can be illustrated through the more dynamic process of an 'institutional loop' between legal and ethical actors connecting the international and national levels, which was instigated with the UDHGHR and the IDHGD. UNESCO, a legal actor at the international level, ensured the participation of all actors in the field of bioethics in the drafting of the UDBHR, which promotes the creation of ethics committees, new and major actors in the implementation machinery.

2.1.1. Emergence of a new Norm-making Process in the Field of International Human Rights Law

Hitherto, UNESCO has always had a strong member-driven approach, and it is significant that it was within this organisation that, under the influence of the Director General, Federico Mayor, a new norm-making process was instigated with the three 'bioethics declarations'.[40] States remain the final decision-makers but the drafting follows a three-step process:[41]

---

[38] Art 1(5) of the Vienna Declaration and Programme of Action, UN World Conference on Human Rights, 14–25 June 1993, UN Doc A/CONF/157/24 (Part I) at 20 (1993).

[39] C Byk, 'Progrès scientifique et droits de l'homme, la rupture?' (2003) 54 *Revue Trimestrielle des Droits de l'Homme*, N°Spécial Bioéthique, 365–70.

[40] It was Federico Mayor who proposed the creation of the IBC. See Lenoir and Mathieu, n 24 above, at 32–3.

[41] De Castro, n 5 above.

consultation with the different stakeholders, production of the preliminary draft by the IBC, finalisation of the draft by government experts and
its submission for adoption by the General Conference. In the traditional
international law-making process, diplomats are in charge of the drafting
of the text which is adopted by states. Here the draft declaration results
from the oral and written contributions of international intergovernmental
organisations, international and domestic non-governmental organisations,
national ethics committees, individual experts and sovereign states. It
ensures transparency and brings the consensus-making process closer to
the reality of interests in issue. However, alongside the elaboration of the
three UNESCO declarations, an evolution has occurred which pulls the
formation process in two directions regarding state control.

On the one hand, the drafting of the UDBHR confirmed the multi-level
norm-making process. Given the aim of the instrument (laying down
universal ethical guidelines), a first international consultation of national
ethics committees, international governmental and non-governmental
organisations was held before the first draft was produced so as to start
from an embryo of consensus.[42] A second was organised further on in the
process, and so was a consultation of authorities of the main religions.[43] An
important number of meetings at the regional and national levels allowed
the consultation process to be completed by means of questionnaire sent
to state and non-state actors. Moreover, two meetings of the UN Inter-
agency Committee on Bioethics[44] ensured dialogue between the different
international organisations, which was undoubtedly more efficient than
written contributions.[45]

However, leaving the drafting of the document in the hands of
independent experts is a double-edged sword. Extensive communication
between the IBC and state representatives is a condition *sine qua non* of
the productivity of the formation process, insofar as it lessens the chance
of further rejection by states of the draft text.[46] It is in response to states,
which asked to be better informed during the drafting of the text, that the

---

[42] The First meeting of the IBC Drafting Group was held on 30 Apr 2004, just after the
Extraordinary Session of the IBC (27–29 Apr 2004). The latter was attended by more than 200
participants from over 70 countries. See ibid.

[43] L De Castro, *Report of the Eleventh Session International Bioethics Committee of UNESCO*
(UNESCO, Paris, 6 Jan 2005).

[44] Created in 2003 under the initiative of the Director-General of UNESCO, the UN Inter-
Agency Committee on Bioethics was established in Mar 2003: it aims to co-ordinate the
activities of the organisations of the UN system and associate members.

[45] The two meetings were respectively held on 24–25 June 2004 and 10 Dec 2004.

[46] M Jean, *UNESCO and Universal Principles in Bioethics: What's Next?*, report of the Twelfth
session of the International Bioethics Committee, Tokyo, Japan, 15–17 Dec 2005, SHS/EST/
CIB-12/05/CONF.509/INF.02, at 8: 'I understand that at the end of the day, member states
are the one who will adopt or reject a text, but this should be done having a full understanding of what is behind the wording of the IBC'.

Intergovernmental Bioethics Committee (IGBC), a body of government experts, was created in 1998 to co-operate with the IBC. This intermediary organ was in the interests of both states and the IBC. Unfortunately, such institutional improvement without any concrete interaction during the drafting process is of little use. The IDHGD suffered from a lack of communication between the two and, in order not to repeat the mistake, a joint meeting was held,[47] and some IBC members attended the IGBC meeting for the drafting of the UDBHR.[48] Some changes made by government experts which will be documented in this chapter illustrate that some room for improvement remains.

2.1.2. Legal and Ethical Implementation Machinery

Implementation of international norms means 'incorporating them in domestic law through legislation, judicial decision, executive decree, or other process'.[49] Therefore the issue of implementation means the margin of discretion states have with regard to the nature of the measures by means of which they intend to implement an international instrument, such as a declaration, which has no direct application. However, especially in the case of a non-binding instrument, it is of paramount importance for the sake of harmonised implementation to provide states with some guidelines. Given the dual nature of the declaration, its authors could favour legal norms (*lato sensu*, namely legislation and regulation) as traditionally found for international instruments, or could pave an alternative way more in conformity with bioethics, which implies institutional changes. According to the WMA, which had the opportunity to set out its views during the first international consultation for the elaboration of the UDBHR, the implementation process of the declaration is unclear. On the one hand, the IBC reported that 'at any level, laws accompanied by effective control should be adopted in order to facilitate personal choices and only a few substantial issues should be regulated through international rules. In other words, the aim should be to maximize moral evolution and to minimize the need for legislation'.[50] On the other hand, the Director General reported that the IBC saw the Declaration 'as a tool to help states wishing to enact laws in that field'.[51]

---

[47] 26–27 Jan 2005.

[48] 24–25 Jan 2005.

[49] D Shelton, *Commitment and Compliance: the Role of Non–binding Norms in the International System* (Oxford/New York, Oxford University Press, 2000), 5.

[50] G Berlinguer and L De Castro, *Report of the IBC on the Possibility of Elaborating a Universal Instrument on Bioethics*, IBC, SHS/EST/02/CIB-9/5 (Rev.3), Paris, 13 June 2003, para 40.

[51] Report by the Director-General on the possibility of Elaborating Universal Norms on Bioethics (32C/59), UNESCO, Paris, 22 Sept 2003, para 27.

In reality, there is no contradiction, but rather a complementarity, between the two forms of implementation, which relate to the adoption of specific legislation and the creation of ethics committees.[52] Such provisions are in line with those adopted in the two previous declarations. The legal approach provides a strong framework, whilst ethics committees allow for the necessary flexibility in a field that is continuously evolving. Ethics committees are presented in UNESCO declarations as the main actors in the regulation of life sciences. Their task is threefold: to promote ethical reflection in the life and health sciences, to fulfil an advisory role for public and private decision-makers and to guide research workers and practitioners in their actions, and to encourage broad public debate. Specific reference to them is made in the UDHGHR and the IDHGD to review research protocols.[53] The conditions of multidisciplinarity, independence and plurality govern their composition.[54] Existing ethics committees are usually established at four different levels: the local (ie, hospital practices), institutional (ie, research funding institutions), professional (ie, medical association) and national (ie, governmental bodies or non-governmental organisations). UNESCO encourages them to develop networks and to co-operate with the IBC.[55]

Ethics committees can also be set up independently of the adoption of specific legislation. In that case, they appeal to the alternative to human rights legislation. In fact, in countries such as China[56] and many African and South American countries, governments are more willing to develop an ethical framework than to enact laws within a human rights framework.

The implementation of the UDBHR is based on vertical (national–international level) and horizontal (state and non-state actors) co-operation. Therefore, we are witnessing a shift from a state to a multi-level process. The plurality of actors—a bioethics feature—affects international

---

[52] In its 2003 Report, the IBC stresses that the UDBHR will 'contribute to a strengthening of the role and the degree of participation of ethics committees at national and institutional levels': Berlinguer and De Castro, n 50 above, para 46.

[53] It results from a combined reading of Art 5d and Art 16 of the UDHGHR. It is provided for in Art 6 of the IDHGD.

[54] 'States should recognize the value of promoting, at various levels, as appropriate, the establishment of independent, multidisciplinary and pluralist ethics committees to assess the ethical, legal and social issues raised by research on the human genome and its application': Art 16 UDHGHR.

[55] In the context of the promotion of the principles set out in the UDHGHR, a special emphasis is put on the need to 'encourage exchanges and networks among independent ethics committees, as they are established, to foster full collaboration: Art 23 UDHGHR. The Guidelines for the Implementation also encourage the 'networking of these institutions so as to facilitate communication and exchanges of experience among them, especially for carrying out joint activities': para 3.3.2 (Implementation of the Universal Declaration on the Human Genome and Human Rights, 16 Nov 1999, UNESDOC Res. 30 C/23).

[56] O Döring, 'Chinese Researchers Promote Biomedical Regulations: What are the Motives of the Biopolitical Dawn in China and Where are they Heading?' (2004) 14 *Kennedy Institutes of Ethics Journal* 39.

human rights law in two ways. From a theoretical perspective, the new norm-making process implies a slight change in the nature of human rights law itself. Professor Van Der Burg illustrated the difference of perspective between ethics and law by the image of an electron that can be seen as a particle ('product') and as a 'wave' ('process'), but not as both at the same time.[57] Schematically presented, bioethics is a 'process' which is based on the assessment and the weighing of benefits, risks and harm in a case-by-case resolution of scientific dilemmas. As such, it is a tool for the resolution of specific problems. In contrast, law may be seen mainly as a 'product' insofar as law is structured as a collection of statutes, judicial decisions and customary rules or a system of rules and principles. The difference between 'law as a product' and 'bioethics as a process' is obviously a matter of emphasis as there is no such clear-cut distinction: to some extent bioethics is a product as it is based on a set of principles and human rights law can be seen as a process. Proclaimed through general principles at the international level, human rights law is implemented at the regional and national levels by states in line with the social and cultural traditions of the country. However, the UDBHR represents a step further in the conceptualisation of human rights law as a 'process' which is strengthened by the involvement of ethics committees in the implementation of this human rights law instrument. From a practical perspective, the recognition of new actors in the implementation would mean that there are potentially new duty-bearers in the field of human rights law.

2.2 The Issue of New Duty-Bearers in a New Legal Context

The determination of the addressees is part of the more general and theoretical issue of the identification of duty-bearers in international law, and more particularly international human rights law. Traditionally, addressees of international human rights instruments can only be states, since they are the only subjects of international law. However a feature of international human rights law instruments is that they seem to impose duties on individuals. To start with, the 'Universal Declaration of Human Rights, proclaimed as a common standard of achievement for all peoples and all nations, seeks to enlist every individual and every organ of society in a universal human rights movement. This document, as a major product of the UN, assumes that the furtherance and realization of human rights is a task to be carried out at all levels in multifarious

---

[57] W Van Der Burg, 'Law and Bioethics' in H Kuhse and P. Singer, n 27 above, at 51.

ways.'[58] In line with the declaration, the UDHGHR, the IDHGD and the Declaration on the Responsibilities of the Present Generations towards Future Generations (DRPGFG) are addressed to state and non-state actors. Similarly, the UDBHR was drafted through the lens of the potential violators (human beings) to provide a benchmark for bioethics at all levels (decision-makers, medical and scientific communities, public and private sectors, patients, research participants and their relatives, the media, any human being), so as to be omni-present.[59] Despite these formulations and laudable claims, positivist theory tells us that there is no intention to recognise the legal personality of the individual and that international norms, in reality, impose upon the state the obligation to prohibit or to sanction individual behaviours or authorise them.[60] The language of human rights instruments remains unchanged, and so does the 'statist approach'. However a significant innovation was proposed by the IBC to give teeth to the recognition of new duty-bearers in the field of bioethics by changing the formulation through the repetitive introduction of every principle with the expression 'any decision or practice'. This expression is defined in Article 2 of the Preliminary Draft, which states:

The principles set out in this Declaration apply as appropriate and relevant:

(i) to decisions or practices made or carried out in the application of medicine, life and social sciences [...]
(ii) to those who make such decisions or carry out such practices, whether they are individuals, professionals groups, public or private institutions, corporations or states.

Government experts instead adopted an expression that was more in line with the traditional human rights approach, since states (direct addressees and duty-bearers) and other entities (possible and subsidiary addressees of guidelines) are not on the same level and are not addressed in the same way:

The Declaration is addressed to States. As appropriate and relevant, it also provides guidance to decisions or practices of individuals, groups, communities, institutions and corporations, public and private [Article 1(2) UDBHR].

---

[58] TC Van Boven, 'United Nations and Human Rights: a Critical Appraisal' in A Cassese, *UN Law, Fundamental Rights, Two Topics in International Law* (Alphen aan den Rijn , Sitjhoff & Noordhoff, 1979), 119.

[59] R Ida, 'Contribution pour une Déclaration Universelle de Bioéthique', report for the Extraordinary session of the IBC, UNESCO, Paris, 27–29 Apr 2004, 3.

[60] Anzilotti, n 26 above, at 134: *'l'obligation de l'individu ne naît pas si l'Etat n'a pas posé la norme qui l'interdit'*.

In reality, there were two different, even if interrelated, issues here. One is the clear identification of those who hold responsibility among the myriad of actors involved in life sciences and their applications, which was the main preoccupation of the IBC. The other is the delicate question which was behind states' rejection and which relates to the possibility for individuals to be addressed as duty-bearers, while they are not subjects in international law.

2.2.1. Distinction between State and Non-state Actors: Clarification of the duty-bearers?

The formulation proposed by the IBC was criticised even before the Intergovernmental Meeting of Experts took place and was the object of extensive debate within the IBC. The WHO had strongly underlined that such a formulation failed to identify the addressees, duty-bearers and right-holders for each provision of the UDBHR.[61] In its opinion, such an omission weakened the enforcement of human rights obligations. To what extent was the criticism justified? It was clear that the formulation of the provisions of the UDBHR constituted a departure from those of the two previous UNESCO declarations and any other human rights instrument, insofar as the expression 'any decision or practice' was used throughout the operational part of the Declaration. However, would the IBC proposal have left the determination of duty-bearers unresolved? Does the formulation adopted adequately clarify the duty-bearers?

The issue was discussed by the IBC Drafting Group during its sixth meeting. The debate centred on the fact that there is no right without an obligation, and no obligation without a duty-bearer. The Preliminary Draft carried the risk that, by referring indistinguishably to all decision-makers rather than states exclusively in each Article, it might not have been possible to hold anyone to account. Therefore, the formulation proposed did not appear to be satisfactory from a legal perspective. Some members were sensitive to the fact that the addressees of each provision were not necessarily the addressees of the entire declaration, and that it could be counter-productive to leave any ambiguity regarding the duty-bearers; however, no modifications were proposed for various reasons. On the one hand, it was stressed by one member of the group that a declaration on bioethics did not aim to recognise pre-existing rights and duties. Bioethics stemmed from human rights, but could not give teeth to their recognition. On the other hand, it was claimed by another member that the addressee of the Declaration was the human species as such, in other words, the international community *lato sensu*, which embodies states and individuals.

---

[61] WHO Response, n 11 above.

The formulation finally adopted by the IBC ('any decision or practice') aims to stress the diversity of actors in the field and calls for accurate identification of duty-bearers on a case-by-case basis. In comparison with the IBC formulation, the text of the UDBHR strengthens the role of states ('[t]he Declaration is addressed to states') and weakens the recognition of new duty-bearers, which takes place 'as appropriate and relevant'. Therefore the main distinction between the two formulations lies in the role recognised to non-state actors. Whereas in the first formulation ('any decision or practice') state and non-state actors are on the same level, the second formulation (dissociation between state and non-state actors) ensures the recognition of states as duty-bearers. It is doubtful whether the latter formulation, which has been adopted in the passive voice used throughout the text, clarifies the duty-bearers for each provision. However, as we have to assume that this was done for the sake of coherence of state action, it implies that states are the duty-bearers of the declaration in any case, in order to ensure that there is no duty without a duty-bearer. This approach is reflected by the use of 'should', which only 'invites' and does not 'order', and which is characteristic of human rights instruments. In the Preliminary Draft, 'shall', which carries a legally binding connotation, was used in the principles addressed to any decision-maker. As soon as states are the only addressees, 'should' reappears in the general principles with a well-known yet ill-defined term: 'is to be'.

Mrs Jean, President of the IBC at that time, attempted to smooth over the change by stressing that, even with the new formulation, 'right at the beginning of the text, there is a will to involve all the actors that can play a role in the application of this Declaration'.[62] However, a departure from the state-centred addressees which remains would have helped to fill the gap between reality and legal positivist theory.

2.2.2. The Gap between Reality and Legal Positivist Theory in the Challenge of Effectiveness

Departing from the traditional discourse on the subjects of international law, the notion of individual responsibility is not absent from the texts, or from the reflection that took place within international organisations.[63]

---

[62] Jean, n 46 above, at 6.

[63] The most thorough and comprehensive works on the subject of individuals and responsibilities are the two following reports: the Report of Ms Erica-Irene A Daes, Special Rapporteur of the Sub-Commission on Prevention of Discrimination and Protection of Minorities, 'The Individual's Duties to the Community and the Limitations on Human Rights and Freedoms under Article 29 of the Universal Declaration of Human Rights, A Contribution to the Freedom of the Individual under Law', E/CN.4/Sub.2/432/Rev.2, New York, July 1982, and the Report of Mr Miguel Alfonso Martinez, Special Rapporteur of the Sub-Commission on the Promotion and Protection of Human Rights to undertake the study requested by the Commission in its resolution 2000/63, E/CN.4/2002/107, 25 Apr 2002.

The gap between the classical approach and the reality of international law-making reminds us of Higgins' comment concerning the fact that the notions of subject and object in international law have no functional purpose,[64] which can be understood in such a way as to challenge rather than serve the effectiveness of international law. This is particularly true in two fields. First, the evolution that has occurred in international criminal law reveals the obvious necessity to adapt the theory of international subjects to the need for implementation of the law. The individual responsibility for international crimes has been recognised as '[c]rime against international law are committed by men, not by abstract entities, and only by punishing individuals who commit such crimes can the provisions of international law be enforced'.[65] Another example relates to trans-national activities, in which control by a state or group of states is ill-suited to finding adequate solutions for the myriad problems which are transnational in scope and where there is a need to consider other addressees of universal obligations.[66] In this regard, ways have been explored of placing obligations on corporations.[67] Bioethics borrows elements from the two examples. Rooted in international criminal law,[68] it aims to regulate scientific research, which undoubtedly has a transnational dimension.[69] Similarly, effectiveness is a strong argument for reconsidering the position of addressees. The open formulation proposed by the IBC ('any decision or practice') was somehow a necessary condition for the effective implementation of the UDBHR. The field of bioethics is regulated in diverse ways throughout the world. The level at which the legislative or regulatory framework is placed depends on the country.[70] The duty-bearers are thus likely to differ from one country to another. States may be actors, just as they may be completely absent from this regulation. By placing the emphasis on the role of states, the UDBHR blurs the responsibility of entities other than states (ie, groups of researchers, pharmaceutical corporations) and overestimates the

[64] Higgins, n 33 above, at 49.

[65] Judgment of 1 Oct 1946 of the International Military Tribunal at Nuremberg, Transcript of proceedings, at 16–878. See Lauterpacht, n 32 above.

[66] Maggio and Lynch, n 33 above.

[67] See n 29 above.

[68] Bioethics emerged from the principles of the 1947 Code of Nuremberg.

[69] While considering the feasibility of a universal instrument on bioethics, the IBC listed the scientific practices that have extended beyond national borders as an illustration of the problems that should be addressed (healthcare, human reproduction and the beginning of life, genetic enhancement, gene therapy and genetic modification, human genetic data and other personal healthcare data, end of life, research involving human subjects, intellectual property rights, human organ and tissue transplantation, the use of embryonic stem cells in therapeutic research, behavioural genetics, genetically modified organisms). See Berlinguer and De Castro, n 50 above, at 4–9.

[70] L Nielsen, 'From Bioethics to Biolaw' in CM Mazzoni (ed), *A Legal Framework for Bioethics* (The Hague, Kluwer 1998), 39–52.

possibility of control of the state, for example in the case of transnational or international research projects. Quite the reverse, the legal construction used in the expression 'any decision and practice' highlighted the plurality of actors at the national level and stressed the departure from a state-centred scheme ('[s]tates should/shall'). On the one hand, by recognising other entities as duty-bearers, it restricted obligations placed on states and took into account that states can have only limited responsibility in bioethical reflection and in the drafting of any legislation that may stem therefrom,[71] insofar as they are not the main actors in the field at the national, trans-national or international level. On the other hand, states were invited to participate in the drafting of the regulations insofar as the declaration imposed positive duties, such as facilitating complementary action between non-state actors, on them.[72]

This debate on the scope of the UDBHR is of paramount importance regarding the implementation of the Declaration. The IBC's formulation would have opened the way to legal recognition of the reality of the plurality of actors involved in the field of bioethics. In this way, it would have contributed to the effectiveness of the Declaration.

The ethical standards which are set out in the UDBHR are to be transposed at the domestic level through a process which involves both ethical and legal actors. The emergence of new actors highlights the dominance of bioethics in the drafting and implementation of the UDBHR. Another form of interaction between the two bodies of norms is found when we consider the content and form of the UDBHR, in which we may talk of convergence and reconciliation of ethical and legal principles.

3. CONVERGENCE AND RECONCILIATION OF ETHICAL AND LEGAL PRINCIPLES

The ambiguity of the normative content of the UDBHR—an ethically-inspired legal instrument—stems from the mandate given to the IBC by the General Conference. The latter considered that 'it [was] opportune and desirable to set universal standards in the field of bioethics *with due*

---

[71] Berlinguer and De Castro, n 50 above, para 13: '[s]tates have a special responsibility not only with respect to bioethical reflection but also in the drafting of any legislation that may stem therefrom'.

[72] With regard to the European Convention on Human Rights and Biomedicine ('Oviedo Convention') 1997, see C Byk, 'La Convention européenne sur la biomédecine et les droits de l'homme et l'ordre juridique international' (2001) 128 *Journal du Droit International* 48: 'the paradox is that while the concept of the welfare state is declining, public authorities are mandated to exercise the control which means the right to an equitable access to new biomedical applications and the full respect of human rights'.

regard for human dignity and *human rights* and freedoms, in the spirit of *cultural pluralism* inherent in bioethics'.[73]

The human rights approach, which was previously adopted in the UDHGHR and the Oviedo Convention, is particularly appropriate in the field in question. This can be justified, as bioethics suffers from the plurality and range of actors involved and the overproduction of divergent norms,[74] whereas human rights law offers a strong framework and a common language, which may constitute a starting point for the development of universal bioethical principles.[75]

On closer examination of the UDBHR, to what extent does it borrow from human rights law? On the one hand, the title refers to human rights: it was proposed by the IBC Drafting Group instead of the original title, the 'Declaration on Universal Norms on Bioethics'. On the other hand, there is no recognition of specific rights ('everyone has the right to') or any absolute prohibition ('any form of … is prohibited'). However, the document imposes negative and positive duties on states. In addition, we find in the 'Final Provisions' the human rights mechanisms of limitations and interpretation of the 'Principles' (Articles 26, 27 and 28 UDBHR). Furthermore, the language of human rights law (such as 'informed consent', 'privacy and confidentiality' and 'should/shall') underpins the entire text. Beside the general provisions devoted to the aims and scope of the declaration (Articles 1 and 2), the UDBHR embodies in its operational parts, on the one hand, the 'principles' and the procedure for the 'application of the principles' which aim to guide state action within its territory, and, on the other hand, the principles governing the 'promotion of the declaration', which relate to the duty of states at the international level. Therefore the declaration lays down only general principles and no specific issue has been settled,[76] owing to the difficulty of reaching consensus.

If we consider the content of the 'principles' of the UDBHR, as highlighted by WHO in its comments on the Third Outline of the Declaration, there is 'a very substantial conflict between the human rights-based orientation of some (such as Article 3 [on human dignity, human rights and justice]) and the bioethics basis of others (such as Article 4 [on benefit and harm])'.[77] In reality, the substantive role played by human rights law or bioethics is rarely that

---

[73] Resolution 32C/24, 32nd Session of the General Conference of UNESCO, 2003, emphasis added.

[74] D Human and SS Fluss, *The World Medical Association's Declaration of Helsinki: Historical and Contemporary perspectives*, (2001), available at www.wma.net/e/ethicsunit/pdf/draft_historical_contemporary_perspectives.pdf.

[75] IBC Drafting Group, n 31 above, at 2.

[76] For the list of the specific issues considered during the preparatory works, see n 69 above.

[77] Letter of Comments from the World Health Organisation on the Third Outline of the Declaration on Universal Norms on Bioethics, 29 Nov 2004, 5.

clear-cut and is often blurred by the overlapping (informed consent,[78] respect for life, integrity and vulnerability,[79] and autonomy[80] are both legal and ethical principles) or intertwining (non-discrimination, which is legally recognised, is associated with non-stigmatisation,[81] which is more ethically grounded) of the two bodies of norms. Regarding the sources of each provision, the IBC Explanatory Memorandum on the Elaboration of the Preliminary Draft Declaration on Universal Norms on Bioethics refers either to the Oviedo Convention and its protocols or to the UDHGHR. There is no need to recall that these instruments are used in the same logic of intertwining between bioethics and human rights law. As a result, this 'self-referential' mechanism organises the autonomy of the ethico-legal normative system, which appears to be indirectly rooted in human rights law and bioethics.

In this intertwining of ethical and legal features, the most challenging aspect of the UDBHR can be found in the two guiding principles, which are mentioned in the IBC mandate. These principles include human dignity and cultural diversity, which underpin the entire text, although their respective understanding in law and bioethics are not necessarily compatible. On the one hand, respect for human dignity is common to international human rights law and Western bioethics; however, it is not necessarily recognised or respected in the same way by all cultures. There is no homogenous understanding of human dignity in bioethics and, therefore, it is not possible to ensure a common approach with human rights law. On the other hand, the principle of cultural diversity is inherent in ethics, which is pluralistic by nature, whereas it is acknowledged with some limitations in international human rights law.

However, in the field of life sciences, we witness a new approach to these two principles which reconciles the divergences in bioethics and law. This also conforms to the evolution of the content and the form of international human rights law. With regard to content, a new understanding of the principle of human dignity, which is oriented towards the safeguarding of humanity, can be seen as the first value shared among all cultures. In relation to form, the acknowledgement of cultural diversity leads to a development towards a more procedural approach in order to reach harmonisation on the substance. These issues will be discussed in more detail below.

---

[78] The principle of informed consent to medical and scientific experimentation, which has been a key principle in bioethics since the 'Nuremberg Code', is set out in Art 7 of the International Covenant on Civil and Political Rights.

[79] The right to life is recognised first in Art 3 UDHR. The rights to physical and spiritual integrity were recognised respectively in Art 3 and Art 5 of the UDHR. The notion of vulnerability has been developed by the human rights bodies.

[80] See below in the discussion regarding the principle of human dignity.

[81] IBC Drafting group, n 31 above, at 8: '[w]hile prohibition of discrimination can be more easily targeted by legal instruments, elimination of the stigma requires a longer process of social transformation in which ethics and ethics teaching can play a significant role'.

3.1. Human Being versus Human Species: Towards a New Balance in International Human Rights Law through the Principle of Human Dignity

The concept of human dignity is not recognised as such by all legal cultures, for example, in the Japanese legal system.[82] In addition, it is not well-defined and its qualification as a normative concept is problematic. However, its strength lies in the fact that it can be understood as an overarching principle in international human rights law and bioethics, or at least in Western bioethics. This section does not aim to define human dignity in bioethics and in human rights law but to explore it, in the context of the study, as a functional notion that aims to protect the interests of different entities (which are categorised as 'human') from possible infringements arising in the application of biotechnology or biomedicine (which can be perceived as a threat to 'dignity'). The emergence of the human species as a new entity in the field of human rights law raises the question of the recognition of the rights of right-holders to protect their interests.

3.1.1. Convergence in the Interpretation of the Principle of Human Dignity in Bioethics and Law

Human dignity is originally based on the idea of the intrinsic value of every human being, who possesses inherent rights as a means of protecting this value. It may be related to or equated with the value of autonomy and expressed through the requirement of informed consent.[83] Individual empowerment also stems from this principle, which is understood as the recognition of the right to control one's actions with regard to the choices that one has made. This constitutes the subjective approach to human dignity that is dominant in both human rights law and Western bioethics.

In Western bioethics, the principle of autonomy was highlighted in reaction to the overly paternalistic behaviour of the medical profession. Respect for autonomy clearly signals that individuals should not be sacrificed for the greater good of medical progress. Patients are empowered: even if doctors claim to know best, it is ultimately patients who make choices and exercise control.[84]

---

[82] Ida, n 59 above, at 5.

[83] JD Rendtorff and P Kemp, *Basic Ethical Principles in European Bioethics and Biolaw*, and Barcelona, centre for Ethics and Law and Institut Barja de Barcelona, 2000, at 31. The traditional correlation between human dignity and informed consent is still undoubtedly an important principle, but needs to be adapted for the child, and now for the foetus or the embryo. In the last case, informed consent of the genitors and human dignity of the embryo do not seem to go hand-in-hand: see the recent ECtHR case, *Evans v UK*, App no 6339/05, Judgment of 7 Mar 2006.

[84] MA Grodin, 'Introduction: the Historical and Philosophical Roots of Bioethics' in MA Grodin, *Meta Medical Ethics: the Philosophical Foundations of Bioethics* (Dordrecht, Kluwer Academic Publishers, 1995), 17.

In the field of human rights, it is claimed that a balance needs to be struck between the conflicting interests at stake: individual interests, on the one hand, and state interests in protecting the general welfare, on the other. However, human rights law has focused until recently on rights, autonomy and individual freedom.[85] The interests of society could be given priority in some circumstances; however, only the concept of human dignity was understood as linked to individuals who could invoke their own part of humanity (through the right to informed consent).

The prevalence of the individual is inherent in Western bioethics, and it can be argued that international human rights law embodies Western values. In fact some cultures give prevalence to societal interests, rather than the individual. Social ethics (or communitarian ethics) is opposed to individual ethics.[86]

The biotechnological revolution has led to a new approach to the principle of human dignity. As biosciences (ie, genomic sciences) affect human existence as a whole, it has been recognised that the human species has a value in itself, beyond the dignity of each of its members. Whereas the debate has so far focussed on the dialectic between individuals and society, the conflict over the position of the individual with regard to society has moved on. In addition to individual and societal interests, the interest of the human species has emerged and, alongside this, the protection of the biosphere. Moreover individual choice is not always contingent on collective interests. From this recognition stems a dual approach to respect for human dignity: in addition to the idea of human dignity as empowerment, human dignity as constraint has also emerged.[87] Whereas dignity as empowerment is the ability to exercise control over one's life, dignity as constraint asserts collective control over the exercise of autonomy when the interest of humankind is at stake.

Considering the interest of humanity, the aim of bioethics is to ensure that science guarantees the safety and wellbeing and the development of the human species, while it pursues its own development.

In human rights law, a new balance has appeared between the interests of the human species and those of individuals. Some commentators refer to the emergence of 'humanity rights' based on belonging to the human species, along with individual-centred human rights.[88]

---

[85] Daes, n 63 above, at 38–47.

[86] Grodin, n 84 above, at 17–18.

[87] D Beyleveld and R Brownsword, *Human Dignity in Bioethics and Biolaw* (Oxford, Oxford University Press, 2001).

[88] M Delmas-Marty, 'Interdire et punir, le clonage reproductif humain' (2003) 54 *Revue Trimestrielle des Droits de l'Homme*, Bioethics Special Edition, at 433.

3.1.2. The Recognition of new Rights-holders?

This section aims to consider how the dual approach of human dignity, as presented above, has found legal expression and the possible consequences of such a turn in the traditional theory of the beneficiaries or 'right-holders' of international human rights law.

It was in 1997 that both at the regional level, through the adoption of the Oviedo Convention, and at the international level, through the UDHGHR, the protection of the human species appeared in human rights law. In the same year, the General Conference of UNESCO adopted the Declaration on the Responsibilities of the Present Generations towards Future Generations, which views the human species through an intra-(spatial) and inter-(temporal) generational dimension. The UDBHR is in line with these instruments.

In the Oviedo Convention, the human species was described for the first time as a new beneficiary of human rights law. In the Explanatory Report, it says:

> It was necessary to take account of the actual developments in medicine and biology, while indicating the need for them to be used solely for the benefit of present and future generations. This concern has been affirmed at three levels: the first is that of individuals ... The second level relates to society ... The third and final concern relates to the human species ... The Convention sets up safeguards starting with the preamble where reference is made to the benefits to future generations and to all humanity, while provision is made throughout the text for the necessary legal guarantees.[89]

Therefore the Oviedo Convention promotes a new generation of human rights, the so-called 'biorights', which involve the three generations of human rights, and the third one (principles of justice and solidarity, right to development and right to a healthy environment) in particular. At the same time, it widens the scope of human rights law by linking human rights to the safeguarding of the human species.

In the UDHGHR, given the specificity of the object of protection—the human genome—it was of paramount importance to find a balance between the protection of the inalienable rights of the individual and common interest of humanity. In fact, considerable emphasis was placed on the dual nature of the human genome as the genetic heritage of each individual, on the one hand, and as the 'heritage of humanity' (Article 1 UDHGHR), on the other. This dual qualification underpins the new balance in international human rights law and reflects the profound

---

[89] Explanatory report to the Convention on human rights and biomedicine, Secretary General of the Council of Europe, CETS n 164, Strasbourg, 17 Dec 1996, para 14.

change that has occurred in the interpretation of the principle of human dignity. 'The aim is not only to protect the individual, in his rights and his freedoms, since dignity concerns the human being as such, in its [*sic*] largest sense'.[90] The dignity of the human species is therefore recognised. The interests of the individual have to be balanced with the interests of future generations, which are protected by the concept of heritage of humanity. This is in line with the preamble to the DRPGFG, which states that "[t]hese responsibilities include, *inter alia*, the maintenance and perpetuation of humankind with due respect for the dignity of the human person'.

In the UDBHR, the different interests protected appear within the 'principles', which 'determine gradually widening obligations and responsibilities in relation to the individual human being itself, to another human being, to human communities, to humankind as a whole; and towards all living beings and their environment'.[91]

At the first level (Articles 3–12), we find the principles which protect the human person as such (human dignity, autonomy, informed consent, personal integrity, privacy and confidentiality) or as a member of society (equality, justice, equity, non-discrimination, non-stigmatisation, cultural diversity and human vulnerability). Article 3 restates the principle of human dignity-empowerment ('[h]uman dignity, human rights and fundamental freedoms are to be respected'). The principle of autonomy (Article 5) is counterbalanced by the recognition of responsibility towards others. The principle of informed consent (Articles 6 and 7) is reaffirmed as the core of the protection of the human person in bioethics, with the possibility of an additional collective agreement (among a group or community).

The second level (Articles 13 to 16) concerns humanity in its entirety, that is present and future generations. Article 13 ('[s]olidarity and cooperation'), Article 14 ('[s]ocial responsibility and health') and Article 15 ('[s]haring of benefits') implicitly refer to the rights of the present and future generations to enjoy the benefits of life sciences and medicine. Furthermore, Article 16 explicitly mentions the protection of future generations.

The third level (Article 17) refers to a collective responsibility towards the biosphere. This recognition is an important development in international human rights law, insofar as it constitutes a step further in the protection of the human species. The idea of a common destiny is accompanied by

---

[90] Report of the Fourth Meeting of the Legal Commission of the IBC, UNESCO, Paris, 27 Apr 1994, in UNESCO document, *Birth of the Universal Declaration on the Human Genome and Human Rights* (Paris, UNESCO, Division of the Ethics of Science and Technology 1999), 53. The '[*sic*]' was in the original quotation.

[91] IBC Drafting Group, n 31 above, at 6.

the idea that the future of the human species is linked to the rest of the biosphere. In light of this interdependence, the impact of human beings on the biosphere cannot be disregarded.

With regard to the beneficiaries of these provisions, the Intergovernmental Meeting of Experts introduced two modifications. First, it retained a narrow understanding of human rights as individual rights as opposed to collective rights. In the Preliminary Draft, the scope of the UDBHR embodied the decisions or practices made or carried out with relation 'to individuals, families, groups and communities' (Article 2(i)). Instead, government experts adopted a formulation focussed on the individual, since human rights were to be held by individuals and not by groups: '[t]he declaration addresses ethical issues ... as applied to human beings, taking into account their social, legal and environmental dimensions' (Article 1(a)). Secondly, government experts agreed by consensus on the insertion of an Article devoted to the protection of future generations (Article 16). It could seem paradoxical that states widen the beneficiaries of protection (future generations) while claiming that human rights are individual rights.

Individuals, the traditional subjective beneficiaries in human rights law, now have duties towards new beneficiaries, 'the present and future generations'. With Article 16 ('Protecting Future Generations'), the UDHGHR strengthens the 'framework of inter-generational responsibility', which was envisaged during the drafting process by the IBC Drafting Group.[92] The interests of present and future generations have been increasingly emphasised since the 1972 Stockholm Declaration (Principle 1 on the duty to protect the environment). This recognition could challenge the traditional implementation of international human rights law if we consider that individual interests and interests of the human species do not necessarily go hand-in-hand. It also questions who has legal standing to speak on behalf of humanity, the biosphere and the present or future generations to ensure their protection.

Traditionally, two systems of control of respect for human rights coexist, one juridical, at the regional level, where individuals and states have legal standing in the European and Inter-American courts, and the other, non-juridical, at the international level, where only states can represent individual interests. The existence of an individual remedy depends on states, as is recognised by the Optional Protocol to the International Covenant on Civil and Political Rights.[93] It consists of the possibility for

[92] Final Report of the Third Meeting of the IBC Drafting Group for the Elaboration of a Declaration on Universal Norms on Bioethics, 2–3 July 2004, SHS/EST/04/CIB-GRED-3/2, UNESCO, Paris, 18 Aug 2004.

[93] In 2004, the Optional Protocol had been ratified by 104 States Parties (Office of the United Nations High Commissioner for Human Rights, Status of ratifications of the principal international human rights treaties, 9 June 2004, available at www.unhchr.ch/pdf/report.pdf).

individuals to send a communication with a complaint to the Human Rights Committee (HRC), but the decisions of the latter have no legal force and do not bind states. The juridical/non-juridical panorama schematically presented will not change drastically; however some refinements may occur regarding the system of protection recognised by the 'bioethics instruments'.

At the regional level, there is no mechanism in place to ensure respect for the Oviedo Convention. The European Court of Human Rights (ECtHR) is responsible only for its interpretation. However the ECtHR can adopt an evolving interpretation of the 1950 European Convention on Human Rights (ECHR) in the light of other conventions. Therefore there could be grounds for widening the scope of Article 2 (right to life), Article 3 (prohibition of torture), Article 6 (right to a fair trial) or Article 8 (right to respect for private and family life) of the ECHR in light of the Oviedo Convention. However, this would not imply that the individual had the right to invoke the interests of present and future generations and the case law of the ECtHR does not seem to encourage this evolution. In considering the application of Article 6 in the case of a licence to operate nuclear power plants, the European Court of Human Rights found that the right to a fair trial was not applicable in a case in which the applicants alleged 'not so much a specific and imminent danger in their personal regard as a general danger in relation to all nuclear power plants; and many of the grounds they relied on [were] related to safety, environmental and technical features inherent in the use of nuclear energy'.[94] The requirement of the proof of a serious, specific and imminent danger obviously runs counter to the interests of future generations. Therefore considering Article 6 ECHR, states are not required to ensure the right of access to justice in order to protect the interests of present and future generations. Another approach to considering future interests could lie in the protected status of the embryo as a member of future generations, but here again the Court does not appear as its guardian.[95]

Thus far, only one domestic court decision has allowed the bringing of an action (in this case, a class action) brought by the plaintiffs acting as representatives for themselves and future generations. The Philippines Supreme Court in 1993 addressed intergenerational equity in the context of state management of public forest land.[96] The Court found that the petitioners had *locus standi* (in other words, they were qualified to sue) on

---

[94] ECtHR, *Athanassoglou and Others v Switzerland*, App no 27644/95, of [2000] ECHR 159, para 52.

[95] See ECtHR, *Vo v France*, App no 53924/00, of [2004] ECHR 326 and ECtHR, *Evans v UK* n 83 above.

[96] Philippines Supreme Court, *Minors Oposa v Secretary of the Department of Environment and Natural Resources* ('DENR'), GR 101083, 224 SCRA 792, 30 July 1993.

behalf of present and future generations in the Philippines to invoke their right to a healthy environment.

At the international level, human rights obligations are traditionally regarded as *erga omnes* obligations, that refer to obligations incumbent upon all states *vis-à-vis* all other states, in any given sphere of legal relationship, which benefit not the subjective interest of the states, but a third party, usually that of individuals, collectivised as 'the international community'.[97] Therefore international human rights law is originally characterised by what we can call a 'normative and conceptual matter':[98] there is no correlation between the right-holders—namely, the international community of states—and the subjective beneficiaries—namely, the individuals or the international community of individuals.

This legal construction could be appropriate in the field of biomedicine and life sciences. The human species, the future generations or the biosphere would be the new subjective beneficiaries, and states would remain the right-holders. The forefront of the logic would be the same and would conform to the dual approach to human dignity relating to the individual or to the human species: any denial of human dignity is the concern of the entire 'international community'.

The ability of the individual to speak on behalf of humanity would represent an innovation in international law. In 1980, the HRC addressed the question whether a communication could be submitted on behalf of 'future generations' and found that this question did not have to be resolved in the circumstances of a case concerning emissions caused by nuclear waste in waste tips.[99] The author of the communication had the standing to submit it both on her own behalf and also on behalf of the residents of Port Hope who had specifically authorised her to do so. As stated by the HRC, '[t]he Committee will treat the author's reference to "future generations" as an expression of concern purporting to put into due perspective the importance of the matter raised in the communication'. Consequently, the question remains and it will be interesting to follow the developments arising from the future application of the UDBHR.

---

[97] *Barcelona Traction, Light and Power Company, Limited (Belgium v Spain)* [1970] ICJ. Rep 3: the ICJ refers to them as 'obligations of a state towards the international community as a whole … all states can be held to have a legal interest in their protection'.

[98] L-A Sicilianos, 'Classification des obligations et dimension multilatérale de la responsabilité internationale' in P-M Dupuy, *Obligations multilatérales, droit impératif et responsabilité internationale des Etats* (Paris, Pedone, 2003), 63.

[99] CCPR/C/17/D/67/1980, *EPH v Canada*, Decision of 27 Oct 1982, at para 8.

3.2. Universal Values versus Cultural Diversity: Towards a Common Procedural Approach to Reach a Harmonisation of the Substance

In its comments on the Third Outline, the WHO emphasised that the text of the Declaration suffered from 'the unresolved tension between declaring "universal norms" and wanting to accept local differences demanded by certain cultures, which may contradict those norms'.[100] Although it could be claimed that the UDBHR is the product of this 'tension', this does not affect its coherence. During its second meeting, the IBC Drafting Group acknowledged that it did not seem possible to find a common position on bioethics; however, there was a great risk of the development of practices contrary to human dignity and harmful to the human species if everyone was left to act according to his or her own ethical considerations. Therefore, the scope of the Declaration was confined to the role of giving teeth to the basic principles of bioethics.[101] In reality, this represents only one side of the coin as the IBC Drafting Group chose to adopt a dualist approach, based on the need for unification—namely a single, shared position on a given subject—and harmonisation—namely guiding principles of a more general nature with a national margin of discretion (while respecting a minimal threshold of compatibility).[102] The two mechanisms will be described in further detail in the following sections.

3.2.1. Unification of the Limitations on Cultural Diversity

The IBC reached a consensus on the limitations placed on the recognition of cultural diversity. This delimitation was of paramount importance considering the need to ensure respect for the principles set out in the UDBHR.

First, the recognition of cultural diversity as the 'common heritage of humanity' in the Preamble can be interpreted as a limitation on individuals freely invoking their cultural background, tradition and religion. This interpretation is compatible with the UNESCO Declaration on Cultural Diversity 2001 and the Convention on Cultural Diversity 2005.

In addition, Article 12 of the UDBHR ('Respect for cultural diversity and pluralism') states that 'such considerations [related to cultural diversity and pluralism] are not to be invoked to infringe upon human dignity, human rights and fundamental freedoms, nor upon the principles set out in this declaration, nor to limit their scope'. This Article is in line with the 2003 Report of the IBC which stresses that the limitation of the

---

[100] WHO Letter, n 77 above, at 3.

[101] L De Castro and G Berlinguer, 'Preliminary Report on the Possibility of Elaborating a Universal Instrument on Bioethics', report of the Working Group of the IBC, SHS/EST/02/CIB-9/5, Paris, 15 Nov 2002, para 36.

[102] De Castro, n 5 above, at 4.

recognition of bioethical pluralism is 'the right of present and future generations'.

3.2.2. *Harmonisation of the Ethical Process*

The IBC Drafting Group was also mandated to find common ground that could act as the starting point for harmonising divergent bioethical positions. There are two factors relating to harmonisation in the final draft.

First, the final draft borrows from the common language of international human rights law, which contributes to harmonising the understanding of the principles set out in the declaration.

Secondly, the UDBHR can be interpreted as a 'codex bioethicus'[103] which seeks to find consensus on the procedure rather than lay down apodictic prescriptions. Two series of provisions ('Application of the Principles' and 'Promotion of the Declaration') are devoted to procedural principles. The underlying idea is to assist states by providing them with guidelines so as to achieve a harmonised approach to the implementation of general principles, and to ensure an ethical framework for scientific and technological progress. A fair process is envisaged to lead to ethically acceptable decisions.

In order to complete these guidelines, the IBC drafted the Explanatory Memorandum on the Elaboration of the Preliminary Draft Declaration on Universal Norms on Bioethics ('Explanatory Memorandum'). The Explanatory Memorandum should have constituted a 'first' interpretation of the principles set out in the UDBHR and later adapted in line with a common and evolving interpretation provided by the IBC. Moreover Member States were requested to submit reports every five years on 'the steps they have taken ... to give effect to this declaration' (Article 27(b) of the Preliminary Draft). On the basis of these reports, the IBC and the IGBC would have been responsible for monitoring and evaluating the implementation of the UDBHR: '[t]he two committees should be responsible in particular for the formulation of any opinion or proposal likely to further the effectiveness of this Declaration' and should have made recommendations addressed to the General Conference (Article 27(a) of the Preliminary Draft). UNESCO committed itself 'to take appropriate measures to examine this Declaration in the light of scientific and technological development and if necessary, to ensure its revision' (Article 28(c)). Finally, it was stated that the Declaration should be further developed through international instruments adopted by the General

---

[103] Contribution of the Pan America Health Organisation (PAHO), 'Towards a Codex Bioethicus, A position paper by the PAHO Bioethics Unit', 'Towards a Declaration on Universal Norms on Bioethics, written contributions', SHS/EST/04/CIB-EXTR/INF 1, Paris, 25 Apr 2003, 27–29.

Conference of UNESCO, in accordance with its statutory procedures (Article 28(d)).

All these provisions (Article 27(a) and (b) and Article 28(c) and (d)) and the Explanatory Memorandum were rejected by the Intergovernmental Meeting of Experts. The rejection of these measures imposed serious limitations on the efforts made to ensure the effectiveness of the Declaration. Effectiveness, which is the question whether the goals of the norms are achieved,[104] relies mainly on compliance, which refers to 'whether countries in fact adhere to the provisions and to the implementing measures that they have instituted'.[105] Non-legally-binding instruments suffer from the absence of legal guarantees to ensure that states comply with agreed standards, since their responsibility cannot be invoked in the event of a violation. However in order to lessen this shortcoming, the first two UNESCO declarations incorporate supervisory mechanisms traditionally found in hard law texts.[106] The rationale of the IBC approach lies in the balance between normative ('soft law') and institutional (monitoring) mechanisms to ensure compliance with the declaration. For the moment, there is no equivalent to the monitoring machinery[107] which was instigated with the UDHGHR.[108] On the one hand, we find the mechanism of reports, which is traditional in human rights law. On the other hand, the IBC is entrusted with giving advice concerning the follow-up of the UDHGHR, organising workshops aimed at providing a standard framework of legislations and regulations in the field of bioethics and identifying practices that could be contrary to human dignity.[109] The same reporting mechanism and mechanism of periodic review is provided for in the IDHGD.[110] There was therefore a strong argument for maintaining the same monitoring mechanism so as to ensure coherence between UNESCO bioethics declarations and to

[104] Shelton, n 49 above, at 5.
[105] HK Jacobson and EB Weiss, 'Compliance with International Accords' (1995) 1 *Global Governance* 119.
[106] It is against the different systems established under the terms of the main international instruments for the protection of human rights that the monitoring mechanism for the UDHGHR was chosen. See Hector Gros Espiell, 'Mechanism for Monitoring the Future Universal Declaration on the Human Genome and Human Rights', 1 July 1996, in UNESCO, *Birth of the Universal Declaration on the Human Genome and Human Rights* (Paris, UNESCO, Division of the Ethics of Science and Technology, 1999), 79.
[107] Shelton, n 49 above, at 5: '[m]onitoring and supervision refer to the procedures and institutions which are used to assess compliance'.
[108] Only the 1978 UNESCO Declaration on race and racial prejudice provides for a system of reports and the Director-General is invited to submit any recommendation deemed necessary to promote its implementation.
[109] Art 24 UDHGHR. See also Implementation of the Universal Declaration on the Human Genome and Human Rights, n 55 above.
[110] Art 25 IDHGD.

strengthen the role of the IBC and IGBC. Moreover, such mechanism allows each national community to adapt these standards to its own system of ethical values and cultural traditions, while respecting a minimum threshold of compatibility through the intervention of the international organs. The creation of the IGBC, at the request of states as we saw above, should have secured state participation in the overall process, so that the states would not have lost control and been completely under experts' supervision . However it seems that the monitoring system was still too constraining for government experts, who rejected it without any alternative argument.[111] As a result, the third UNESCO declaration does not incorporate any monitoring machinery.

The major contribution of the UDBHR will undoubtedly be found in the innovative provisions for its implementation through procedural principles. However, there is clearly a risk that the procedural guidelines will be insufficient if they are completed without the co-operation of the IBC.

4. CONCLUSION

The 'normative spectrum' presented in this chapter leads to a two-fold conclusion. On the one hand, it would be misleading to consider the 'translation of ethical concepts into legal terms' undertaken in the UDBHR as a simple incorporation of these concepts into a legal instrument without any possibility of a backlash. For this reason, the above analysis of the UDBHR reveals the influence of the 'source of inspiration' (bioethics) in the norms of regulation (human rights law). There appears to be a dominance of bioethics, in the sense that the new approach to the field of life sciences is framed within human rights law, yet it is based on the idea of a process. The interdependence between human rights law and bioethics leads us to move from a 'static image of existing or positive law in favour of a more dynamic concept in which views on law as it is and views on law as it should be are continually merging into views on law as it is becoming'.[112] The diversification of sources of human rights, which stems from the diversification of actors at the national, regional, and international level responsible for the regulation of bioethics, has raised

---

[111] The Final Report of the Intergovernmental Meeting of Experts does not provide any justification for the rejection; one can assume it represents a reluctance by states to ensure compliance with the Declaration. See B Ouoba, 'Final Report of the Second Session of the Intergovernmental Meeting of Experts aimed at Finalizing a Draft Declaration on Universal Norms on Bioethics', SHS/EST/05/CONF.204/6, UNESCO Headquarters, Paris, 27 July 2005, at 7.

[112] Van der Burg, n 57 above, at 51.

concerns among some commentators of a conflict with basic human rights principles. It has been stressed that the multiplication of actors involved in the human rights discourse runs counter to the principles of universality, indivisibility and interrelatedness of human rights.[113] However, the acknowledgement of pluralism and the quest for universality, contradictory statements at first glance, enable the UDBHR to come closer to the reality at stake. Given the difficulty in adapting human rights law to diverse apprehensions about scientific progress, the bioethical-oriented approach provides a unique opportunity to confront issues relating to the implementation of human rights. In this sense, the UDBHR may be useful as a means of stimulating new debate.

On the other hand, state experts were sensitive to the need for coherence with previous human rights instruments but they did not take into account all of the consequences that may arise from the drafting of an international instrument in the field of bioethics. Therefore numerous uncertainties remain which compromise the implementation of the declaration: the increased number of duty-bearers, the determination of the right-holders, the harmonisation of the procedure and, finally, the absence of any provision for follow-up by the IBC.

It is clear that the UDBHR illustrates the way forward: the normative (ethical and legal) and institutional (co-operation and communication of legal and ethical actors) loops are without doubt the keystone of the efficient interaction between bioethics and human rights law to deal with scientific dilemmas.

---

[113] Byk, n 39 above, at 364–74.

Part III

Economic, Social and Cultural Rights

7

Agricultural Biotechnology and the Right to Food

KERSTIN MECHLEM* AND TERRI RANEY†

1. INTRODUCTION

B IOTECHNOLOGY CAN BE defined as any technological application that uses biological systems, living organisms or derivatives thereof to make or modify products or processes for specific use.[1] Agricultural biotechnology encompasses a range of research tools that enable scientists to understand and manipulate the genetic make-up of organisms for use in agriculture: crops, livestock, forestry and fisheries.[2] Most discussions of biotechnology focus on transgenic crops and other genetically modified organisms (GMOs). GMOs result from genetic engineering, which involves transferring genetic material from one organism to another without sexual mating. The cultivation of transgenic crops is rapidly increasing. In 2004, nine years after the beginning of the commercialisation of GM crops, they were grown on 81.0 million hectares of land in a total of 17 countries, and their global area continued to grow at a growth rate of 20 per cent *per annum*.[3]

---

* Legal Officer, Food and Agriculture Organization of the United Nations (FAO). Email: Kerstin.mechlem@fao.org. The views expressed in this article are personal and do not represent the views of the Food and Agricultural Organization of the United Nations (FAO).
† Senior Economist, FAO, and editor of *The State of Food and Agriculture*. Email: Terri. Raney@fao.org. The views expressed in this article are personal and do not engage the FAO.
[1] Art. 2 of the Convention on Biological Diversity (1992) 31 ILM 818.
[2] FAO, *The State of Food and Agriculture 2003–04—Agricultural Biotechnology: Meeting the Needs of the Poor?* (Rome, FAO, 2004), 4. In a narrow sense, biotechnology can be understood as a range of different molecular technologies, such as gene manipulation and gene transfer, DNA typing and cloning of plants and animals: A Zaid, HG Hughes, E Porceddu, and F Nicholas, 'Glossary of Biotechnology for Food and Agriculture', FAO Research and Technology Paper No. 9 (Rome, FAO, 2001), 35.
[3] C James, 'Preview: Global Status of Commercialized Biotech/GM Crops: 2004', International Service for the Acquisition of Agri-biotech Applications (ISAAA) Briefs No. 32 (Ithaca, NY, ISAAA, 2005), 3.

The spread of agricultural biotechnologies and the conditions conducive to their development have economic, social and environmental implications. This chapter explores these implications for the realisation of the right to food, emphasising genetic engineering as the most controversial biotechnological application. First, this chapter will examine the risks and opportunities of transgenic crops to realise the different components of the right to food, namely the availability, accessibility, safety, nutritious quality and acceptability of food, as well as the degree of sustainability with which that food is grown (see section 4 below). The opportunities seem boundless: higher productivity on the same amount of land, saving virgin soils and increasing the overall availability of food; improved nutritional values; the development of crops for saline, dry or other marginalised soils, etc. At the same time uncertainties about the long-term health and environmental risks give rise to legitimate concern.

Secondly, the analysis broadens to examine the effects on the right to food of the manner in which transgenic crops are developed, protected and marketed (see section 5 below). This section will address issues such as the consequences of marked privatisation and concentration of research and market share in the hands of a small number of powerful transnational corporations; the effects that increased protection of intellectual property rights has on further research, on the access of developing countries to the new technologies and on farmers' traditional practices of saving and re-using seeds from their own harvest; and the lack of mechanisms to adequately acknowledge the important contributions of farmers and developing countries to biotechnological inventions.

2. AGRICULTURAL BIOTECHNOLOGIES: CONCEPT AND APPLICATIONS

Biotechnology is multisectoral. The same tools are used in medicine, industry and agriculture. Industrial as well as some medical uses attract relatively little controversy; agricultural applications are more sensitive. Within agriculture, biotechnology includes simple technologies like micropropagation,[4] and more complex methods such as marker assisted selection,[5] genetic engineering[6] and genomics[7] that are based on genomic analysis.

---

[4] Miniaturised *in vitro* multiplication and/or regeneration of plant material under aseptic and controlled environmental conditions: A Zaid, *et al*, above n 2, 184.

[5] Marker assisted selection (MAS) is the use of DNA markers to improve response to selection in a population: ibid, 177.

[6] Modifying genotype, and hence phenotype by transgenesis, ie, by the introduction of a gene or genes into animal or plant cells, which leads to the transmission of the input gene (transgene) to successive generations: ibid, 123, 288.

[7] Genomics is the research strategy that uses molecular characterisation and cloning of whole genomes to understand the structure, function and evolution of genes and to answer fundamental biological questions: ibid, 125.

Biotechnology allows the characterisation of plants and animals at the genomic level, so that the specific gene responsible for a desirable trait can be targeted in breeding and conservation programmes. Conventional breeding, in contrast, relies on the physical appearance of a specimen, an often imperfect guide to its value in breeding. By integrating biotechnology into their agricultural research programmes, countries can speed up breeding programmes and tackle challenges that are not tractable with conventional methods. For example, micropropagation techniques can generate disease-free planting materials for clonally propagated species, such as potato and banana, on which many subsistence farmers rely. Biotechnology is also used in diagnosing plant and animal diseases before the host is badly damaged, while treatment is still possible. By distinguishing vaccinated animals from infected ones, biotechnology can facilitate vaccination programmes without disrupting trade.

The most contentious agricultural biotechnologies are transgenic crops and other GMOs. GMOs are the result of genetic engineering, which involves transferring genetic material from one organism to another without sexual mating. This ability of genetic engineering to move genes across species barriers gives it its tremendous potential and makes it so controversial. Genetic engineering is at once a more precise extension of, and a radical departure from, conventional breeding methods. It can meet some challenges that other biotechnologies cannot address, but in many cases it complements other research approaches.

Only six countries, four crops and two traits accounted for 99 per cent of the global area planted with transgenic crops in 2003. The countries are: the USA, Argentina, Canada, China, Brazil and South Africa; the crops are soybean, maize, cotton and canola; and the traits are insect resistance (Bt[8]) and herbicide tolerance.[9] These same crops and traits are the subject of most of the transgenic crop research underway in both developed and developing countries and the public and private sectors.

A much broader range of crops and traits is under research, although the needs of poor farmers in developing countries are relatively neglected (see section 5.1 below). Some of the products in the pipeline that could be especially relevant for developing countries and an improved realisation of the right to food include nutritionally enhanced foods such as 'Golden Rice' (vitamin A), higher protein potatoes, iron-fortified rice, virus resistant sweet potato and maize, and striga-resistant herbicide-tolerant maize.[10]

---

[8] *Bacillus thuringiensis*, a bacterium that produces a toxin against certain insects, particularly *Coloeptera* and *Lepidoptera* and a major insecticide approved for use in organic farming.

[9] FAO, above n 2, 5.

[10] Due to limitations of scope, the use of biotechnology in livestock and fisheries will not be addressed in this chapter.

3. THE RIGHT TO FOOD

The use of agricultural biotechnologies can have a number of implications for the right to food. The right to food, or aspects of it, is recognised in a large number of binding and non-binding instruments.[11] Most prominent among these are Article 25(1) of the Universal Declaration of Human Rights (UDHR), which states that '[e]veryone has the right to a standard of living adequate for the health and well-being of himself and his family, including food, clothing, housing',[12] and Article 11 of the International Covenant on Economic, Social and Cultural Rights (ICESCR).[13] In Article 11(1), 'States Parties ... recognise the right of everyone to an adequate standard of living ... including adequate food'; in Article 11(2), 'States Parties ... [recognise] the fundamental right of everyone to be free from hunger'.

The right to food is realised when every individual 'alone or in community with others, has physical and economic access at all times to adequate food or means for its procurement'.[14] It implies the *'availability* of food in a quantity and quality sufficient to satisfy the *dietary needs* of individuals, *free from adverse substances,* and *acceptabl*e within a given culture [and] the *accessibility* of such food in ways that are *sustainable* and that do not interfere with the enjoyment of other human rights' (emphasis added).[15] This definition from General Comment No 12 on the Right to Adequate Food was adopted by the Committee on Economic, Social and Cultural Rights (CESCR) in 1999. For the 152 states parties to the ICESCR this definition is of high authoritative value.[16] Under the ICESCR states have to use all

---

[11] See also Art 12(2) of the Convention on the Elimination of All Forms of Discrimination against Women (CEDAW) of 1979; Art 24(2)(c) and (e), as well as Art 27(3) of the Convention on the Rights of the Child (CRC) of 1989; the Protocol to the African Charter on Human and Peoples' Rights on the Rights of Women in Africa (2003). Among non-binding instruments, see the 1992 World Declaration on Nutrition; the 1996 Rome Declaration on World Food Security, and Plan of Action; the 2002 Declaration of the World Food Summit Five Years Later; the 2002 Plan of Implementation of the World Summit on Sustainable Development; and, in particular, the 2004 Voluntary Guidelines for the Progressive Realisation of the Right to Adequate Food.

[12] Adopted by UN General Assembly Resolution 217 A (III) of 10 Dec 1948.

[13] 993 UNTS 3.

[14] Committee on Economic, Social and Cultural Rights (CESCR), General Comment No 12 on The Right to Adequate Food, UN Doc E/C.12/1999/5 of 12 May 1999, para 6.

[15] Ibid, para 8.

[16] *Cf* Updated Study on the Right to Food, Submitted by Mr Asbjørn Eide in Accordance with Sub-Commission Decision 1998/106, UN Doc E/CN.4/Sub.2/1999/12 of 28 June 1999, para 45; *The Right to Food,* Report Submitted by the Special Rapporteur on the Right to Food, Jean Ziegler, in Accordance with Commission on Human Rights Resolution 2002/25, UN Doc. E/CN.4/2003/54 of 10 Jan 2003, paras 17, 23. On the legal relevance of General Comments of the CESCR, and other treaty bodies in general, see also P Alston, 'The Historical Origins of the Concept of "General Comments" in Human Rights Law' in: L Boisson de Chazournes and V Gowlland-Debbas (eds), *The International Legal System in Quest of Equity and Universality, Liber Amicorum Georges Abi-Saab* (The Hague/London,

appropriate means progressively to realise the right to food (Article 2(1)), including measures 'to improve methods of production, conservation and distribution of food by making full use of technical and scientific knowledge' (Article 11(2)(a) of the ICESCR). Also, Article 12 of the Protocol of San Salvador requires improved methods of food production.[17] Article 24(2)(c) of the Convention on the Rights of the Child lists the application of readily available technology as an obligatory measure to combat malnutrition. Hence, human rights law recognises the important contribution technological advances can make to the realisation of the right to food.

A commonly used analytical framework of states' human rights obligations distinguishes between three types of obligations, *viz* the obligations to respect, protect and fulfil the right to adequate food.[18] The obligation to *respect* requires states to refrain from interfering directly or indirectly with the enjoyment of the right.[19] They must refrain from denying or limiting access to food or interfering arbitrarily with existing arrangements, eg, by destroying existing, functioning market systems. The obligation to *protect* requires states to take measures to ensure that third parties such as individuals, groups, corporations or other entities do not interfere in any way with the enjoyment of the right.[20] States must enact and enforce effective legislation and take other measures—such as food safety measures—to control and restrain third parties' activities. The obligation to *fulfil* means that states must take positive measures to *facilitate* and *provide* for individuals' enjoyment of their rights.[21] States must develop comprehensive national

Boston, Mass, Martinus Nijhoff, 2001), 763; M Craven, *The International Covenant on Economic, Social and Cultural Rights—A Perspective on its Development* (Oxford, OUP 1995) 90 ff; M Scheinin, 'International Mechanisms and Procedures for Implementation', in R Hanski and M Suksi (eds), *An Introduction to the International Protection of Human Rights.* (Turku/Åbo, Åbo Akademi University, 1999) 429, 444; T Meron, *Human Rights Law-Making in the United Nations* (Oxford, Clarendon, 1986), 10; C Tomuschat, 'National Implementation of International Standards on Human Rights', (1984) 85 *Canadian Human Rights Yearbook*, 31, 36.

It is noteworthy that this definition of the right to food bears considerable resemblance to the definition of food security adopted at the World Food Summit in 1996: '[f]ood security, at the individual, household, national, regional and global levels is achieved when all people, at all times, have physical and economic access to sufficient, safe and nutritious food to meet their dietary needs and food preferences for an active and healthy life'. 'World Food Summit Plan of Action', in FAO, *Report of the World Food Summit, Rome 13 to 17 November 1996*, Part I, Appendix (Rome, FAO, 1996). For the development of the concepts of food security and the right to food, and an analysis of the distinctiveness of rights-based food security policies from other approaches, see K Mechlem, 'Food Security and the Right to Food in the Discourse of the United Nations' (2004) 10 *European Law Journal*, 631.

[17] OAS, Treaty Series No 69.

[18] This framework of obligations is increasingly used by UN and regional human rights actors, academia and national courts, in particular for economic, social and cultural rights. It is based on a concept which was originally proposed by Henry Shue and later developed by Asbjørn Eide: see A Eide, *Right to Adequate Food as a Human Right* (New York, United Nations, 1989).

[19] CESCR, above n 4, para 15.

[20] Ibid.

[21] Ibid.

right to food strategies and policies, repeal legislation that impairs the progressive realisation of the right, and enact necessary laws. In short, to facilitate the realisation of the right to food means to create an enabling framework in which as many individuals as possible can provide for their own food. Finally, states have the obligation to *provide directly* for the fulfilment of the right in those cases in which individuals are unable, for reasons beyond their control, to realise their rights themselves.[22] Food safety nets and food interventions targeted towards vulnerable groups fall within the 'provide' dimension. The use of genetically modified (GM) crops can facilitate or interfere with the fulfilment of each of the obligations outlined above. Additional state obligations stem from crosscutting human rights principles, which comprise participation, non-discrimination and the right to redress in cases of a right's violation.

The right to food requires protection of both physical and economic access to food. While in developed countries few people depend directly on agriculture for the realisation of their right to food, in developing countries the importance of agriculture, not only for directly accessing food but also as a source of income and employment and as a basis for livelihoods, cannot be overestimated. About three quarters of the world's poor live in rural areas and rely on agriculture for their survival.[23] Any changes in the agricultural sector can immediately and dramatically affect the realisation of their right to food.

4. THE IMPLICATIONS OF THE USE OF GENETICALLY MODIFIED CROPS FOR THE DIFFERENT COMPONENTS OF THE RIGHT TO FOOD

4.1. Availability of Food

The first prerequisite of the fulfilment of the right to food is the availability of food. At present the amount of food produced globally is sufficient to feed the world.[24] However, this macro view conceals the fact that some parts of the world have food surpluses, whereas others have food shortages. Food availability in the latter must be increased. In addition, overall food shortages are likely to increase as the global population, mainly in developing countries, grows from six billion to an estimated 9–10 billion within the next 50 years. With this growth food demand will rise proportionately. The demand can be fulfilled only by increasing cultivated land, by intensifying productivity, or by a combination of both.

---

[22] Ibid.

[23] IFAD, *Rural Poverty Report 2001—the Challenge of Ending Rural Poverty* (Rome; IFAD, 2001), 1.

[24] FAO, *The State of Food Insecurity in the World 2003* (Rome; FAO, 2003), 6.

The Green Revolution demonstrated that technological innovation—higher yielding seeds and the inputs required to make them grow—can play an important role in increasing food availability. Between 1960 and 2000, global cereal production more than doubled on virtually the same amount of land. As a consequence, an additional two billion people are being fed, and fed much better. The share of the world's chronically hungry population fell by half during the same period, from 34 per cent to 17 per cent. Although the number of undernourished people remains stubbornly high—852 million at present[25]—without the yield gains of the Green Revolution many more people would be hungry today and much more land would be under cultivation.[26]

High-yielding GM crops could offer similar opportunities to grow more food on the same amount of cultivated land. Until now, however, biotechnological innovations have focused mainly on crops grown with large-scale farming methods in developed countries, not on those of greatest importance to developing countries. GM crops are under research that would provide particular benefits in marginalised areas, such as drought or salinity resistance. Examples include striga-resistant maize in Africa, aluminium-tolerant wheat and salt-tolerant rice. Such crops would improve the livelihoods of people depending on poor soils.

Biotechnology can be a means of increasing the availability of food, not as an alternative but in addition to other methods.[27] It needs to be combined with other approaches, such as improvements in conventional plant breeding, integrated pest management, organic farming, better farming methods,[28] and, where necessary, area expansion. Converting virgin land for agricultural production has, however, clear negative environmental effects, as it exacerbates water scarcity and biodiversity loss. Furthermore, land conversion is limited by the fact that pushing crops further into marginal land decreases returns.

---

[25] FAO, *The State of Food Insecurity in the World 2004* (Rome; FAO, 2004), 6.

[26] The environmental record of the Green Revolution is more controversial. Critics argue that the Green Revolution high-yielding varieties required the use of irrigation and fertilisers, and that they were often sold 'bundled' with chemical insecticides that gave rise to pest resistance and pollution. Most scientists concede that insecticides were over-used, especially in the early years of the Green Revolution, but many dispute the environmental evidence regarding irrigation and fertiliser use, and argue that claims of environmental harm ignore that by reducing the need to expand cultivated area, the Green Revolution saved large tracts of virgin land from the plough.

[27] Nuffield Council on Bioethics, *Genetically Modified Crops: The Ethical and Social Issues* (London; Nuffield Council on Bioethics, 1999), 62 ff available at www.nuffieldbioethics. org/go/ourwork/gmcrops/introduction (last accessed 14 Jan 2005).

[28] Modern irrigation technologies, such as drip irrigation, can improve efficiency, but their potential is limited by increasing scarcity of water resources and competition between household, agricultural and industrial uses, as well as between urban and rural uses.

4.2. Accessibility of Food

In human rights terms, the notion of *accessibility* encompasses economic accessibility:[29] Personal or household financial costs associated with acquiring food for an adequate diet should not compromise the satisfaction of other basic needs.[30] Lack of economic accessibility of food due to overall poverty is the most common cause of hunger in developing countries, as well as in the hunger pockets that remain in developed countries.

In developing countries, the majority of the poor are farmers or landless people living in rural areas who depend on agriculture for their livelihoods. GM seeds can benefit small-scale and resource-poor farmers. As the technology is embodied in the seed, it is scale-neutral and easily transferable.[31] Evidence suggests that some transgenic crops, especially insect-resistant cotton, yield significant economic gains to small farmers.[32] While prices for GM seeds may be higher, they can be compensated for by superior effective yields and lower pesticide costs. Bt cotton is a case in point. More than four million small farmers grow Bt cotton in China. These farmers have seen their costs of chemicals decline and their yields increase because their crops suffer less damage from pests.[33] Small farmers in several other developing countries are also benefiting from Bt cotton, Bt maize and herbicide-tolerant soybeans.

The positive effects extend beyond the farming community. When many farmers adopt a productivity-enhancing technology, they collectively produce more output, and this increased supply on the world market causes consumer prices to fall. Falling food prices were the most important avenue through which the Green Revolution improved the nutritional status of the poor. Lower prices mean that consumers can buy more for the same amount of money. Hence, the economic accessibility of food increases both for consumers in countries where the technology is adopted as well as elsewhere as markets are linked.[34]

Despite fears of corporate control of the sector, farmers and consumers so far are reaping a larger share of the economic benefits of transgenic crops than the companies that develop and market them.[35] Experience with transgenic cotton suggests that small farmers are as likely as large farmers to benefit from the adoption of the new crop. It must, however,

---

[29] CESR, above n 14, para 13.
[30] Ibid.
[31] FAO, above n 2, 104.
[32] Ibid, 51.
[33] Ibid.
[34] Farmers that do not adopt GMOs face falling prices for their produce and, unlike their competitors, they do not enjoy lower production costs, so their incomes fall. In the case of transgenic cotton, farmers in countries where transgenic cotton is not grown have experienced small economic losses as a result of lower cotton prices.
[35] FAO, above n 2, 104.

be borne in mind that this economic evidence is based on only two or three years of data for a relatively small number of farmers in just a few countries. The short-term farm-level gains may not be sustained over time as larger numbers of farmers adopt the technologies. More evidence is required to determine what the level and distribution of benefits from transgenic crops will be in the longer term, and to assess the risk that the costs of access to modern crops might put their potential benefits beyond the reach of poor farmers.

Also, the wider socio-economic effects need to be examined. In different circumstances, the same transgenic crop can have negative or positive effects depending on the labour constraints facing farm households and the nature of rural labour markets. Herbicide-resistant crops that render hand weeding superfluous provide an example. Reducing the demand for farm labour could free workers for higher value activities, such as off-farm wage employment or, in the case of children, for schooling. If alternative employment opportunities do not exist, however, reducing farm labour will detrimentally affect those displaced workers' ability to afford sufficient food.[36]

4.3. Food Safety

The right to food is realised only when individuals have access to safe food. Many concerns have been voiced about the safety of GM foods and their short and long-term effects. The main specific food safety concerns regarding GM crops and foods derived from them relate to allergens and toxins,[37] antibiotic resistance[38] and other unintentional changes in food composition.

---

[36] *Cf* Nuffield Council, above n 29, 72.

[37] Gene technology—like traditional breeding—may increase or decrease levels of naturally occurring proteins, toxins or other harmful compounds in foods.

[38] Antibiotic resistance is a food safety concern because many first generation GM crops were created using antibiotic resistant marker genes. If these genes could be transferred from a food product into the cells of the body, or to bacteria in the gastrointestinal tract, this transfer could lead to the development of antibiotic resistant strains of bacteria, with adverse health consequences. Although scientists believe the probability of transfer is extremely low (GM Science Review Panel, *GM Science Review: First Report—an Open Review of the Science Relevant to GM Crops and Food Based on the Interests and Concerns of the Public* (2003), available at www.gmsciencedebate.org.uk/report/pdf/gmsci-report1-full.pdf (last accessed 20 Feb 2005), 96 ff), the use of antibiotic resistance genes has been discouraged by an FAO and WHO expert panel (FAO/WHO, 'Safety Aspects of Genetically Modified Foods of Plant Origin, Report of a Joint FAO/WHO Expert Consultation on Foods Derived from Biotechnology, Geneva, Switzerland, 29 May–2 June 2000, available at ftp://ftp.fao.org/es/esn/food/gmreport.pdf (last accessed 20 Feb 2005), 13) and other bodies. Researchers have developed methods to eliminate antibiotic resistance markers from genetically engineered plants.

The foods derived from the transgenic crops that are currently being grown commercially have been evaluated by the national food safety authorities of several countries using procedures that are consistent with internationally agreed principles.[39] They have been judged safe to eat, and the methods used to evaluate their safety have been deemed appropriate.[40] To date, no verifiable untoward toxic or nutritionally deleterious effects resulting from the consumption of foods derived from GM crops have been discovered anywhere in the world.[41] Although potential risks remain, up to now it appears that GM foods are as safe as their conventional counterparts.[42]

However, absence of the evidence of harm is not evidence of the absence of potential harm. Little is known about the long-term food safety effects of foods derived from transgenic crops (or any foods), and continued monitoring is recommended. In addition, some critics argue that risk analysis based on the concept of substantial equivalence[43] is not sufficient for GM foods because complex, unexpected differences arising from genetic modification may be missed. Consensus does exist that foods derived from emerging, more complex genetic transformations may require additional food safety procedures.[44]

One aspect that has received little attention up to now is the fact that agricultural biotechnologies may have the potential to *improve* the safety of some foods. This has not yet been clearly established in the literature, but there are several opportunities that arise from GM crops currently on the market. For example, Bt crops typically involve lower applications of chemical insecticides, so the resulting foods may contain fewer chemical

---

[39] International Council for Science (ICSU), *New Genetics for Food and Agriculture: Scientific Discoveries—Societal Dilemmas* (2003), available at www.icsu.org/Gestion/img/ICSU_DOC_DOWNLOAD/90_DD_FILE_ICSU_GMO%20report_May%202003.pdf (last accessed 21 Feb 2005), 19.

[40] FAO/WHO, above n 38.

[41] ICSU, above n 39, 13.

[42] This is a result of a review carried out by FAO in 2004 of scientific evidence contained in several recent major reports on biotechnology authored by institutions as diverse as the International Council for Science (ICSU), which represents 101 national academies of science and 27 international scientific unions, The Nuffield Council on Bioethics, The UK GM Science Review Panel, The Royal Society, DANIDA, and the FAO/WHO Codex Alimentarius Commission: FAO, above n 2, 58 ff.

[43] Risk analysis of foods derived from GMs is conducted using principles similar to those used for food additives: ie, differences from the conventional food are identified and those differences are tested. If no harmful effects are found, the food is deemed to be as safe as its conventional counterpart. The concept of 'substantial equivalence' is the cornerstone of this approach endorsed by the FAO/WHO Codex Alimentarius Commission. See Codex Alimentarius Commission, 'Principles for the Risk Analysis of Foods Derived from Modern Biotechnology' (Doc CAC/GL 44-2003), 'Guideline for the Conduct of Food Safety Assessment of Foods Derived from Recombinant-DNA Plants' (Doc CAC/GL 45-2003), and 'Guideline for the Conduct of Food Safety Assessment of Foods Produced Using Recombinant-DNA Micro-organisms' (Doc CAC/GL 46-2003).

[44] ICSU, above n 39, 33, GM Science Review Panel, above n 38, 12.

residues. Bt crops also suffer less insect damage so they may have lower concentrations of mycotoxins, such as aflatoxin, which can cause serious liver damage at low exposures and even acute toxicity. Such occurred in Kenya where over 100 people died from eating contaminated grain in 2004.[45] More directly, GM crops with reduced allergens (wheat, soybeans and peanuts) and less toxic compounds (eg, cassava with less cyanide) are under development. These benefits need, however, to be better documented.[46]

4.4. Nutritious Food

The right to food can only be fully realised with nutritious food. Individuals should have access to a diet that contains an adequate mix of nutrients for physical and mental growth, development and activity. It should meet physiological needs at all stages throughout the life cycle and according to gender and occupation.[47] In particular, micronutrient deficiencies, the so-called hidden hunger, which is mainly caused by a lack of vitamin A, iron and iodine, can have devastating effects. Vitamin A deficiency, for instance, affects more than 200 million people worldwide and causes an estimated 2.8 million cases of blindness in children under the age of 5.[48]

Biotechnology can improve the nutritional profile of certain food stuffs, eg, vitamin A enriched mustard seeds, high-protein potatoes and iron-enriched rice. A well-known example is Golden Rice, a biotechnology derived rice which contains large amounts of beta carotene (a precursor of vitamin A).[49] Golden Rice is a sustainable and low-cost alternative to food supplements and vitamin fortification,[50] and has been proposed for people that depend on rice for the bulk of their diet.[51] In addition, anti-nutrients, such as cyanide in cassava, can be reduced through genetic engineering. All of these innovations have been developed, although they have not been approved for commercial cultivation, and it remains to be seen whether their potential will be realised.

Critics argue that Golden Rice is a simplistic high-tech solution to a problem that should instead be solved by increasing access to more balanced diets rich in fruit and vegetables. While this criticism is

---

[45] See www.cdc.gov/mmwr/preview/mmwrhtml/mm5334a4.htm (last accessed 22 Feb 2005).

[46] FAO, above n 2, 61, ICSU, above n 39, 25.

[47] CESCR, above n 14, para 9.

[48] FAO, above n 2, 42.

[49] The high beta carotene content was achieved by inserting two genes from a daffodil and one from the bacterium *Erwinia uredovora*.

[50] R Zimmermann and M Qaim 'Projecting the Benefits of Golden Rice in the Philippines', Discussion Paper on Development Policy No 51 (Bonn, Centre for Development Research, 2002).

[51] FAO, above n 2, 42.

well-founded in principle, it needs to be taken into account that hidden hunger is typically not due to a lack of nutritious food on offer but, rather, to a lack of financial means to access higher quality foods. The poor already spend a large proportion of their income on food and typically cannot allocate more resources to food consumption. Therefore, nutrition can only be improved by either increasing incomes or the nutritional value of a subsistence diet.

4.5. Acceptability

Realising the right to food requires accounting for consumer acceptability, ie, the non-nutrient-based values attached to food and food consumption, and informed consumer concerns regarding the nature of accessible food supplies.[52]

The acceptability of GM crops and foods varies considerably among and within countries. What is readily embraced in the US may meet sharp protest and deep suspicion in Europe.[53] Lack of acceptability is more pronounced when advantages or benefits of GM food cannot be perceived because the products are neither cheaper or healthier, nor superior in taste.

Consumers resist genetic engineering for many reasons. Among these are ethical, religious, health and environmental ones. Some oppose the technology itself as being 'unnatural' and as interfering with nature or God's will. Others may refrain from using specific applications, eg, Muslims may avoid food produced with swine genes or vegetarians abstain from fruit and vegetables modified with animal genes. The most common reason is the already addressed fear of potential health risks.

From a human rights point of view, consumers' resistance to GM food has to be taken seriously. Labelling allows consumers to make a choice between GM and non-GM foods. Labelling has, however, two counteracting effects. One the one hand, it increases transparency, consumer information and choice. One the other hand, establishing and monitoring a labelling system also raises the costs of food production. In developing countries, the costs of labelling should not create access hurdles to food for the poor.

Acceptability is also an issue in inter-state relations. Zambia rejected food aid from GM products in 2002 on a number of grounds, amongst

---

[52] CESCR, above n 14, para 11.

[53] In general, Europe and Japan are less accepting than North and South America and developing Asia. In the least accepting European countries—France and Greece—22% approve of GMOs; in the most accepting Asian countries—Indonesia and China—only 20–30% have reservations: FAO, above n 2, 77 ff.

which were health concerns and fear of gene flow[54] from GM imports to local crops, which could prevent access to external markets.[55] In situations of famine, states must generally accept food aid as part of their obligation to fulfil the right to food.[56] When GM food is offered, the case becomes more complicated as states need to strike a difficult balance between contradictory short-term (fulfilling the core content of the right to food through famine relief) and long-term (eg, potential loss of access to foreign markets) effects on the realisation of the right to food.[57] If possible, donors should offer food aid that does not create such a dilemma for a recipient country.[58]

4.6. Environmental Sustainability

Sustainability in the context of the right to food means that food should be available and accessible for both present and future generations.[59] Sustainability in this sense is linked to environmental sustainability.

The environmental sustainability implications of GMOs are highly contested and give the most reason for concern. Many of the potential environmental impacts of transgenic crops are not yet fully clear. Agreement exists on the types of hazards and potential benefits that exist, but not on their likelihood and potential consequences. Possible types of harm include gene flow, direct harm to non-target organisms,[60] emergence of

---

[54] Gene flow, ie, the spread of transgenes to related crops (conventional, organic or landraces) or wild relatives can occur when transgenic crops are grown in proximity to related plants. Transgenes will persist and spread in these circumstances only if they convey a competitive advantage on the recipient plant. This is not likely to be the case for herbicide tolerance since this trait is advantageous only in the presence of the herbicide. Insect and disease resistance could provide an advantage, however.

[55] Nuffield Council on Bioethics, The Use of Genetically Modified Crops in Developing Countries—A Follow-up Discussion Paper (2004), available at www.nuffieldbioethics.org/go/ourwork/gmcrops/page_218.html (last accessed 21 Feb 2005), 78.

[56] *Cf* CESCR, above n 14, para 19.

[57] Some concerns can be met by milling GM crops.

[58] FAO, FAO Expert Consultation on Food Safety: Science and Ethics, 3–5 Sept 2002, Rome; FAO, Readings in Ethics No 1 (Rome; FAO, 2004), 32.

[59] CESCR, above n 14, para 7.

[60] Insect resistant crops could harm non-target insects and other organisms, potentially disrupting food chains and soil microbial communities, and thereby having negative effects on biodiversity. In the famous monarch butterfly case, pollen from some Bt maize plants was found to harm monarch caterpillars when they were force-fed under laboratory conditions: JE Losey, LS Rayner and ME Carter, 'Transgenic Pollen Harms Monarch Larvae, (1999) 399 *Nature*, 214. Follow-up studies found the same effect to be highly unlikely in field conditions for all Bt maize varieties except one, which was subsequently removed from the market: AJ Connor, TR Glare, and J-P Nap, 'The Release of Genetically Modified Crops into the Environment: Part II. Overview of Ecological Risk Assessment,' (2003) 33 *Plant Journal*, 19. Bt maize has also been found to be less harmful to non-target insects than maize produced using conventional pesticides.

Bt resistant pests,[61] and indirect environmental effects. Conversely, potential benefits include replacing more toxic insecticides and herbicides with less toxic ones[62] and, in the case of herbicide resistant crops, a decreased need for tillage, leading to better soil conservation, improved water retention and less soil erosion.

Scientific evidence concerning transgenic crops is still emerging. In the countries where transgenic crops are grown commercially, there have been no verifiable reports that they cause any significant environmental harm, but some benefits have been observed. However, the lack of negative impacts so far does not mean that they cannot occur, and the current understanding of ecological processes is incomplete. A cautious case-by-case approach is needed, taking into consideration the crop, the trait and the agro-ecosystem in which it is to be released. A number of legal instruments, such as the Convention on Biological Diversity, the Cartagena Protocol on Biosafety and European Community instruments, refer to the precautionary principle in this context.[63]

5. THE GOVERNANCE OF GENETICALLY MODIFIED CROPS AND THE RIGHT TO FOOD

The preceding brief discussion of the actual and potential future implications of using GM crops for each component of the right to food has revealed a complex set of effects depending on the specific GMO, the right to food component in focus and the particular context. Taking this interim conclusion as a starting point, this part will assess the right to food implications of agricultural biotechnology by looking at the effects of the wider socio-economic and legal governance framework for GM crops, which determines their development, marketing and use.

---

[61] Insect resistant Bt crops may lead to the emergence of pests that are resistant to Bt, as those pests not killed survive and breed, passing resistance to their progeny. Crop management strategies, such as the creation of refugia, are recommended to avoid or delay that occurrence. Newer generations of insect resistant crops contain two Bt genes, rather than one, significantly reducing the likelihood of resistance developing. So far, no evidence of Bt resistance has been observed in the field.

[62] Herbicide tolerant crops are associated with decreased use of the most highly toxic herbicides, but an overall increase in herbicide use of lower toxicity. Replacing more toxic herbicides with less toxic ones is generally considered as an environmental benefit.

[63] Preamble to the Convention on Biological Diversity, above n 1; Preamble to and Arts 1, 10(6) and 11(8) of the Cartagena Protocol on Biosafety to the Convention on Biological Diversity, (2000) 39 ILM 1027; Art 1 of Directive 2001/18/EC on the Deliberate Release into the Environment of Genetically Modified Organisms [2001] OJ L 106/1. On biosafety see R Mackenzie, 'The International Regulation of Modern Biotechnology' (2002) 13 *Yearbook of International Environmental Law*, 97.

5.1. The Privatisation of Research

In contrast to the Green Revolution that was led by public sector research, agricultural biotechnological research has become increasingly private and dominated by a few large corporations. It is supported by private funds, carried out in private institutions, and protected by intellectual property rights that restrict the utilisation of its findings.

The world's top 10 transnational bioscience corporations spend about US$3 billion per year on agricultural biotechnology research and development.[64] The CGIAR[65] system has a total crop improvement budget of one-tenth that amount—about US $300 million—of which not more than one-tenth is devoted to biotechnology.[66] Among developing countries, the three largest national agricultural research programmes (those of Brazil, China and India) have total budgets of less than US $500 million each, of which about 5 to 10 per cent goes to biotechnology research.[67] China is the only developing country that has developed transgenic crop technologies independently of the international private sector. India and Brazil may develop this capacity, but few other developing countries will be able to do so.

This private sector dominance of research has two consequences: knowledge is accumulated in the private and not the public domain, and the needs of the poor are systematically neglected.[68] Private sector biotechnology research naturally focuses on developing technologies suitable for the major commercial agricultural input markets in the temperate-zone production environments of North America and Europe. Some farmers in developing countries have been able to take advantage of 'spillover' benefits from private sector research aimed at farmers in the developed world. These farmers are located primarily in temperate production zones in South America, South Africa and China. Barring a few initiatives, no major public sector or private sector programmes are tackling the critical problems of the poor, or target crops and animals on which they rely. This includes the crops that provide the bulk of their food supply and livelihoods—rice and wheat—but also a variety of 'orphan crops' such as sorghum, pearl millet, pigeon pea, chickpea and groundnut that are largely neglected in conventional or biotechnology research

---

[64] FAO, above n 2, 32.

[65] Consultative Group on International Agricultural Research.

[66] FAO, above n 2, 32.

[67] D Byerlee and K Fischer, 'Accessing Modern Science: Policy and Institutional Options for Agricultural Biotechnology in Developing Countries', (2002) 30 *World Development*, 931 ff.

[68] P Pingali and G Traxler, 'Changing Locus of Agricultural Research: Will the Poor Benefit form Biotechnology and Privatisation Trends?' (2002) 27 *Food Policy* 223. Some of the CGIAR centres are working with national research systems and the private sector to develop transgenic crops for developing countries, but these programmes are small and poorly funded.

programmes. Under-researched traits of particular interest to the poor include resistance to production stresses like drought, salinity, disease and pests, as well as nutritional enhancement.

To what extent do these developments point to shortcomings in the fulfilment of right-to-food obligations? Under the ICESCR states have to use 'all appropriate means' to progressively realise the rights recognised in the ICESCR.[69] As part of the obligation 'facilitate' the realisation of the right to food, states should promote research and development of technological advances, including research on better seeds.[70] However, each state has broad discretion on how to promote human rights.[71] When allocating resources, states must take all reasonable options into account and adequately consider the potential of biotechnological research in designing their research programmes. Given the often prohibitive costs involved[72] and the uncertain beneficial effects of biotechnological inventions, a duty of developing countries to carry out biotechnological research cannot be assumed.

There are also no obligations for developed countries to carry out or promote research for the benefit of individuals in other states. Each state has an obligation to respect, protect and fulfil the human rights of individuals under its jurisdiction, not under the jurisdiction of other states.[73] While some assert 'extra-territorial obligations', ie, human rights obligations of one state versus individuals under the jurisdiction of another,[74] these obligations do not yet reflect existing law. Also state-to-state cooperation duties that are mentioned in a number of binding and non-binding human

---

[69] Art 2(1) of the ICESCR.

[70] For the right to health, the CESCR has explicitly mentioned an obligation to promote research: General Comment No. 14 The Right to the Highest Attainable Standard of Health, UN Doc. E/C.12/2000/4 of 11 Aug 2000, paras 36 and 37. While for the right to food research might not play the same role as for the development of new medicines, crops adapted to the needs of developing countries, with improved nutritious value or with higher productivity, can play a vital role in improving its realisation. See also Art 15(2) of the ICESCR.

[71] CESCR, General Comment No. 3 on The Nature of States Parties Obligations of 14 Dec 1990, reprinted in 'Compilation of General Comments and General Recommendations Adopted by Human Rights Treaty Bodies', UN Doc HRI/GEN/1/Rev.7 of 12 May 2004, 15.

[72] For many states research in the field of biotechnology will exceed the 'maximum … available resources' requirement of Art 2(1) of the ICESCR.

[73] This is made explicit in Art 1 of the ECHR: '[t]he High Contracting Parties shall secure to everyone within their jurisdiction the rights and freedoms defined in Section 1 of this Convention', Rome, 4 Nov 1950, 213 UNTS 221, as amended by Protocol No 11 of 5 May 1994, ETS No 155.

[74] See, eg, CESCR, above n 14, para 36 ff; J Ziegler, 'Second Submission of Jean Ziegler, Special Rapporteur on the Right to Food of the United Nations Commission on Human Rights to the Intergovernmental Working Group (IGWG) for the Voluntary Guidelines on the Right To Adequate Food' para 13 ff available at www.righttofood.org/SECOND%20S UBMISSION%20SR%20RIGHT%20TO%20FOOD.htm#_ednref (last accessed 20 Feb 2005); R Künnemann, 'The Extraterritorial Scope of the International Covenant on Economic, Social and Cultural Rights, FIAN', available at www.fian.org/fian/index.php?option=cont ent&task=view&id=76&Itemid=61 (last accessed on 20 Feb 2005); S Skogly, 'Extra-national Obligations Towards Economic and Social Rights, International Council on Human Rights Policy', (2002), available at www.ichrp.org/ac/excerpts/92.doc (last accessed 20 Feb 2005).

rights and other instruments[75] do not oblige states to undertake specific positive measures, such as carrying out or directing research towards the needs of other states. Although frequently mentioned, cooperation obligations remain notoriously vague and their content is difficult to grasp in clear legal terms.

Privatising research with the potential to contribute significantly to the better realisation of the right to food, amongst other rights, is problematic. The technology's potential in assisting the poor to realise their right to food has remained underused, thus preventing the poor from enjoying their right to benefit from scientific progress and its application (Article 15(1)(b) ICESCR, see below). Concerted international efforts, including public/private partnerships and contributions by international organisations, are required to ensure that the technology needs of developing countries are addressed and that barriers to access are overcome.[76]

5.2. The Role of Transnational Corporations

Where GMOs are introduced, concerns arise whether transnational biotech corporations will misuse their commercial and political clout.[77] These concerns are aggravated by major consolidation in the global seed and agricultural input industries aimed at vertical and horizontal integration to optimise investment through better control of distribution channels, including those of complementary agricultural inputs such as herbicides and pesticides.[78]

---

[75] Eg, Arts 2(1), 11(2) and 23 of the ICESCR, Arts. 55 and 56 of the UN Charter, Arts 16 and 19 of the Convention on Biological Diversity (above n 1), Arts 16 and 19, or Art 22 of the Cartagena Protocol on Biosafety to the Convention on Biological Diversity (above n 63).

[76] See Nuffield Council, above n 27, 63.

[77] Sub-Commission on the Promotion and Protection of Human Rights, Globalisation and its Impact on the Full Enjoyment of Human Rights, 'Progress Report Submitted by J Oloka-Onyango and Deepika Udagama, in Accordance with Sub Commission Resolution 1999/8 and Commission on Human Rights Decision 2000/102, UN Doc. E/CN.4/Sub.2/2001/10 of 2 Aug 2001, para 32. For an NGO view on the practices of global agrifood businesses, see ActionAid, 'Power Hungry—Six Reasons to Regulate Global Food Corporations', available at www.global-policy.org/socecon/tncs/2005/01powerhungry.pdf (last accessed 12 Jan 2005). The fact that 85% of all fines imposed on global price-fixing operations in the past several years were paid by food and agricultural cartels shows the dangers of market concentration in this field; cf RA Levins, 'Dwindling Competition Will Mean Higher Food Costs', available at www.apec.umn.edu/faculty/dlevins/Dwindling Competition.pdf (last accessed 22 Feb 2005).

[78] Commission on Intellectual Property Rights, *Integrating Intellectual Property Rights and Development Policy* (London, Commission on Intellectual Property Rights, 2002), 65, available at www.iprcommission.org/graphic/documents/final_report.htm (last accessed 12 Jan 2005). Also, in developing countries there is evidence of rapidly growing control of transnational corporations. In Brazil, eg, following the introduction of plant variety protection in 1997, and presumably also related to the expected permission to grow GM crops, Monsanto increased its share of the maize seed market from 0% to 60% between 1997 and 1999, through mergers and acquisitions. Dow and Cargill and Agrevo (now Aventis) also increased their market shares by acquisition. Only one Brazilian-owned firm remained with a 5% share in the maize market: ibid.

States have to protect the right to food against infringements by private actors.[79] They must shield individuals against market practices that rely on structural differences in knowledge and experience between the buyers and sellers, and work systematically and in an unfair manner to the disadvantage of one party. A state must, for example, ensure that adequate information is available about the consequences of using a new product. If patented GM seeds are introduced, which farmers are not allowed to reuse, states have to make sure that farmers have access to clear information about this consequence. This is particularly relevant where this prohibition interferes with traditional seed uses. States may also need to regulate certain marketing practices, such as aggressive and misleading promotional offers of patented products through special loans and grants, which are tied to designated seed and chemical packages,[80] and may lead to indebtedness and loss of livelihood. Conversely, overregulation that impedes access to technology needs to be avoided, in particular until present farmers and consumers have reaped most of the benefits of GM crops (see section 5.2 above). Where to draw the line between individual responsibility and state duty depends on the circumstances of each case and the concept of the role of the state in each society. The greater the structural imbalance in knowledge and influence, the more pronounced the state's obligations. The limits of legitimate market power are normally regulated by competition and consumer protection law which may be weak in many developing countries.

The role of the state is particularly important because corporations—as private actors—are not directly bound by human rights law. While increasingly the scope of international legal personality of transnational corporations is reconsidered in order to hold such corporations directly accountable,[81] such developments are still in their early stages. The recent adoption of the '[n]orms on the responsibilities of transnational corporations and other business enterprises with regard to human rights' by the Sub-Commission on the Protection and Promotion of Human Rights is an

[79] *Cf* CESCR, above n 14, para 15.

[80] A Chapman, 'Core Obligations Related to ICESCR Article 15(1)(c)', in: A Chapman and S Russell, *Core Obligations: Building a Framework for Economic, Social and Cultural Rights* (Antwerp, Intersentia, 2002), 305, 327.

[81] Coomans, eg, sees a trend towards increased recognition of the applicability of human rights standards to corporations: F Coomans, 'The Ogoni-Case Before the African Commission on Human and Peoples' Rights' (2003) 52 *International and Comparative Law Quarterly*, 749, 760, n 47; N Jägers, *Corporate Human Rights Obligations: In Search of Accountability* (Antwerp, Intersentia, 2002); D Weissbrodt, and M Kruger, 'Business and Human Rights' in Mars, Bergsmo (ed), *Human Rights and Criminal Justice for the Downtrodden—Essays in Honour of Asbjørn Eide* (Leiden/Boston, Mass, Marinus Nijhoff, 2003), 421.

interesting step in this direction.[82] Other initiatives comprise the Global Compact[83] and codes of conduct developed by industries or businesses themselves[84]. With regard to biotechnology corporations, further research on the standards such corporations could be required to abide by, and the kind of human rights duties they could be expected to fulfil could serve as a useful basis for a human rights dialogue with this specific industry.[85]

5.3. Intellectual Property Rights

The increase in intellectual property rights protection since the early 1980s has stimulated the rapid growth of agricultural biotechnology research and product development by the private transnational sector.[86] It has also facilitated the concentration of agricultural biotechnological research capacity, knowledge and market share in products developed with such knowledge in the developed world. In particular, the entry into force of the Agreement on Trade-Related Aspects of Intellectual Property Rights (TRIPS Agreement) in 1995 has strengthened the protection of intellectual property rights by setting minimum standards binding upon all 148 WTO Members and enforceable through the WTO dispute settlement mechanism.

A number of developments in the field of intellectual property rights have raised human rights concerns. Human rights and other actors have reacted by paying increased attention to intellectual property rights' effects on human rights since 1999:[87] The 1999 UNDP Human Development Report warned against the negative consequences of the

---

[82] UN Doc E/CN.4/Sub.2/2003/12/Rev.2 of 26 Aug 2003. For a discussion of the norms, see D Weissbrodt, 'Norms on the Responsibilities of Transnational Corporations and Other Business Enterprises with Regard to Human Rights' (2003) 97 AJIL, 901. Similar efforts were made earlier on. Noteworthy in this context are, eg, the Code of Conduct on the Marketing of Breastmilk Substitutes (Infant Formula Code) reprinted in S Shubber, *The International Code of Marketing of Breastmilk Substitutes—An International Measure to Protect and Promote Breast-Feeding* (Dordrecht, Kluwer, 1998), or the OECD Guidelines for Multinational Enterprises, available at: www.oecd.org/dataoecd/56/36/1922428.pdf (last accessed 20 Feb 2005), or the self-commitment of certain industries or firms to their code of conducts.

[83] See www.unglobalcompact.org/Portal/Default.asp? (last accessed 20 Feb 2005).

[84] See, eg, OECD Working Party of the Trade Committee, Codes of Corporate Conduct: An Inventory, 1999, OECD Doc No TD/TC/WP(98)74/FINAL, available at www.olis.oecd.org/olis/1998doc.nsf/LinkTo/td-tc-wp(98)74-final (last accessed 20 Feb 2005).

[85] *Cf* the suggestion by the Special Rapporteur on the Right to Health to explore the right to health duties of the pharmaceutical sector by expert mechanisms composed of representatives from the human rights and pharmaceutical sectors: P Hunt, 'The Right to Health—What Do I Expect from a Pharmaceutical Company?' paper presented at *The Right to Health: A Duty for Whom?*, International Symposium, 2 Dec 2004, Basel, Switzerland, available at www.novartisfoundation.com/pdf/NFSD_Symp_04_Speech_Paul_Hunt.pdf (last accessed 13 Jan 2005).

[86] FAO, above n 2, 31.

[87] See also the related FAO Report of the Panel of Eminent Experts in Ethics in Food and Agriculture, Second Session 18–20 Mar 2002 (Rome; FAO 2003), 15.

TRIPS Agreement, particularly in relation to food security, indigenous knowledge, bio-safety and access to health care.[88] The Sub-Commission on the Promotion and Protection of Human Rights adopted resolutions on 'Intellectual Property Rights and Human Rights' in 2000[89] and in 2001.[90] At the request of the Sub-Commission, the High Commissioner for Human Rights prepared a report on 'The Impact of the Agreement on Trade-Related Aspects of Intellectual Property Rights on Human Rights'.[91] In 1999, the CESCR issued a Statement to the Third Ministerial Conference of the WTO in Seattle;[92] in 2000, it held a day of general discussion on intellectual property rights and human rights; and in 2001 it adopted a Statement on Human Rights and Intellectual Property.[93] It is now discussing Draft General Comment No 18 on the Right of Everyone to Benefit from the Protection of the Moral and Material Interests Resulting from any Scientific, Literary or Artistic Production of Which He Is the Author (Article 15 ICESCR). The Special Rapporteurs on the Right to Health (Paul Hunt) and the Right to Food (Jean Ziegler) have addressed intellectual property rights and, in particular, the TRIPS Agreement in some of their reports.[94] In the words of the United Nations Sub-Commission on the Promotion and Protection of Human Rights, 'actual or potential conflicts exist between the implementation of the TRIPS Agreement and the realisation of economic, social and cultural rights in relation to, inter alia, impediments to the transfer of technology to developing countries, the consequences for the enjoyment of the right to food of plant variety rights and the patenting of genetically modified organisms, "bio-piracy" and the reduction of communities' (especially indigenous communities')

[88] UNDP, *Human Development Report 1999* (New York, Oxford, Oxford University Press, 1999), 68.

[89] Sub-Commission on the Promotion and Protection of Human Rights, Intellectual Property Rights and Human Rights, Resolution 2000/7 of 17 Aug 2000 (adopted without a vote). On this resolution see D Weissbrodt and K Schoff, 'Human Rights Approach to Intellectual Property Protection: The Genesis and Application of Sub-Commission Resolution 2000/7' (2003) 5 *Minnesota Intellectual Property Review,* 1.

[90] Sub-commission on the Promotion and Protection of Human Rights, above n 89.

[91] High Commissioner for Human Rights, 'The Impact of the Agreement on Trade-Related Aspects of Intellectual Property Rights on Human Rights', UN Doc. E/CN.4/Sub.2/2001/13 of 27 June 2001.

[92] UN Doc. E/C.12/1999/9 of 26 Nov 1999.

[93] Statement by the CESCR on Human Rights and Intellectual Property, CESCR, Report of the Twenty-fifth, Twenty-sixth and Twenty-seventh session CESCR Official Records, 2002, Supplement No. 2, UN Doc E/2002/2, E/C.12/2001/17, Annex XIII.

[94] Eg, 'The Right of Everyone to the Enjoyment of the Highest Attainable Standard of Physical and Mental Health', Report of the Special Rapporteur, Paul Hunt, Addendum— 'Mission to the World Trade Organisation', UN Doc E/CN.4/2004/49/Add.1 of 1 Mar 2004; 'The Right to Food', Report by the Special Rapporteur on the Right to Food, Mr Jean Ziegler, submitted in accordance with Commission on Human Rights resolution 2000/10, UN Doc E/CN.4/2001/53 of 7 Feb 2001, para 73.

control over their own genetic and natural resources and cultural values'.[95] A less nuanced position is held by the Special Rapporteur on the Right to Food, Jean Ziegler, who views 'developments in biotechnology, including genetically modified plants, ownership of international patents by agribusinesses from the North and worldwide protection of those patents, hampering access to food and the availability of food' as one of seven major economic obstacles that hinder or prevent the realisation of the right to food.[96] He calls for a moratorium on GM crops.[97]

Beyond the general concerns mentioned by the Sub-Commission, a number of developments more specifically concern the right to food. Amongst these developments are: the potential for stifling further research by a rapidly increasing number of often broad patents[98] concentrated in few hands, the difficulties of developing countries to access the technology; the interference of patents and other intellectual property rights with farmers' rights to use and exchange farm-saved seeds; and the lack of recognition and protection of farmers' contributions to the development of the available genetic resource pool.

5.3.1 The Protection of Creativity in Human Rights Law and the Need to Balance Private and Public Interests

Article 15 of the ICESCR, developed from the similarly worded Article 27 of the UDHR, recognises the right of everyone 'to benefit from the protection of the moral and material interests resulting from any scientific, literary or artistic production of which he is the author' (Article 15(1)(c)). While not human rights in themselves, intellectual property rights safeguard the right to protection of one's production. Article 15(1) of the ICESCR, however, recognises not only a right of the inventor to benefit from his or her production (Article 15(1)(c)), but also a right of everyone 'to enjoy the benefits of scientific progress and its applications' (Article 15(1)(b)) and

---

[95] Sub-Commission on the Promotion and Protection of Human Rights, above n 89, 11th preambular para.

[96] 'The Right to Food', above n 94, summary and paras 69 and 73.

[97] Report of the Special Rapporteur of the Commission on Human Rights on the Right to Food, UN Doc. A/57/356 of 27 Aug 2002, paras 19 and 20.

[98] Broad patents claim the gene, the vector or carrier for effecting the transformation and so on, which may cover a number of potential varieties or crops incorporating the gene. In fact, the patent on a product containing or consisting of genetic material normally extends to all material in which the product is incorporated and in which the genetic information is contained and performs its function: see, eg, Art 9 of EC Directive 98/44/EC on the Legal Protection of Biotechnological Inventions of 6 July 1998 [1998] OJ L 213/13. The granting of broad patents has given rise to calls that patents be given only when there has been a genuine invention that has created a biological product significantly different from any that existed before, and the patent should cover only the inventive step itself, nothing beyond it: FAO, above n 87, 15.

a right 'to take part in cultural life' (Article 15(1)(a)). These three components—the right to protection of one's production, the right to benefit from scientific advancement and the right to take part in cultural life—are intrinsically interrelated with one another, not only in the ICESCR, but also in the UDHR and the American Declaration of the Rights and Duties of Man.[99] States must strike a balance between public and private interests in knowledge that does not unduly advantage private interests and gives due consideration to public interest in enjoying broad access to new knowledge.[100] Protecting creators' interests must not interfere with the realisation of other human rights including the right to food, and the protection should be compatible with overall human rights principles.[101]

Any system of intellectual property rights should serve a double function. It should protect the interests of both inventors and society at large. With the right to reap the exclusive benefits from their invention for a limited period, inventors are provided with incentives to research and develop. Society at large, in turn, should profit from such new developments. While, in fact, patents, plant breeders' rights, and other types of intellectual property rights have greatly stimulated the growth of private agricultural research, the present balance between private and public interests may need to be reconsidered.

5.3.2 Patents: Incentive or Stifling of Research?

One question to resolve is whether the rapid expansion of intellectual property protection in level and scope and through higher global minimum standards is undermining the research incentive function of intellectual property rights.[102]

The patent is the most relevant intellectual property right to agricultural biotechnology. Plant variety rights play a smaller role as they can protect only new plant varieties[103] and because the title holder's rights are more

---

[99] Chapman, above n 80, 314.

[100] CESCR, above n 93, para 17. The CESCR also points to the Declaration on the TRIPS Agreement and Public Health adopted at the WTO Fourth Ministerial Conference in Doha in Nov 2001 as an example of the need to strike a balance between public and private interests: WT/MIN(01)/DEC/W/2.

[101] Cf CESCR, above n 93 and A Chapman, 'The Human Rights Implications of Intellectual Property Protection' (2002) 5 *Journal of International Economic Law*, 861, 866 ff.

[102] There is some evidence suggesting such a trend: ibid, 16 ff.; Nuffield Council, above n 27, 50 ff. There are, eg, several hundred overlapping patent rights for the Bt technology, and at least 4 companies obtained patents that cover Bt-transformed maize; Commission on Intellectual Property Rights: above n 78, 6. In addition, there is evidence that patents are contributing to the rapid market concentration in the agricultural biotechnology field, which has adverse effects on the degree of competition: Commission on Intellectual Property Rights, above n 78, 17.

[103] Plant variety can be defined as a plant grouping within a single botanical taxon of the lowest known rank, which grouping can be defined by the expression of the characteristics

limited. In contrast, patents have been granted for living organisms,[104] biological molecules (DNA), proteins and other biochemical compounds, and on various biotechnologies.[105]

In the field of agricultural biotechnology, the proliferation of biotechnology patents—in particular, that of broad patents on the enabling technologies[106] concentrated in the hands of five major industrial groups[107]—has led to widespread concern that patents no longer stimulate research but, on the contrary, hinder the generation of further knowledge. Patent holders are not in all cases willing to license their patents, as they may wish to retain a property advantage over their competitors and enjoy the benefits that patent monopoly brings.[108] Even if licences are given, researchers must still acquire agreements from many patent holders, often at prohibitive prices. Such developments have been termed the 'tragedy of the anticommons'.[109] It remains to be seen whether corporations will be willing to grant licences on favourable terms to develop crops for the developing world. If this is not the case, tensions will arise with the right of everyone to benefit from scientific progress and its application. To avoid such negative effects, intellectual property rights must be regulated in a manner that impedes research as little as possible. For instance, if patents for genes are allowed, they should cover only uses set out in them, not other uses of the same invention which

that result from a given genotype or a combination of genotypes, distinguished from any other plant grouping by the expression of at least one of those characteristics, and considered as a unit with regard to its suitability for being propagated unchanged: EC Council Regulation 2100/94 on Community Plant Variety Rights [1994] OJ L 227/1, Art 5. Under the International Convention for the Protection of New Varieties of Plants (UPOV Convention) signed in 1961, subsequently revised in 1972, 1978, and 1991, a plant variety is eligible for protection if it is distinct, stable, uniform, and novel (Art 6 UPOV 1978).

[104] Patents on living organisms are not unique to GM and biotechnology. Micro-organisms in particular have commonly been patented. As far back as the 19th century, Louis Pasteur was granted a patent on a strain of yeast in both France and the US. Plant patents were occasionally granted prior to the first UPOV Convention in 1961 which specifically excluded the granting of both patents and plant variety rights for the same plant variety: (Art 2(1)) Nuffield Council, above n 27, 45, n 16.

[105] Ibid, 45.

[106] There are two main types of patents: those which enable the technology, and application patents which cover specific traits for improving plants but which are dependent on the enabling patents for their implementation: Nuffield Council, above n 27, 51.

[107] Formerly, the six main groups were: AstraZeneca, Aventis, Dow, DuPont, Monsanto and Novartis. With the merger of the agricultural arms of AstraZeneca and Novartis to form Syngenta, there are now five left: Commission on Intellectual Property Rights, above n 78, 65, n 35.

[108] Monsanto, eg, has intellectual property rights over several key technologies relating to one crop, transgenic cotton, solely under its control. It has been unwilling to license its broad patent for the technology that produces transgenic cotton: Nuffield Council, above n 27, 50. It should be noted that researchers at the Chinese Academy of Agricultural Science independently developed an alternative method for the transformation of cotton: FAO, above n 2, 44.

[109] MA Heller, and RS Eisenberg, 'Can Patents Deter Innovation? The Anticommons in Biomedical Research' (1998) 280 *Science*, 698.

others may discover.[110] In addition, reforms of the patent laws of developed countries or alternative practices by patent holders (eg, patent pooling and open source models) may be needed to address the challenges set out above. Such changes may be necessary both from a utilitarian point of view, which would focus on maintaining incentives for broad-based research, and from a human rights point of view, which would stress that the current situation may impede research essential for the better realisation of human rights.

5.3.3 Access to Technology

Ninety-seven per cent of all patents worldwide are held in industrial countries.[111] More than 80 per cent of the patents granted in developing countries belong to residents of industrial countries, usually multinational corporations.[112] Hence, the direct beneficiaries of patent protection are located in the developed world. Current evidence suggests that the extension of the protection of intellectual property rights worldwide will result in substantial additional net transfers from developing to developed countries.[113] Licence fees and royalties paid for using patented technology may increase the costs of essential agricultural inputs, thus putting available technologies out of reach of farmers in developing countries.[114] Unlike during the Green Revolution when technologies were developed and disseminated freely as public goods, based on an explicit strategy for international technology transfer, developing countries might not be able to access privately held biotechnologies useful for their development.[115] When designing their national patent systems, developing countries should therefore, within the framework of the TRIPS Agreement and other international instruments, try to mitigate such negative effects. They should 'provide a pro-competitive patent system that limits the scope of subject matter that can be patented; applies strict standards of patentability; facilitates competition; includes extensive safeguards against abuses of patent rights; and encourages local innovation'.[116]

With regard to the right to food, states have a duty to prevent unreasonably high licence fees or royalties either for seeds or other means of food

---

[110] Ibid.

[111] World Bank, *World Development Report 2000/2001* (Washington, DC, World Bank, 2001), 184.

[112] UNDP, above n 88, 68.

[113] Commission on Intellectual Property Rights, above n 78, 11. The focus here is only on the flow of royalties, not on the wider set of economic advantages or disadvantages that may derive from the use of a patented product.

[114] On access to technology by developing countries see J Ntambirweki, 'Biotechnology and International Law Within the North-South Context' (2001) 14 *The Transnational Lawyer*, 103.

[115] Commission on Intellectual Property Rights, above n 78, 12.

[116] Ibid, 23. For technologically more advanced developing countries, however, systems that provide more extensive intellectual property rights protection may be more advantageous, as they may provide better incentives for research: ibid, 22.

production which would interfere with access to food for large segments of the population. As with pharmaceuticals, one could debate whether countries should intervene to improve access to seeds that promise important benefits, eg, through compulsory licensing. In the pharmaceutical sector spectacular achievements have been made, such as the 50 per cent reduction in AIDS deaths in Brazil following four years of making patented medication affordable, amongst other measures.[117] Such results are unlikely to result from compulsory licensing of improved seeds because seeds are environment-specific, unlike medication which can be used globally. In addition, any similar measure must be handled carefully because longer-term effects, such as decreased interest of corporations to enter the national market, may outweigh short-term advantages.

5.3.4 The TRIPS Agreement and the 'Farmers' Privilege'

The TRIPS Agreement now obliges countries to offer patent protection in all fields of technology to an invention that is new, involves an inventive step and is capable of industrial application (Article 27(1)).[118] States may exclude from patentability 'plants and animals other than micro-organisms, and essentially biological processes for the production of plants or animals other than non-biological and microbiological processes. However, Members shall provide for the protection of plant varieties either by patents or by an effective sui generis system or by any combination thereof' (Article 27(3)(b)). Under the TRIPS Agreement, states can thus allow patents for all of the above. Alternatively, they can exempt plants and animals, but not micro-organisms, from patentability. However, for plant varieties some kind of intellectual property right protection must be offered, through patents, or a *sui generis* system, or a combination of both.[119] The TRIPS Agreement does not prescribe whether genes should be patentable or not. Whether they need to be patentable hinges on the question of what constitutes an invention in the sense of Article 27(1) of the TRIPS Agreement.

---

[117] The Brazilian example is mentioned in a Report of the High Commissioner on 'The Impact of Trade-Related Intellectual Property Rights on Human Rights', above n 91, paras 51–58. Brazil considered using compulsory licences for two patented drugs. For one drug however, it achieved the negotiation of a reduced price, for the other negotiations were still continuing when the report was finalised.

[118] Patents shall be available and patent rights enjoyable without discrimination as to the place of invention, the field of technology and whether products are imported or locally produced: Art 27(1) of the TRIPS Agreement.

[119] One option for *sui generis* protection of plant varieties is through the recognition of plant variety rights or plant breeders' rights under the International Convention for the Protection of New Varieties of Plants (the UPOV Convention; signed in 1961, entry into force in 1968. Subsequently revised in 1972, 1978, and 1991).

Genetically modified seeds are typically protected by patents. A patent confers on its owner the exclusive right to make, use, offer for sale and sell the patented product.[120] Planting, harvesting, saving, re-planting and exchanging seeds of patented plants, or of plants containing patented cells and genes, constitutes use in the sense of Article 28(1) of the TRIPS Agreement.[121] Unless exceptions are provided for, the product of the harvest of patented seeds may not be re-sown or exchanged against other seeds. This interferes with traditional farming practices. Farmers have always saved seeds from their harvest to replant and exchange informally with other farmers to improve both the quality and quantity of their yields (the so-called 'brownbagging'). In developing countries this practice prevails, and it also occurs, to some extent, in developed countries. For poor farmers in developing countries, the right to use their own harvest is essential to secure their livelihoods. They may have the means to buy seeds one year, but may need to rely on farm-saved seed the next for propagation or multiplication on their farms and in exchange for other seeds. These practices were implicitly allowed under the 1978 International Convention for the Protection of New Varieties of Plants (UPOV Convention). This non-prohibition is commonly termed the 'farmers' privilege'. It is—with the exclusion of seed exchange practices—still allowed under UPOV 1991.[122] Also, the International Treaty on Plant Genetic Resources for Food and Agriculture[123] mentions the right

---

[120] Art 28(1)(a) of the TRIPS Agreement.

[121] This question was, *inter alia*, discussed in the Canadian case of *Monsanto Inc v Schmeiser* 2004 SCC 34.

[122] The scope of protection under UPOV extends to the production for purposes of commercial marketing, to the offering for sale, the marketing of the reproductive or vegetative propagating material, as such, of a variety: Art 5(1) UPOV 1978. UPOV provides for exemptions from the prior authorisation requirement for breeders: they have the right to use protected varieties as an initial source of variation for the purpose of creating other varieties or for the marketing of such varieties without authorisation from the original breeder: Art 5(3) UPOV 1978. In UPOV 1978, farmers' re-sowing of seed harvested from protected varieties was not listed as an activity requiring prior authorisation from the breeder and was therefore implicitly allowed. Under UPOV 1991, the scope of breeders' rights was extended to include production or reproduction and conditioning for the purpose of propagation: Art 14(1). UPOV 1991, however, provides explicitly that countries may restrict the breeders' right in order to permit farmers to use for propagating purposes, on their own holdings, the product of the harvest which they have obtained by planting the protected variety: Art 15(2). UPOV 1991 does not provide for any restriction of the breeders' right to permit farmers to exchange seeds. Therefore, the farmers' privilege has become more restricted.

[123] Art 9(3), the text of the treaty is available at ftp://ext-ftp.fao.org/ag/cgrfa/it/ITPGRe. pdf (last accessed 20 Feb 2005). On the International Treaty on Plant Genetic Resources for Food and Agriculture, see, for instance, G Moore and W Tymowski, 'Explanatory Guide to the International Treaty on Plant Genetic Resources for Food and Agriculture', IUCN Environmental Policy and Law Paper No. 57 (Gland, and Cambridge, IUCN, 2005); D Cooper, 'The International Treaty on Plant Genetic Resources for Food and Agriculture', (2002) 11 *Review of the European Community and International Environmental Law*, 1; A Mekouar, 'Treaty Agreed on Agrobiodiversity—the International Treaty on Plant Genetic Resources for Food

of farmers to save, use, exchange and sell farm-saved seed/propagating material, in the broader context of farmers' rights.

Patent laws can provide for at least some components of the farmers' privilege, as TRIPS allows limited exceptions to the exclusive rights conferred by a patent (Article 30). In the European Community (EC) some elements of the farmers' privilege have been recognised in Article 11(1) of EC Directive 98/44/EC on the Legal Protection of Biotechnological Inventions.[124] From a right to food perspective, it is important that developing countries consider whether and how best to use the flexibility of Articles 27(3)(b) and 30 of the TRIPS Agreement to protect small-scale and subsistence farmers, which are among the poorest worldwide. The recognition of a farmers' privilege in national laws may also be viewed in the broader context of Article 8(1) of the TRIPS Agreement that allows states to 'adopt measures necessary to protect public health and nutrition, and to promote the public interest in sectors of vital importance to their socio-economic and technological development, provided that such measures are consistent with the provisions of this Agreement'.

5.3.5 Protecting the Contribution of Farmers, Local Communities and Developing Countries to Biotechnological Inventions

The development of a biotechnological invention can be based on traditional knowledge or genetic resources, including traditional crop varieties, from developing countries where most biodiversity is concentrated.[125] For hundreds of years farmers have developed and improved crops. They have selected, developed and conserved landraces[126] and traditional varieties, thereby making an invaluable contribution to the development of the modern crop varieties that are now the basis of biotechnological research.

---

and Agriculture', (2002) 32 Environmental Policy and Law, 20; G Rose, 'The International Law of Sustainable Agriculture in the 21st Century: The International Treaty on Plant Genetic Resources for Food and Agriculture', (2003) 15 Georgetown International Environmental Law Review, 583; SM Ruby, 'The International Treaty on Plant Genetic Resources for Food and Agriculture: Friend or Foe of the International Farmer', (2004) 2 *Oklahoma Journal of Law & Technology*, 23.

[124] Above n 98. Art 11(1) states that 'the sale or other form of commercialisation of plant propagating material to a farmer by the holder of the patent or with his consent for agricultural use implies authorisation for the farmer to use the product of his harvest for propagation or multiplication by him on his own farm'. Hence, farmers can re-use farm-saved seed, but not exchange and sell it.

[125] It should be noted that while plant breeders tend to use modern varieties as resources rather than landraces in their breeding programmes, exotic germplasm may be used when particular traits are sought, when new breeding programmes are being started, or for long-term genetic enhancement, and also in the breeding of certain crops (eg, potatoes): G Dutfield, *Intellectual Property Rights, Trade and Biodiversity* (London, Earthscan, 2000), 4 ff.

[126] A crop cultivar or animal breed that evolved with and has been genetically improved by traditional agriculturalists, but has not been influenced by modern breeding practices; Commission on Intellectual Property Rights, above n 78, 191.

These services by farmers are little recognised, and their contributions are only weakly protected. Local varieties, for example, cannot be protected by intellectual property rights as they are genetically too heterogeneous, unstable and typically not attributable to specific rights-holders.[127] Conversely, biotechnological inventions derived from such genetic resources or traditional knowledge may be patented in other countries. All royalties from such a patent will accrue only to the patent holder—a highly inequitable result. Patenting can occur without the prior informed consent of the genetic resources' country of origin and without an adequate sharing of the benefits derived from the commercialisation of these resources.[128] A number of non-biotech cases of such so-called 'biopiracy'[129], such as the tumeric and neem plant, the ayahuasca ceremonial drink and the Hoodia cactus, have been widely documented.[130]

Several instruments address these concerns. The International Treaty on Plant Genetic Resources for Food and Agriculture is one attempt to protect and recognise farmers' contributions. It recognises the enormous contributions local and indigenous communities and farmers of all regions of the world, particularly those in the centres of origin and crop diversity, have made for the conservation and development of plant genetic resources, which constitute the basis of food and agriculture production throughout the world (Article 9(1)). It provides for a farmers' right equitably to participate in sharing benefits arising from the utilisation of plant genetic resources for food and agriculture (Article 9(2)(b)) and the establishment of a multilateral system of access and benefit sharing (Articles 10–13). Access and benefit sharing and prior informed

---

[127] Cf the criteria for plant variety protection, above n 103.

[128] Commission on Intellectual Property Rights, above n 78, 9.

[129] There is no accepted definition of the term 'biopiracy'. It has been defined as the unauthorised commercial exploitation of the knowledge and biological resources of indigenous peoples or traditional communities: Dutfield, above n 125, 41. The ETC Group defines it as 'the appropriation of the knowledge and genetic resources of farming and indigenous communities by individuals or institutions seeking exclusive monopoly control (usually patents or plant breeders' rights) over these resources and knowledge': ETC Group, 'Ag Biotech Countdown: Vital Statistics and GM Crops', Update June 2002, available at www.etcgroup.org/search2. asp?srch=Ag+Biotech+Countdown (last accessed 22 Feb 2005). The following situations have been called biopiracy: the granting of patents for inventions that are either not novel or not inventive with regard to traditional knowledge already in the public domain; the granting of patents on inventions derived from a community's traditional knowledge or genetic resources, on the basis of low patenting standards, where patents are allowed, for instance, for inventions which amount to little more than discoveries, or where the national patent regime (eg, as in the US) does not recognise some forms of public disclosure of traditional knowledge as prior art; or cases in which the patent represents a genuine invention, but no arrangements have been made to obtain the prior informed consent of the countries/communities providing the knowledge or resource, and for sharing the benefits of commercialisation to reward them appropriately in accordance with the principles of the Convention on Biological Diversity; Commission on Intellectual Property Rights, above n 78, 74.

[130] Commission on Intellectual Property Rights, above n 78, 76 ff; Dutfield, above n 125, 65 ff. Some of the patents were successfully challenged subsequently.

consent are also addressed in the Convention on Biological Diversity, which mentions particularly access to the results and benefits arising from biotechnologies based upon genetic resources provided.[131] Other relevant instruments include the Draft United Nations Declaration on Indigenous Rights,[132] and the African Convention on the Conservation of Nature and Natural Resources.[133] The relationship between the TRIPS Agreement and the Convention on Biological Diversity is problematic and disputed. One suggestion to better align patenting procedures, on the one hand, and access and benefit sharing and prior informed consent requirements, on the other, is to include in national patent laws, or possibly in international instruments, requirements to disclose the genetic resources' geographical origin and to provide proof that the resource was acquired with the prior informed consent of the country of origin.[134] Some countries have already enacted laws to that effect.[135]

In the context of the current review of the TRIPS Agreement, the Council for TRIPS has been instructed by the Doha Ministerial Declaration to examine, inter alia, 'the relationship between the TRIPS Agreement and the Convention on Biological Diversity'[136] within the review of Article 27(3)(b).

6. CONCLUSIONS

The use of GM crops has manifold implications for the different components of the right to food, namely for the availability, accessibility, safety, nutritional quality and acceptability of food, and the environmental sustainability with which it is grown. The implications depend on the

---

[131] Arts 1, 15 and 19(2) of the Convention on Biological Diversity, above n 1: see also the Bonn Guidelines on Access and Benefit Sharing, Decision VI/24, available at www.biodiv. org/decisions/default.aspx?m=cop-06&d=24 (last accessed 22 Feb 2005).

[132] Art 29 of the Draft United Nations Declaration on Rights of Indigenous Peoples states 'indigenous peoples are entitled to the recognition of the full ownership, control and protection of their cultural and intellectual property. They have the right to special measures to control, develop and protect their sciences, technologies and cultural manifestations, including human and other genetic resources, seeds, medicines, knowledge of the properties of fauna and flora': Sub-Commission on Prevention of Discrimination and Protection of Minorities, Resolution 1994/45, Annex, of 26 Aug 1994, reprinted in: UN Doc E/CN.4/Sub.2/1994/56 of 28 Oct 1994.

[133] Art XVII of the Traditional Rights of Local Communities and Indigenous Knowledge states: '[t]he Parties shall take legislative and other measures to ensure that traditional rights and intellectual property rights of local communities including farmers' rights are respected in accordance with the provisions of this Convention' (para 1); and '[t]he Parties shall require that access to indigenous knowledge and its use be subject to the prior informed consent of the concerned communities and to specific regulations recognizing their rights to, and appropriate economic value of, such knowledge' (para 2).

[134] Commission on Intellectual Property Rights, above n 78, 85 ff.

[135] See examples in ibid, 86.

[136] Para 19, WT/MIN(01)/DEC/1.

specific context, the kind of GM crop used, and the wider socio-economic and legal framework. GM crops promise a number of important opportunities to improve the realisation of the right to food, such as increased food production, improved nutritional values, crops suitable for difficult production environments, and lower environmental costs. According to present knowledge regarding consumption, GM foods are as safe as conventional foods. However, there may be long-term risks to health and the environment, the extent of which may not be fully clear as yet. Evidence suggests that biotechnology, including GM crops, can play a role in integrated and comprehensive agricultural research and development programmes in developed and developing countries alike. At this point, however, nine years after the commercialisation of GM crops, sufficient experience has not yet been gained to judge the overall long-term effect of GMOs on the right to food and on economic and social development in general.

A number of challenges exist with respect to the overall governance of the development, marketing and use of agricultural biotechnologies, such as the private–public and developed–developing country divides in research and access to the technology. Areas of concern include the privatisation of knowledge, which leads to a bypassing of the needs of the poor; the increasing risk of research being stifled by the manner in which intellectual property rights are granted; tensions between intellectual property rights with farmers' practice of using and exchanging farm-saved seed; and the inadequate recognition of the contribution of farmers and developing countries to agricultural biotechnological research.

Whether, and how, to use or allow biotechnology is part of states' wide range of policy options when determining how best to fulfil their obligation progressively to realise the right to food and other human rights. They need to take into account all opportunities and risks when making decisions about agricultural biotechnologies. They should make use of available policy space in creating appropriate national frameworks that promote a better realisation of the right to food while protecting those individuals that may be harmed. States need to strike an adequate balance between various legitimate interests, namely the rights and hopes of the hungry and malnourished who may profit from some GMO crops; the interest of everyone in a safe and healthy environment and in being reasonably well protected against unknown risks (including the risks associated with current technologies); the interests of agribusiness in its return for investment in research; and the right of everyone to benefit from scientific progress. A human rights approach to this balancing task will ensure that the interests of all individuals, particularly the poorest and most vulnerable, are given due weight.

8

A Case Study of the European Union's Regulation of GMOs: Environment, Health, Consumer Rights and Economic Freedom

ELISABETTA RIGHINI[*]

1. INTRODUCTION

Biotechnologies and human rights are two concepts which are very broad, both in scope and content. In its narrowest sense, 'biotechnology' or, more appropriately, 'modern biotechnology' refers, in fact, to that branch of biological science which deals with genetic engineering and recombinant DNA technology.[1] Biotechnology is applied in a number of different areas, from the development of new diagnostic and therapeutic tools, to DNA profiling and cloning, to the production of genetically modified organisms (hereinafter 'GMOs'). The notion of 'human rights', on the other hand, can vary in space and time and covers very diverse concepts, of both a collective and individual nature, which range from the respect for the rule of law and democracy, to the protection of health and the environment, to fundamental freedoms, economic freedoms, etc. Thus, when related to one another, these two concepts give rise to a wide variety of different combinations in terms both of ethical concerns and of legal issues.

Numerous examples of these concerns and issues are contained elsewhere in this book. This contribution will limit its analysis to that part

---

[*] Member of the Legal Service of the European Commission. Email: Elisabetta.Righini@ ec.europa.eu. The views expressed are the personal opinions of the author.
[1] In legal terms, the most authoritative definition is that of the Convention on Biological Diversity ('CBD'), concluded in Rio de Janeiro on 5 June 1992 and entered into force on 29 Dec 1993, available at www.biodiv.org/convention/articles.asp (last visited 26 Sept 2005), which, in its Art 2, defines biotechnology as meaning any technological application that uses biological systems, living organisms, or derivatives thereof, to make or modify products or processes for specific use'.

of biotechnology that leads to the production of and trade in GMOs and to the geographical and regulatory space of the European Union ('EU'). The premise on which this analysis is conducted is the customary rule of international law according to which the world is divided into national spheres of jurisdiction coinciding with the territory of every state, in this case of 25 states which have attributed part of their sovereignty to a supranational entity, the EU. Within that territory, states have the sovereign and exclusive right to exercise their government authority; they are free to exploit their resources and to exercise their jurisdiction over the people and goods which are present on their territories. They also have the responsibility to ensure that their activities do not damage other states or areas beyond the limits of national jurisdiction.[2]

2. THE DEBATE AROUND GMOS

The introduction of GMOs into the environment and the food chain has given raise to a highly controversial debate. The positive view is that GMOs will contribute to optimising agriculture, thus reducing hunger in the world, and to revolutionising medicine. Yet, the strong growth in GM crop production worldwide has given rise to fervent reactions, ranging from outrage to uneasiness.

2.1 What are GMOs

A GMO is an organism (ie, any biological entity capable of replication or of transferring genetic material[3]) in which the genetic material has been altered in a way that does not occur naturally by mating and/or natural recombination.[4] GMOs are also referred to as 'living modified organisms' (LMOs),[5] or 'genetically engineered organisms'.

Irrespective of the term used, GMOs are obtained through the transfer of foreign genes from one organism into another or the deletion or silencing of certain genes. Genes are pieces of deoxyribonucleic acid ('DNA') that lead to the production of a particular protein. Proteins are the basic 'building blocks' of any organism and they determine the

---

[2] See, in this sense, ibid, Art 3.

[3] See Art 2(1) of Directive 2001/18 of the European Parliament and of the Council of 12 Mar 2001 on the deliberate release into the environment of genetically modified organisms and repealing Council Directive 90/220 [2001] OJ L/106/1.

[4] Ibid, Art 2(2); or the 'Glossary of Biotechnology and Genetic Engineering' FAO Research and Technology Paper No 7 (Rome, Dec 1999).

[5] See Art 3(g) of the Cartagena Protocol on Biosafety to the CBD (hereinafter 'Biosafety Protocol'), signed in Montreal on 29 Jan 2000, available at www.biodiv.org/doc/legal/carta-gena-protocol-en.pdf (last visited 27 Sept 2005).

organism's physical characteristics and development. When genes are modified, the cell may produce a different set of proteins that may lead to changed characteristics in the plant or fruit. Thus, a potato may be given different starch content, a tomato may deliver vaccines and antibodies, and some plants may become capable of generating a new toxin that kills only certain insects.

The qualitative difference between genetic modification and conventional breeding practices is that the latter do not allow for the crossing of natural species barriers (as, for instance, from a plant into a bacterium or an animal), or for the transfer of single or few genes instead of the whole of an organism's genetic material (genome).[6]

2.2 Potential Benefits of GMOs

GMOs are developed because they have the potential to provide significant benefits, such as increased agricultural output, enhanced durability, added nutritional value to foods, and certain environmental benefits such as reductions in the use of pesticides.

The first GM plants were developed to improve crop protection. GM crops were modified to be insect-pest resistant or herbicide tolerant, thus minimising crop losses and maximising yield.

The second generation of GM crops went beyond on-farm benefits to provide value-enhancing traits (offering the potential ability to change the agronomic characteristics of a product) as nutritional values, colour, texture, flavour or processing properties, such as potatoes with less moisture, and thus absorbing less fat during frying, and with anti-browning properties for better storage.

Other, more recent, GMOs have been developed for medical purposes, such as the so-called 'functional foods', where crops contain micronutrients capable of reducing some of the risk factors for diseases. Examples are rice modified to produce pro-vitamin A, and canola oil with high beta carotene content.

In terms of volume, the most common GMOs remain herbicide tolerant crops, accounting for some 73 per cent of the commercially planted area worldwide, followed by insect resistance (18 per cent), and stacked genes (ie, both herbicide tolerant and insect resistant) 8 per cent. Virus

---

[6] Ibid, Art 3(i), in fact, defines 'modern biotechnology' as:
'the application of:
a. in vitro nucleic acid techniques, including recombinant deoxyribonucleic acid (DNA) and direct injection of nucleic acid into cells or organelles, or
b. fusion of cells beyond the taxonomic family,
that overcome natural physiological reproductive or recombination barriers and that are not techniques used in traditional breeding and selection'. (emphasis added)

resistant and quality traits are in less than 1 per cent of GM crops grown worldwide.[7] This notwithstanding, the promotion and application of this new technology are supported by the potential GMOs have in alleviating illness or hunger, as well as by the willingness to safeguard the principle of freedom of research.

2.3 Concerns and Objections

The development of GMOs has also caused growing concerns and prompted strong objections. These are mainly related to the actual or potential risks to human and animal health, to the environment, to the reduction of biodiversity and the modification of traditional agricultural practices.

From a scientific point of view, harmful effects may be generated by the fact that, despite huge and fast progress, the various techniques for inserting foreign DNA still do not control exactly where the insertion takes place, the number of copies inserted or their level of expression, nor do they guarantee that the foreign gene is stably integrated into the host genome. Further unintended effects may take place in the process of creating a GMO: there may be too much foreign or unwanted extraneous DNA unintentionally inserted, multiple rearranged integration events may occur, or the foreign DNA may have been contaminated during laboratory manipulations.

2.3.1 Hazards to Human Health

With regard to human health, the principal hazards identified are accrued toxicity, allergenicity and horizontal gene transfer, in particular as far as development of antibiotic resistance is concerned.

A number of plants produce toxins as a protection against insect and fungal pests. These are parts of their innate defence systems and, as such, are important to maintain. They are generally present at such low levels that humans and animals are able to tolerate them. Through genetic modification, however, plants which do not naturally contain toxins may become toxic or increase their levels of toxins.

Similarly, a GM food containing DNA derived from a species that has known allergenic effects may acquire allergenicity.[8] An example of this kind is that of the soyabean variety genetically modified to contain

---

[7] International Service for the Acquisition of Agri-Biotech Applications, 'Global Status of Commercialised Transgenic Crops: 2003' (2003) 30 *ISAAA briefs*.

[8] See, eg, the Report of the 2nd Joint FAO/WHO Expert Consultation on Foods Derived from Biotechnology, Evaluation of Allergenicity of Genetically Modified Foods (FAO, Rome, 22–25 Jan 2001), available at www.fao.org/es/ESN/food/pdf/allergygm.pdf (last visited 29 Sept 2005).

a certain protein from the Brazil nut in order to increase its nutritional value. The modified soyabean turned out to have acquired the same allergenic qualities as the parental crop, the Brazil nut.[9]

Allergenicity may also be caused by novel proteins that the organism produces, or produces in increased quantities, as a result of the genetic modification. This was the case with the famous *StarLink* maize, genetically modified to contain a plant pesticide protein which kills certain insects but which was believed to be potentially allergenic. This maize, initially registered by the US government for animal feed and industrial purposes only, had to be withdrawn from production altogether when it turned out that maize for human consumption had been contaminated.[10]

Another hazard to human health comes from the possibility, through genetic manipulation, of horizontal gene transfer, that is to say the transfer of genetic material from one organism to another cell that is not its descendant. Horizontal gene transfer can be contrasted with vertical gene transfer, which occurs in normal reproduction techniques when an organism receives genetic material directly from its ancestor. Horizontal gene transfer may take place through ingestion of GM food products and the transfer and integration of the DNA into the resident gut microflora.

Such hypothesis represents a particular risk to human health, especially in the case of horizontal transfer of antibiotic resistance marker genes. During the creation of GMOs, antibiotic resistance genes are placed into the vectors which carry the inserted gene of interest as 'markers' so as to determine whether the gene of interest has been successfully inserted into the genome of a plant cell: the cells are treated with the antibiotic in question, and only those with the correct inserted antibiotic resistance gene survive. Even though used only as markers, the antibiotic resistance genes may remain in the GMO and if, when ingested, fragments of that DNA are taken up by gastrointestinal bacteria, they could potentially result in the development of antibiotic resistance of human bacteria against known antibiotic medication, thus rendering ineffective important medical treatments. The use of technology without antibiotic resistance genes has been recently recommended by a FAO/WHO expert panel.[11]

---

[9] J Nordlee, S Taylor, J Townsend, L Thomas and R Bush, 'Identification of a Brazil-nut Allergen in Transgenic Soybeans' (1996) 334 *The New England Journal of Medicine* 688.

[10] EPA, *A Set of Scientific Issues Being Considered by the Environmental Protection Agency Regarding Assessment of Scientific Information Concerning StarLink Corn* (FIFRA Scientific Advisory Panel Meeting, Environmental Protection Agency of the United States, 2000); W Lin, GK. Price and E Allen, 'StarLink: Impacts on the U.S. Corn Market and World Trade' [2001] *Economic Research Service USDA Feed Yearbook* 40, available at www.ers.usda.gov/Briefing/Biotechnology/starlinkarticle.pdf (last visited 29 Sept 2005).

[11] *Report of the Joint FAO/WHO Expert Consultation on the safety assessment of foods derived from genetically modified animals* (Rome, Nov 2003), available at www.who.int/foodsafety/biotech/meetings/en/gmanimal_reportnov03_en.pdf (last visited 29 Sept 2005) point 5.2.2.4.

2.3.2 Risks for the Environment

When released into the environment, some GMOs may also constitute major hazards because of the dispersal of their modified genes by pollen, seeds or other means—what is commonly referred to as 'genetic pollution'.

Thus, pest- or insect-resistant GMOs may have adverse effects on other, non-target organisms, such as insects, which fly on to the GM plant and are not pests for it, or birds, which feed on the GM plant.

GM plants with herbicidal or insecticidal genes may cause a problem of invasiveness and persistence in the environment, by crossing with other plants surrounding the crop such as wild relatives, neighbouring non-GM crops of the same species or volunteers (ie, re-growth of a previous crop in a subsequent crop), and thus transferring the genetic modification. In Canada, for instance, cross-pollination between three distinct varieties of oilseed rape led to the emergence of volunteers expressing resistance to three distinct herbicides.[12]

Such phenomena pose considerable challenges to traditional agricultural management practices. The *StarLink* maize case showed how difficult it is to prevent unwanted genes from reappearing year after year. After *StarLink* was taken off the market, its Bt gene continued to be present in trace amounts in US grain supply three years later.[13]

Furthermore, when these plants become invasive of and persistent in natural habitats, they may have detrimental effects on biodiversity. Biodiversity, or 'biological diversity', 'means the variability among living organisms from all sources including, inter alia, terrestrial, marine and other aquatic ecosystems and the ecological complexes of which they are part; this includes diversity within species, between species and of ecosystems'.[14] The conservation of biodiversity is considered of the outmost importance, to the point of deserving a dedicated international treaty, the CBD, with a specific protocol on the impact of GMOs on biodiversity, the Biosafety Protocol. Biodiversity is important primarily because it keeps an ecosystem intact and functioning and makes it resilient to external perturbations. The diversity that exists in micro-organisms, plants and animals is required for essential functions such as decomposition, nutrient transformations, soil formation, detoxification, etc.

In Europe, the problem of preservation of biodiversity in relation to GMOs is increased by the fact that valuable biodiversity is considered to

---

[12] L Hall, K Topinka, J Huffman, L Davis and A Good, 'Pollen Flow between Herbicide-resistant Brassica Napus is the Cause of Multiple-resistant B. Napus Volunteers' (2000) 48 *Weed Science* 688.

[13] 'USDA continues to investigate StarLink situation', *USDA-GIPSA News*, 6 Jan 2003, available at http://151.121.3.117/newsroom/release/2003/0103.htm (last visited 29 Sept 2005).

[14] See Art 2 of the CBD, n 1 above.

include not only natural and semi-natural areas (mountains, extensive forests, etc), but also farmland. In fact, much of the non-urban landscape in Europe consists of farmland, which is thus the primary habitat for native insects, birds and other animals.

In certain instances, GM plants may also constitute an irreversible hazard for their own centre of origin, ie, the location or locations in the world where the oldest cultivation of a particular crop has been identified and where there are the highest observable levels of genetic variability. This can all be put in danger if the GM crop cross-pollinates with its wild relatives. A recent example of this phenomenon occurred in Mexico, where GM corn was found to have out-crossed with wild relatives, even over very great distances. Mexico is the centre of origin of maize and, in order to preserve it, the relevant advisory committee of the North American Free Trade Agreement ('NAFTA') recommended to the Environment Ministers of Canada, Mexico and the United States that a moratorium on imports of transgenic corn to Mexico be put in place until the risks to human health, cultural integrity of maize producers in Mexico and the environment in general were better understood and appropriate long-term decisions made.[15]

2.3.3 Unequal Distribution of Potential Benefits

Objections have also been raised with regard to the distribution of potential benefits of GMOs.

The costs necessary to develop, produce and trade GMOs are such that only large agrochemical and seeds enterprises may have the resources to incur them. Large companies naturally carry out the research and the commercial strategy that is in their interest. Some fear that the exploitation of GMOs will thus be based on commercial considerations only and will end up benefiting a small group of companies rather than the interests of the wider public (consumers, farmers, researchers, in the developed as well as in the developing world).[16]

3. THE HUMAN RIGHTS AND FUNDAMENTAL FREEDOMS RELEVANT TO THE GMOs DEBATE

In the context of the EU, any legal analysis related to human rights has to take as its primary point of reference the Charter of Fundamental

---

[15] See letter of 13 Apr 2004 from the Joint Public Advisory Committee of the North American Commission for Environmental Cooperation, available at www.cec.org/files/PDF/JPAC/JPAC-Letter-Maize-13-Apr-04_en.pdf (last visited 29 Sept 2005).

[16] See, for instance, the considerations developed in *Genetically Modified Crops: the Ethical and Social Issues* (London, Nuffield Council on Bioethics, May 1999), point 1.21 and ff, available at www.nuffieldbioethics.org/fileLibrary/pdf/gmcrop.pdf (last visited 29 Sept 2005).

Rights of the EU (hereinafter 'The Charter'),[17] which is now also part of the Treaty establishing a Constitution for Europe, signed in Rome on 29 October 2004 (hereinafter 'the Constitution').[18]

The original three European Communities Treaties (hereinafter 'EC Treaties') signed in the 1950s contained no reference to human rights and values. It was the European Court of Justice (hereinafter 'the Court') that, a few years later, declared that the 'general principles of EC law' included the protection of fundamental rights which were part of the common constitutional traditions of the Member States.[19] Over the years, the Court has then progressively developed a sort of unwritten catalogue of fundamental rights and freedoms that was first given express recognition in Articles 6 and 7 of the Treaty Establishing the European Union ('TEU').[20] The purpose of the Charter was to make the existing rights more visible and to 'deepen and strengthen the culture of rights and responsibilities in the EU'.[21]

3.1 Freedom of the Sciences and to Conduct a Business

One of the central elements of the debate concerning GMO development and production is whether and how the freedom of scientific research and the freedom to conduct a business are guaranteed.

In Europe, economic rights such as the right to purse a trade or a profession were among the first fundamental rights protected by the Court.[22] Their respect is now sanctioned by Articles 13 and 16 of the Charter. Article 13, consolidated in Article II–73 of the Constitution on the 'Freedom of the arts and the sciences', provides that:

> The arts and scientific research shall be free of constraint. Academic freedom shall be respected.

Similarly, Article 16 of the Charter, now Article II–76 of the Constitution entitled 'Freedom to conduct a business', establishes that:

> The freedom to conduct a business in accordance with Union law and national laws and practices is recognised.

The first freedom is the natural consequence and corollary of the freedom of thought and expression (Articles 10 and 11 of the Charter).

---

[17] Done at Nice on 7 Dec 2000 [2000] OJ C/364/1.

[18] [2004] OJ C/310/1.

[19] Case 29/69, *Stauder v City of Ulm* [1969] ECR 419.

[20] Treaty on European Union (TEU) (consolidated text) [2002] OJ C/325/1.

[21] Lord Goldsmith, 'A Charter of Rights Freedoms and Principles' (2001) 38 *Common Market Law Review* 1201. See also K Lenaerts and E De Smijter, 'A "Bill of Rights" for the European Union' (2001) 38 *Common Market Law Review* 273.

[22] See Case 11/70, *Internationale Handelsgesellschaft mbH v Einfuhr– und Vovratstelle für Getreide und Futtermittel* [1970] ECR 1125.

Such freedom finds its limit in Article 1 of the Charter, now Article II–66 of the Constitution, which declares human dignity inviolable and to be protected and respected.

The freedom to conduct a business, on the other hand, is based on the case law of the Court which has recognised freedom to exercise an economic or commercial activity[23] and freedom of contract.[24]

It is interesting to note, however, that already in the very early case law of the Court the protection of these rights was limited by considerations related to public interest. Thus, in *Nold*, the Court held that:

> If rights of ownership are protected by the Constitutional laws of all the Member States and if similar guarantees are given in respect of their right freely to choose and practice their trade or profession, the rights thereby guaranteed, far from constituting unfettered prerogatives, must be viewed in the light of the social function of the property and activities protected there under. For this reason, rights of this nature are protected by law subject always to limitations laid down in accordance with the public interest. Within the Community legal order it likewise seems legitimate that these rights should, if necessary, be subject to certain limits justified by the overall objectives pursued by the Community, on condition that the substance of these rights is left untouched.[25]

3.2 Environmental and Health Protection

Environmental protection as an area of Community activity was first introduced into the EC Treaties by the Single European Act in 1986.[26] This was further developed by the TEU which made a balanced and sustainable development one of the objectives of the EU.[27] Now Article 37 of the Charter, consolidated in Article II–97 of the Constitution entitled 'Environmental protection', provides that:

> A high level of environmental protection and the improvement of the quality of the environment must be integrated into the policies of the Union and ensured in accordance with the principle of sustainable development

The principle set out in this Article draws on the provisions of some national constitutions and it is clearly based on Articles 2, 6 and 174 of

---

[23] See Case 230/78, *Eridiana SPA and others v Ministry of Agriculture and Forestry and others* [1979] ECR 2749, para 20.
[24] See Case C–240/97 *Spain v Commission* [1999] ECR 1–6571, para 99.
[25] Case 4/73, *J Nold Kohlen– und Baustoffgrosshandlung v Ruhrkohle AG* [1974] ECR 491, para 14.
[26] Single European Act (1986) [1987] OJ L/169/1.
[27] Art 2 TEU.

the Treaty Establishing the European Community (hereinafter 'TEC').[28] Its content can thus be better appreciated by reference to the latter. Article 174, in particular, includes among the objectives of the EU policy on the environment the protection of human health, as well as the prudent and rational utilisation of natural resources. Paragraph (2) of Article 174 further establishes that:

> Community policy on the environment shall aim at a high level of protection taking into account the diversity of situations in the various regions of the Community. It shall be based on the precautionary principle and on the principles that preventive action should be taken, that environmental damage should as a priority be rectified at source and that the polluter should pay.
>
> In this context, harmonisation measures answering environmental protection requirements shall include, where appropriate, a safeguard clause allowing Member States to take provisional measures, for non-economic environmental reasons, subject to a Community inspection procedure.

3.3 Consumer Protection

Another fundamental right of relevance to the GMO regulatory framework is that of 'consumer protection' enshrined in Article 38 of the Charter, now Article II–98 of the Constitution, which provides that:

> Union policies shall ensure a high level of consumer protection.

The notion of consumer protection was absent in the original EC Treaties and was given some attention only by the modifications brought about by the TEU. Even today, Article 3(t) of the TEC only provides for, as one of the activities of the Community, 'a contribution to the strengthening of consumer protection', clearly a less far-reaching activity than the policy in the sphere of environment.

As far as the content of this right is concerned, Article 153 of the TEC establishes that the objects of consumer protection are the protection of the 'health, safety and economic interests of consumers', as well as the promotion of 'their right to information, education'.

4. THE RULE OF LAW AS THE DEMOCRATIC MEANS TO ADJUDICATE ON CONFLICT

In the previous sections the issues surrounding GMOs, as well as the fundamental freedoms and rights relevant to this debate were outlined.

---

[28] TEU, n 20 above, 33.

Now it remains to be seen how these two worlds interact with each other. Talking about 'the impact of biotechnology on human rights', in fact, suggests that it is the fast moving scientific and industrial evolution of biotechnology that has influenced the scope and interpretation of human rights. Such a suggestion can be true from a strictly 'chronological' viewpoint, in the sense that, generally, it is the law that struggles to catch up with scientific and technological developments. From a substantial point of view, however, the contrary axiom, ie, talking of an 'impact of human rights on biotechnologies', appears to be the correct one.

Far from being a word game, this distinction reflects a very different policy approach towards biotechnologies and human rights. The two different ways of formulating a title imply very different conceptions of the role of public authorities in the regulation of the society we live in and of its economic system. Talking of an 'impact of biotechnologies on human rights' may be taken, in fact, to presume a situation where biotech research and industry are allowed to develop without regulatory boundaries (or very limited regulatory boundaries), so that choices related to common goods, such as environment and health, are left to industry and researchers. On the contrary, a regulatory system that is based on the existence of fundamental rights that inform and shape research as well as economic activity, thus balancing the different interests and benefits at stake, leads rather to talk about an 'impact of human rights on biotechnologies'.

This chapter will show that within the limited scope of its analysis, the regulation of GMOs in the EU, the second approach appears to be the more appropriate.

4.1 The European Approach

In Europe, the approach followed in the regulation of GMOs is based on the belief that 'our democratic societies should offer the necessary safeguards to ensure that the development and application of life sciences and biotechnology takes place respecting the fundamental values recognised by the EU in the Charter of Fundamental Rights'.[29] In other words, it is considered the task of the Community and of the Member States to enact rules that, in full respect for fundamental constitutional rights and freedoms, can reconcile the potential benefits offered by modern biotechnology with the risks linked to its impact on the environment and human health.

---

[29] See Communication of the Commission to the Council, the European Parliament, the Economic and Social Committee and the Committee of the Regions, *Life Science and Biotechnology—A Strategy for Europe*, COM(2002)27 of 23 Jan 2002; available at http://europa. eu.int/eur-lex/en/com/cnc/2002/com2002_0027en01.pdf (last visited 26 Sept 2005).

This is consistent with the primacy of human rights and fundamental freedoms as founding values on which the EU is built, sanctioned in Article 6 of TEU. The second Article of the Constitution, Article I–2 entitled 'The Union's values', which refines this principle, establishes, in fact, that:

> The Union is founded on the values of respect for human dignity, freedom, democracy, equality, the rule of law and respect for human rights, including the rights of persons belonging to minorities. These values are common to the Member States in a society in which pluralism, non-discrimination, tolerance, justice, solidarity and equality between women and men prevail.

The Biosafety Protocol summarises well the core of the matter in its Preamble, in which it recognises 'the growing public concern over [modern biotechnology] potential adverse effects on biological diversity, taking also into account risks to human health', and affirms that 'modern biotechnology has great potential for human well-being if developed and used with adequate safety measures for the environment and human health'.

Notwithstanding the international dimension of the debate concerning GMOs, the regulatory responses have been far from homogeneous. At one extreme, marketing bans have been imposed, in different forms and for different durations, by countries like Algeria,[30] El Salvador,[31] New Zealand,[32] Thailand,[33] or by regions such as the Australian States and Territories.[34] The United States, on the other hand, is an example of a

---

[30] Ministère de l'agriculture et du developpement rural, Arrêté ministériel no 910 du 24 décembre 2000 (Ministerial order by the Minister of agriculture and development of 24 Dec 2000), quoted in Bureau of Economic and Business Affairs, *2001 Country Reports on Economic Policy and Trade Practices* (Washington, DC, US Department of State, Feb 2002), available at www.state.gov/documents/organization/8182.pdf (last visited 26 Sept 2005).

[31] See Art 30 of the *Ley de Semillas*—Decreto Legislativo 530 de 5 de septiembre 2001, *Diario Oficial*, Tomo N. 352, available at www.glin.gov/view.do?documentID=76541&summaryLang=es&fromSearch=trae (last visited 30 Sept 2005).

[32] See Art 4(b) of the Hazardous Substances and New Organisms (Genetically Modified Organisms) Amendment Act 2002, which imposed a temporary ban on GMOs from 29 Oct 2001 to 29 Oct 2003; available at www.legislation.govt.nz/browse_vw.asp?content-set=pal_statutes&clientid=2310743439&viewtype=contents (last visited 30 Sept 2005).

[33] *Plant Quarantine Act*, BE 2507 (1964), BE 2546 (2003), available at www.thaifloriade.thaigov.net/hort_cd/html/PLANT%20_QUARANTINE%20%20ACT%20%201.htm (last visited 26 Sept 2005). S 6 gives the minister in charge the power to declare any plant, pest or medium prohibited; s 8 prohibits import of prohibited items without permission from the Department of Agriculture. According to some sources, this ban was lifted in Sept 2004; http://www.afaa.com.au/news/news-1507.asp (last visited 26 Sept 2005).

[34] The Australian government recognised GM crop free areas as they were designated by several Australian states and territories, in the policy principle, *Gene Technology (Recognition of Designated Areas) Principle 2003*, Commonwealth of Australia Special Gazette No 340, 5 Sept 2003, available at www.tga.gov.au/gene/policy/gtrdap03.pdf (last visited 26 Sept 2005). For examples of such regional laws see South Australia Genetically Modified Crop Management Act 2004, available at www.parliament.sa.gov.au/Catalog/legislation/Acts/G/2004.8.un.htm (last visited 13 May 2004).

country that has adopted, at least in part, an approach that comes close to a model of regulatory *laissez faire*. Its position with regard to GM food is contained in a policy statement issued in 1992 by the Food and Drug Administration ('FDA') which establishes that GM foods 'are regulated within the existing framework of the act, FDA's implementing regulations, and current practice, utilizing an approach identical in principle to that applied to foods developed by traditional plant breeding'[35]. Thus, the only means available to developers of GMOs to address safety and regulatory questions prior to commercial distribution is a voluntary consultation process which FDA has set up.[36]

4.2 The EU Regulatory Framework for GMOs

The EU legislator was the first to regulate GMOs, and has established the most complete regulatory system worldwide.[37] As early as the mid-1980s, it established a general policy approach to GMOs, following the recommendations of the OECD.[38] In 1990, it adopted the first Directive on the deliberate release into the environment and the placing on the market of GMOs.[39] As an instrument of so-called 'horizontal' legislation, it covered all GMOs across all sectors.

Since then, the EU regulatory framework has undergone several revisions in order to come into line with the development of scientific understanding and of international regulation, as well as with growing awareness and interest on the part of the public at large. EU regulation currently governs such diverse aspects as the patenting of biotechnological inventions, the authorisation of pharmaceutical products, contained use of genetically modified micro-organisms, and release and marketing of products consisting of or derived from GMOs, including foods, feeds and seeds.

---

[35] *FDA Statement of Policy: Foods Derived From New Plant Varieties*, US Fed Reg 22984, 29 May 1992, available at www.cfsan.fda.gov/~lrd/bio1992.html (last visited 30 Sept 2005).

[36] FDA, *Guidance on Consultation Procedures. Foods Derived from New Plant Varieties*, as last amended in Oct 1997; available at www.cfsan.fda.gov/~lrd/consulpr.html (last visited 30 Sept 2005).

[37] For an in-depth analysis of the EU regulatory framework for GMOs see T Christoforou, 'The Regulation of Genetically Modified Organisms in the European Union: the Interplay of Science, Law and Politics' (2004) 41 *Common Market Law Review* 637.

[38] See Communication from the Commission to the Council, 'A Community Framework for the Regulation of Biotechnology', COM(1986)057.

[39] Council Directive 90/220/EEC of 23 Apr 1990 on the deliberate release of genetically modified organisms [1990] OJ L/117/15. A second Directive adopted on the same day specifically covered the so-called 'contained use' of GMOs, ie, use of genetically modified micro-organisms for research and industrial purposes under circumstances limiting the contact of these organisms with the public and the environment: see Council Directive 90/219/EEC of 23 Apr 1990 on the contained use of genetically modified micro-organisms [1990] OJ L/117/1.

As under EU law once a product is placed on the market in one Member State it can freely circulate in all other 25 Member States, the EU system for authorisation of GMOs is drafted to permit each and every Member State to participate in the decision-making process and, thus, to have its differences in ecosystems, agricultural practices, as well as in public perceptions and attitudes, taken into account.

4.2.1 Deliberate Release into the Environment of GMOs

Currently, the main legislative instrument of the EU framework regulating GMOs is Directive 2001/18.[40] The purpose of the Directive is to harmonise the laws of the Member States on the deliberate release of GMOs so as to guarantee the protection of human health and the environment through both control of the risks that this can generate and safe development of industrial products utilising GMOs. In recital (4) of its Preamble, Directive 2001/18 recognises that:

> Living organisms, whether released into the environment in large or small amounts for experimental purposes or as commercial products, may reproduce in the environment and cross national frontiers thereby affecting other Member States. The effects of such releases on the environment may be irreversible.

The basis of Directive 2001/18 is, thus, a precautionary approach, and the precautionary principle must be taken into account when implementing it.[41] To achieve its objectives, Directive 2001/18 requires a case-by-case evaluation of the potential risks to human health and the environment before any GMO, or product consisting of or containing GMOs, can be placed on the market or in any other way released into the environment within the Community's territory. On the basis of that risk assessment, which is conducted at the national, Member State and Community level, a market authorisation is either granted or refused.

4.2.2 GM Food and Feed

Until the adoption of specific sectoral rules, food and feed, too, to the extent that they consisted of or contained GMOs, were covered by Directive 90/220. Instead, GM food derived from GMOs but no longer consisting of or containing them could be marketed freely across the Community. In 1997, in line with the beginning of a broader debate on food safety in Europe, specific legislation on food, whether GM or not, was adopted.[42]

---

[40] See n 3 above.
[41] See recital (8).
[42] Regulation 258/97 of the European Parliament and of the Council of 27 Jan 1997 concerning novel foods and novel food ingredients [1997] OJ L/43/1.

This legislation has now been replaced, as far as GM food and feed are concerned, by Regulation 1829/2003.[43] The purpose of the new Regulation is to streamline and make more transparent the authorisation procedure which was established by the previous Regulation, thus eliminating all differences between national laws, regulations and administrative provisions concerning the safety assessment and authorisation of GM food and feed, and ensuring a high level of protection of human life and health. To this end, it establishes an authorisation process which is very similar to that provided for by Directive 2001/18 which ensures close cooperation between the EU and the Member States through a comitology procedure[44] and a single risk management process. As for Directive 2001/18, a scientific evaluation has to be undertaken under the responsibility of the European Food Safety Authority (hereinafter 'EFSA')[45] of any risks which GM food and feed present for human and animal health and, as the case may be, for the environment.

Regulation 1829/2003 also eliminates the possibility of avoiding the authorisation procedure and its risk assessment by notifying GM food as substantially equivalent to existing food. Under Regulation 258/97, in fact, products which were produced from, but did not contain, GMOs could be placed on the market on the basis of a so-called 'simplified procedure' if they were considered by a Member State's competent authority as 'substantially equivalent' to a food or food ingredient already present on the market.[46] The Commission had only to forward a copy of that notification to the other Member States indicating that the product could be placed on the market. All 13 GM food products placed on the market on the basis of Regulation 258/97 benefited from the application of this simplified procedure. In light of the international debate on the concept of substantial equivalence,[47] this procedure has now been abandoned.

---

[43] Regulation 1829/2003 of the European Parliament and of the Council of 22 Sept 2003 on genetically modified food and feed [2003] OJ L/268/1.

[44] The so-called 'comitology procedure' is established by Council Decision 1999/468/EC of 28 June 1999 laying down the procedures for the exercise of implementing powers conferred on the Commission [1999] OJ L/184/23.

[45] EFSA was established by Regulation 178/2002 of the European Parliament and of the Council of 28 Jan 2002 laying down the general principles and requirements of food law, establishing the European Food Safety Authority and laying down procedures in matters of food safety [2002] OJ L/31/1.

[46] See Art 5 of Regulation 258/97, n 42 above.

[47] See, for instance, *Report of the FAO/WHO Expert Consultation on Safety Aspects of Genetically Modified Foods of Plant Origin* (Rome, 2000), available at www.fao.org/es/esn/food/risk_biotech_aspects_en.stm (last visited 29 Sept 2005). The Consultation recognised that the concept of substantial equivalence can be used as a comparative approach focusing on the similarities and differences between GM food and its conventional counterpart. However, it also indicated that the concept of substantial equivalence is not a safety assessment in itself nor an endpoint but just a starting point of the safety assessment.

Regulation 1829/2003 also extends the labelling provisions to all genetically modified food and feed irrespective of the detectability of DNA and protein. The applicability of this requirement is ensured through the provisions laid down in Regulation 1830/2003 on traceability and labelling.[48]

4.2.3 Traceability and Labelling

Regulation 1830/2003 is an instrument of so-called 'horizontal' legislation, establishing uniform rules on the traceability and labelling of GMOs and of food and feed produced from GMOs.

According to the Preamble to the Regulation, the purpose of the traceability requirements is to 'facilitate both the withdrawal of products where unforeseen adverse effects on human health, animal health or the environment, including ecosystems, are established, and the targeting of monitoring to examine potential effects on, in particular, the environment'.

Traceability is also necessary to implement risk management measures in accordance with the precautionary principle as well as to ensure accurate labelling of GMOs. Labelling is, in turn, essential to guarantee that accurate information is available to operators and consumers so as to enable them to exercise their freedom of choice in an effective manner.

4.2.4 Coexistence

At present, as seen above, the EU legislative measures in force form a coherent, although complex, legislative framework, which authorises the scientific and industrial use of GMOs, but only after a high level of protection of human health and the environment has been guaranteed. The only area in which harmonised and uniform regulation is painfully lacking is that of coexistence.

As different modes of agriculture are not really compartmentalised, the question has arisen how to ensure coexistence between GM and non-GM (organic or conventional farming) crops. The issue is clearly crucial to any real choice between GM and non-GM crops and, consequently, food and feed, to farmers, industry, retailers and consumers, as well as to the effective ability to protect biodiversity.

Unfortunately, in this area, the absence of political consent has precluded the creation of an EU-wide instrument. In a Recommendation

---

[48] Regulation 1830/2003 of the European Parliament and of the Council of 22 Sept 2003 concerning the traceability and labelling of genetically modified organisms and the traceability of food and feed products produced from genetically modified organisms [2003] OJ L/268/24.

dated July 2003,[49] the Commission considered the issue of coexistence as relating only to 'the ability of farmers to make a practical choice between conventional, organic and GM-crop production, in compliance with the legal obligations for labelling and/or purity standards',[50] and thus to consumer choice. Having limited itself to considering this issue as concerning 'the potential economic loss and impact of the admixture of GM and non-GM crops, and the most appropriate management measures that can be taken to minimise admixture', the Commission has then confined itself to taking stock of the extremely diverse farm structures and farming systems, as well as economic and natural conditions under which farmers in the EU operate. Therefore, it has considered that measures of coexistence should be better developed and implemented by the Member States, and that it could limit itself to providing a list of general principles and elements for the development of national strategies and best practice in the form of guidelines.

The last couple of years have shown that the reality is much more complex and that coexistence is not only a key factor for ensuring consumer choice but it is also at the heart of the fierce debate on how to reconcile GMOs and non-GMOs 'in the fields'. The application of the principle of subsidiarity and the resulting inability to balance divergent fundamental freedoms and rights have thus led to a steadily growing number of regions and local entities declaring themselves 'GMOs free', out of fear of irreversible harm to their citizens' health, their environment or their biodiversity.[51]

4.3 Guaranteeing Freedom of Research and Conducting a Business

The EU regulatory framework for GMOs has been developed on the assumption that research must be guaranteed and industry should remain free and competitive. Both 'the strengthening of the competitiveness of Community industry' and 'the promotion of research and technological development' are, in fact, among the activities the Community is mandated to carry out.[52] Both industry and research also have a dedicated title in the TEC[53] and ensuing Community policies.

---

[49] Commission Recommendation of 23 July 2003 on guidelines for the development of national strategies and best practices to ensure the coexistence of genetically modified crops with conventional and organic farming, C (2003)2624 [2003] OJ L/189/36.

[50] Ibid, recital (3).

[51] See, eg, the Charter of the Regions and Local Authorities of Europe on the Subject of Coexistence of Genetically Modified Crops with Traditional and Organic Farming, signed in Florence (Italy) on 4 Feb 2005, available at www.cor.eu.int/document/Highlight/Charter_of_Regions_Feb_05.pdf (last visited 25 Sept 2005).

[52] See Art 3 (m) and (n) of the TEC, n 3 above.

[53] See Titles XVI and XVIII.

As far as research in the field of biotechnology is concerned, the EU has also provided significant financial support, amounting to some €4 billion since 1982. Of this amount, over €70 million euros has been dedicated to research on the potential effects of GMOs on human health and the environment.

In the Sixth Framework Programme on Research (hereinafter 'FP6'), the EU's main instrument for the funding of research in Europe covering the years 2002–2006, the Commission has enhanced its support for biotech research. Implementation of this programme has in fact increased the financial contribution to research in this area by some 20 per cent. Around, €2,700 million out of a €17.5 billion budget have in fact been allocated to the two key thematic areas of 'life science genomics and biotechnology for health' and 'food quality and safety'.[54]

Also the freedom to conduct a business is guaranteed in the GMOs area, although subordinated to the granting of an authorisation. Thus, any person intending to commercialise, or else introduce into the environment, a GMO must first submit an application to the competent national authority of the Member State in which the product is to be first placed on the market. The application must include a scientific assessment which has to be carried out by the applicant and must contain certain information specified in the legislation.[55] The national authorities of the other Member States and EFSA will then carry out their own risk assessments and, on this basis, the Council and the Commission will decide, through the comitology procedure, whether to grant or to refuse the authorisation.

Similarly, in the case of a GM food, the person intending to place it on the Community market for the first time must submit a request to the Member State in which the product is first to be placed on the market.[56] The competent national authority informs EFSA. EFSA informs the Commission and the other Member States. Where the application is about food and feed containing or consisting of a GMO (rather than food and feed produced from a GMO), an applicant can file a single application to obtain both the authorisation for the deliberate release of a GMO into the environment under the criteria laid down in Directive 2001/18/EC, and the authorisation for use of this GMO in food and/or feed under the criteria laid down in Regulation 1829/2003 (the so-called 'one door–one key' principle).[57] The authorisation thus obtained is valid throughout the Community.

---

[54] All information concerning the FP6 is available at http://europa.eu.int/comm/research/fp6/index_en.cfm (last visited 30 Sept 2005).
[55] See Art 13(1) (1), of Directive 2001/18, n 3 above.
[56] See Art 5 of Regulation 1829/2003, n 43 above.
[57] Ibid, Art 5(5).

Argentina, Canada and the United States have complained before the WTO that an alleged six-year *de facto* moratorium on the granting of authorisations for GMOs[58] has hindered the ability of the GMOs industry to develop in Europe. In their complaints, however, the fault of the EU appears to lie more in the way the system has been applied than in the texts of its legislation. These countries have, in fact, called into question the legitimacy, from a WTO law point of view, only of certain specific approval procedures and safeguard measures and not of the EU legislation as such. The EU, on the other hand, has defended the case as being:

> a case about regulators' choices of the appropriate level of protection of public health and the environment in the face of scientific complexity and uncertainty and in respect of which there is great public interest. It is a case essentially about time. The time allowed to a prudent government to set up and apply a process for effective risk assessment of products which are novel for its territory and ecsytems, and that have the potential of causing irreversible harm to public health and the environment.[59]

According to some commentators:

> The outcome of [this case] carries profound implications for the balance between state and global power and the relationship of science to democracy. WTO adjudicators will define the extent to which particular conceptions of sound science can be used to set boundaries on members' precautionary health and environmental measures.[60]

4.4 Protecting the Environment and Human Health

The EU regulatory framework for GMOs is built around the notion of a high level of environmental and health protection to be ensured through a thorough assessment of the risks, post-marketing requirements and national or regional measures.

4.4.1 Through Risk Assessment

Under Directive 2001/18, the competent authority which has received the notification must assess the potential adverse effects on human health

---

[58] See disputes WT/DS291, WT/DS292, WT/DS293, *EC—Measures Affecting the Approval and Marketing of Biotech Products*. The Panel is expected to publish its report at the end of Sept 2006.

[59] See ibid, *Oral statement of the European Communities at the first meeting of the Panel with the Parties* (Geneva, 2 June 2004), 1, available at http://trade-info.cec.eu.int/wtodispute/show.cfm?id=190&code=2 (last visited 30 Sept 2005).

[60] D Winickoff *et al*, 'Adjudicating the GM Food Wars: Science, Risk, and Democracy in World Trade Law' (2005) 30 *Yale Journal of International Law* 81, at 84.

and the environment of the GMO in question and, within 90 days of the receipt of the notification, it must prepare an opinion indicating whether the GMO should or should not be placed on the market. During the preparation of this report, the leading competent authority normally addresses a number of requests for additional information to the notifier. If the leading competent authority concludes that consent cannot be granted, the application is rejected and the procedure ends. If, instead, it concludes that authorisation for release should be granted, the procedure moves on to the Community level.[61]

At that point, the Commission forwards the application to the competent authorities of all the other Member States, which may then, within a deadline of normally 60 days, ask for further information, make comments or present reasoned objections to the placing on the market of the GMO in question.[62] If there are no objections from other Member States or the Commission, the leading competent authority grants consent to the placing on the market of the product.[63] The product may then be placed on the market throughout the EU in conformity with any conditions required in that consent. If objections are raised and maintained by other Member States or by the Commission, or if the leading competent authority delivers an unfavourable opinion, applications for authorisation are referred to the relevant Scientific Committee of EFSA.[64] EFSA then carries out a further risk assessment.

Similarly, under the regulation for GM food and feed, the applicant must submit a request which has to contain certain information and include material which demonstrates that the product complies with the requirements that: (1) the food does not have adverse effects on human or animal health and the environment; (2) it does not mislead the consumer; and (3) it does not differ from foods or food ingredients which it is intended to replace to such an extent that its normal consumption would be nutritionally disadvantageous for the consumer.[65]

Under this new regime (and contrary to the previous practice that required an opinion of the Scientific Committee only in the event of objections by a Member State or the Commission), EFSA will monitor all applications for the placing on the market of food products containing GMOs and will issue an opinion on each of them, normally within six months from the receipt of a valid application. In order to prepare this opinion, EFSA can rely on the assistance of the national competent

---

[61] See Art 14 of Directive 2001/18, n 3 above. Annex II of Directive 2001/18 contains a detailed set of objectives, principles and methodologies to be followed by the competent authorities when performing the environmental risk assessment.

[62] Ibid, Art 15(1).

[63] Ibid, Art 15(3).

[64] See ibid, Art 28.

[65] See Arts 5(3) and 4(1) of Regulation 1829/2003, n 43 above.

authorities and has to consult the Community reference laboratory on the detection and identification method.[66]

The EFSA opinion is forwarded to the Commission, the Member States and the applicant. The Commission then adopts its decision based on the EFSA opinion by means of the comitology procedure.[67]

It has to be noted that in the EC regulatory framework for GMOs the responsibility for risk assessment is clearly separated from that for risk management. While EFSA advises on possible scientific risks, the responsibility for risk management lies with the EU institutions. It is their role, taking into account EFSA's advice as well as other considerations, to propose and adopt the appropriate regulatory and control measures.

In practice, the carrying out of several risk assessment at national, Member State and Community level has strongly contributed to guaranteeing real scientific progress in the field of GMOs. The studies carried out by individual Member States and their ensuing comments and scientific objections have permitted a deeper understanding of the potential and actual risks of GMOs to the environment and human health. Thus, for instance, the issue of antibiotic resistance marker genes was first raised by the UK competent authority in connection with a request for authorisation of a GM maize variety, Maize Bt176, by Ciba-Geigy (now Syngenta). The Scientific Committees regarded the probability of horizontal gene transfer as very small, and therefore considered the product safe[68] and, on this basis, permission was given for the marketing of the seed in January 1997. Several years later, Article 4 of Directive 2001/18 called for particular consideration of GMOs containing antibiotic resistance marker genes when carrying out an environmental risk assessment and, in 2004, EFSA issued a new opinion on the issue in which it classified antibiotic resistance genes in three categories. Whilst for some of them no restrictions were considered necessary either for field experimentation or for placing on the market, EFSA found that other antibiotic resistance genes should either be restricted to GMOs used in field trials only or not be present in GM plants at all.[69]

---

[66] See Art 18 of Regulation 178/2002, n 45 above. The European Community reference laboratory has been established at the European Commission's Joint Research Centre in Ispra, Italy.

[67] See Art 7 and 35 of Regulation 1829/2003, n 43 above.

[68] See Scientific Committee on Foods, *Opinion On The Potential For Adverse Health Effects From The Consumption Of Genetically Modified Maize (Zea Mays L)* (13 Dec 1996), available at http://europa.eu.int/comm/food/fs/sc/oldcomm7/out02_en.html (last visited 30 Sept 2005).

[69] See *Opinion of the Scientific Panel on Genetically Modified Organisms on the use of antibiotic resistance genes as marker genes in genetically modified plants*, 2 Apr 2004 (2004) 48 *The EFSA Journal* available at www.efsa.eu.int/science/gmo/gmo_opinions/384/opinion_gmo_05_en1.pdf (last visited 30 Sept 2005).

4.4.2 *Through Post-marketing Requirements*

The effective protection of the environment and of human health does not end, however, with the authorisation of the placing on the market of the GMO. Both Directive 2001/18 and Regulation 1829/2003 establish post-marketing requirements and surveillance. In particular, post-marketing monitoring plans are required to investigate the risks identified in the risk assessment and to test any assumption that was included in such risk assessment.[70] The EU Scientific Committees have often recommended such plans in order to verify the accuracy of their assumptions. Thus, for example, in the case of a glufosinate tolerant swede rape notified by Bayer, the Scientific Committee on Plant considered that one of the issues submitted for its attention, the potential transfer of the herbicide resistance gene to wild relatives, was a new issue in Europe given the limited scale of release to date. It thus agreed to the introduction into the environment of the swede rape on condition that the authorisation be accompanied by both an agreed code of practice for field management, also involving the active participation of the notifier in promoting best practice by farmers, and a research programme with an agreed design and implementation plan to detect the occurrence and the establishment of herbicide tolerant volunteers and weeds under field conditions in the EU.[71]

Directive 2001/18 also imposes an obligation on the applicant and the national competent authorities to make known any new information which may become available on the risks to human health and the environment or on the safety risks of GM food. It further requires the applicant to take all measures necessary to protect human health and the environment.[72]

Traceability is also a powerful tool for allowing the withdrawal of GMOs in the event of unforeseen effects on human or animal health, or the environment, as well as for the implementation of risk management measures.

Finally, all authorisations are granted for a limited period of time. After 10 years, in fact, all conditions on the basis of which an authorisation was first granted, as well as the results of monitoring, are verified and assessed.[73]

---

[70] See Annex VII to Directive 2001/18, n 3 above, and Arts 5(3)(k), 5(5)(b), 6(5)(e) and (g), and 9 of Regulation 1829/2003, n 43 above.

[71] See *Opinion of the Scientific Committee on Plants regarding submission for placing on the market of Glufosinate tolerant swede rape transformation event GS 40/90 notified by the agrevo company* (*notification C/DE/96/05*), 14 July 1998, available at http://europa.eu.int/comm/food/fs/sc/scp/out15_en.html (last visited 30 Sept 2005).

[72] See Art 20 of Directive 2001/18, n 3 above.

[73] Ibid. Art 17(6), and Arts 10 and 11 of Regulation 1829/2003, n 43 above.

4.4.3 Allowing National or Regional Measures

As mentioned above, the GMO authorisations granted are valid throughout the Community. However, in line with the requirements of Article 174 TEC (and apart from the role assigned to Member States in the risk assessment and decision-making process leading to the granting or refusal of the authorisations), the EU regulatory framework provides for a number of instruments that allow Member States to govern the use of GMOs on the basis of considerations linked to the specificities of their local ecosystems or agricultural practices. The two main ones are contained in Directive 2001/18 and in Article 95 TEC.

Directive 2001/18 includes a safeguard clause which enables Member States, under certain specific conditions, provisionally to prohibit the marketing within their territory of authorised GMOs. This safeguard clause can be invoked if, as a result of new or additional information affecting the risk assessment or of new or additional scientific knowledge, there is reason to believe that the authorised GMO constitutes a risk to human health or the environment. This measure then has to be reviewed at Community level, where a final decision on whether or not the authorisation granted should continue, be amended or repealed (and hence whether the safeguard measure is justified or not) is taken. Such a decision is, once again, taken by means of the comitology procedure.[74]

Nine safeguard measures are currently in force in certain EU Member States.[75] The reasons for adopting them are varied and range from gene flow, to effects of BT-toxins on non-target organisms and the development of resistance to toxins, to risks associated with antibiotic resistance marker genes. The relevant Scientific Committees, and most recently EFSA, have examined these measures and concluded that there is no new scientific evidence, in terms of risk to human health and the environment, that would invalidate the original risk assessments.[76] On this basis, the Commission has thus proposed to the Council the withdrawal of the measures. On 24 June 2005, however, the Environment Council rejected

---

[74] Art 23 of Directive 2001/18, n 3 above. A similar instrument was already provided for in Art 16 of Directive 90/220/EEC n 39 above, on the basis of which a number of the safeguard measures still in force were adopted.

[75] A complete listing is available at http://europa.eu.int/comm/environment/biotechnology/safeguard_clauses.htm (last visited 2 Oct 2005).

[76] See 'Opinion of the Scientific Panel on Genetically Modified Organisms on a request from the Commission related to the Austrian invoke of Article 23 of Directive 2001/18/EC' (2004) 78. *The EFSA* Journal available at www.efsa.eu.int/science/gmo/gmo_opinions/507/opinion_gmo_safeguard_clauses_austria_en1.pdf (last visited 2 Oct 2005); and 'Opinion of the Scientific Panel on Genetically Modified Organisms on a request from the Commission related to the Greek invoke of Article 23 of Directive 2001/18/EC' (2004) 79 *The EFSA Journal* available at http://www.efsa.eu.int/science/gmo/gmo_opinions/506/opinion_gmo_safeguard_clauses_greek_en1.pdf (last visited 2 Oct 2005).

all eight proposals for withdrawal of the safeguard measures with an unprecedented qualified majority against a Commission proposal on GMOs. This example clearly illustrates the tensions that in reality may exist between different degrees of acceptable risks at EU or Member State level and the concrete difficulty that the Commission, upon which now rests the task of formulating new proposals, has in finding the correct balance between somehow divergent scientific and policy objectives.

Another instrument that allows Member States, under certain specific conditions, to enforce a higher level of protection than that chosen at EU level, is Article 95 TEC. On the basis of that Article, in fact, a Member State is allowed to introduce national provisions of a general nature (ie, in the case of GMOs, not linked to a specific authorisation but to the regulatory regime in general), provided that they are based on new scientific evidence relating to the protection of the environment, on grounds of a problem specific to that Member State that arose after the adoption of the harmonisation measure (ie, in the case of GMOs, Directive 2001/18). Such measures have to be notified to the Commission, which, within six months, has to approve or reject the national provisions 'after having verified whether or not they are a means of arbitrary discrimination or a disguised restriction on trade between Member States' and whether or not they 'constitute an obstacle to the functioning of the internal market'.[77]

In the field of GMOs, Article 95 TEC was recently invoked by Austria as the legal basis for banning the use of GMOs in the region of Upper Austria. The Commission, however, on the basis of an opinion by EFSA, considered that there was no new evidence that could justify this ban.[78] Both Austria and Upper Austria brought annulment proceedings against the Commission's decision to the European of Justice Court[79] and to the Court of First Instance (hereinafter 'CFI'), respectively. The first judgment in these cases has just been delivered by the CFI.[80] In it, the CFI found the Commission's decision to be correct, most importantly because Austria was not able to demonstrate that Upper Austria 'contained unusual or unique ecosystems that required a separate risk assessment from those conducted for Austria as a whole or in other similar areas in Europe'.[81]

---

[77] Art 95(5) and (6) of the TEC.
[78] Commission Decision 2003/653/EC of 2 Sept 2003, relating to national provisions on banning the use of genetically modified organisms in the region of upper Austria notified by the Republic of Austria pursuant to Art 95(5) EC Treaty [2003] OJ L/230/34.
[79] See Case C–492/03, *Austria v Commission*, still pending.
[80] See Judgment of the CFI of 5 Oct 2005 in Case T–366/03, *Land Oberösterreich v Commission*, not yet reported.
[81] Ibid, para 67.

4.5 Protecting Consumers and their Right of Participation

Another notion which is central to the EU regulatory framework for GMOs is that of consumer protection. Under the legislation on GMOs, 'consumers' are protected in many different ways. First and foremost, they are best protected by the fact that the whole regulatory system is aimed at ensuring that the products they consume, or that are present in the environment they live in, are 'safe'. Not only, that but they are also given the opportunity to follow the development of the authorisation procedure and to comment on the scientific opinions of EFSA, in other words to be informed. Thus, both under Directive 2001/18 and under Regulation 1829/2003, a summary of the publicly available data contained in the application dossier (called 'SNIF' in the Directive) is made available to the public.[82] Later in the procedure, when an opinion by EFSA is available, the Commission opens a consultation on its website and the public has the opportunity to submit comments within 30 days of that publication.[83] Once an authorisation is granted, Regulation 1829/2003 requires that it be entered in the Community Register of GM Food and Feed.[84] This Register is intended to provide product information, such as the name of the authorisation holder, the exact scope of the authorisation, links to relevant risk assessments and the date of entry on the EU market, etc. In this way, the scope and content of the authorisations should become accessible and transparent to everybody. The Community Register also contains GM food and feed products that were placed on the market before the date of application of Regulation 1829/2003, or that were notified to the Commission before 18 October 2004.

Under EU legislation, consumers are also given the chance to choose whether or not to make use of or to come into contact with GM products, even though they have been considered 'safe'. Through traceability and labelling, in fact, the current regulatory framework ensures a high level of transparency in the marketing phase of GM products which should guarantee the possibility for consumers to make an informed choice. Norms on traceability requirements, including labelling, are contained in Directive 2001/18 as well as in Regulation 1829/2003.[85] In particular, the

---

[82] See Art 24 of Directive 2001/18, n 3 above. SNIFs can be consulted on-line at http://gmoinfo.jrc.it. See also Art 25 of Regulation 1829/2003, n 43 above.

[83] See ibid, Art 6. For an example see comments received on the EFSA opinion related to the application for the placing on the market of foods consisting of or containing the genetically modified maize '1507' (eg, fresh or canned sweet maize) or food produced from this maize (eg, flour, oil): see http://europa.eu.int/comm/food/food/biotechnology/authorisation/comments_1507.pdf (last visited 30 Sept 2005).

[84] See ibid, Art 28. The Register can be consulted online at http://europa.eu.int/comm/food/dyna/gm_register/index_en.cfm (last visited 30 Sept 2005).

[85] See Art 4(6) and 21 of, as well as Annex IV to Directive 2001/18, n 3 above, and Arts 5(3)(k) and 6(5)(e) and (g) of Regulation 1829/2003, n 43 above.

latter establishes that the words 'genetically modified' or 'produced from genetically modified' shall appear either in the list of ingredients or on the package or on the display of any food authorised under this Regulation. The GM food label shall also mention any differences with respect to its conventional counterpart in regard to composition, nutritional value, intended use and implications for the health of certain sections of the society.[86] A threshold of 0.9 per cent of adventitious or technically unavoidable presence of GMOs is set for the purposes of labelling.[87]

Furthermore, on the basis of Regulation 1830/2003, traceability is required throughout the production and distribution chains and implies that (1) each product must have a unique identifier (ie, a number through which both the transformation event and the developer of the GMO are identifiable);[88] (2) the information to be transmitted is specified; and (3) that all points in the production and distribution chains are reliably linked. As far as GMOs released into the environment are concerned, this should help their withdrawal in the event of unforeseen effects on human or animal health, or the environment, as well as the implementation of risk management measures. In the case of GM food or feed, traceability will also facilitate accurate labelling.[89] The new regulation is also stricter than the sectoral legislation on labelling. It includes all foods produced from GMOs, without drawing a distinction between those containing DNA or protein resulting from genetic modification and those which no longer contain any actual trace. For each product 'consisting of or containing GMOs', any operator in the production and distribution chains shall ensure that the indication that the product contains GMOs appears either on the label or on the display.[90]

4.6 Right to an Effective Remedy

Last but not least, it should be mentioned that respect for the rule of law does not exist if there is no right to an effective remedy. In other words, no fundamental right or freedom really exists if there is no possibility of enforcing it in front of an independent judiciary. The possibility of judicial review constitutes one of the most important democratic guarantees that the rights of individuals are respected.

---

[86] See ibid, Art 13 and 6(5)(d).

[87] See ibid, Art 12(2).

[88] See Commission Regulation 65/2004 of 14 Jan 2004 establishing a system for the development and assignment of unique identifiers for genetically modified organisms, [2004] OJ L/10/5.

[89] See Art 4(A) as well as recitals 3–5 of Regulation 1830/2003, n 43 above.

[90] See ibid, Art 4(B).

Article 47 of the Charter endorses this 'right to an effective remedy' in the following terms:

> Everyone whose rights and freedoms guaranteed by the law of the Union are violated has the right to an effective remedy before a tribunal in compliance with the conditions laid down in this Article.

This first paragraph of Article 47 is based on Article 13 of the European Convention on Human Rights,[91] which recognises that:

> Everyone whose rights and freedoms as set forth in this Convention are violated shall have an effective remedy before a national authority notwithstanding that the violation has been committed by persons acting in an official capacity.

In EU law the protection is more extensive, since it guarantees the right to an effective remedy not just before any public authority but before a court. The Court has been protecting this principle in its case law since 1986.[92] According to the Court, this principle also applies to the Member States when they are implementing EU law. However, the inclusion of this right in the Charter is not intended to change the judicial system laid down by the TEC, and particularly the rules relating to admissibility. This principle is therefore to be implemented in accordance with the procedures laid down in the TEC. It applies to the institutions of the Union and of Member States when they are implementing EU law, and does so for all rights guaranteed by it.

With regard to GMOs, this means that all actors involved should have a means of redress if their rights and freedoms have been infringed. Thus, for instance, if the applicants for authorisations under Directive 2001/18 or Regulation 1289/2003 are dissatisfied with any act or failure to act of the national authority of a Member State or of an EU institution they are free to bring proceedings for judicial review of such acts. In particular, under Articles 230 and 232 of the TEC, the Court has jurisdiction to review the legality of acts of the EU institutions, including the European Commission. So far, however, no such case has ever been brought. On the basis of the same provisions too Member States can contest the actions or omissions of the EU institutions. An example of such proceedings is those brought by Austria against the Commission.[93]

---

[91] Convention for the Protection of Human Rights and Fundamental Freedoms, signed in Rome on 4 Nov 1950.

[92] Case 222/84, *Johnston, v Chief Constable of the RUC* [1986] ECR 1651; see also Case 222/86, *Union nationale des entraineurs et codres techniques professionnels du football (Unectef) v Heylens* [1987] ECR 4097 and case C–97/91, *Oleificio Borelli SpA v EC Commission*, [1992] ECR I–6313.

[93] See n 79 above.

Furthermore, the law of each Member State provides for administrative and/or judicial review of acts or omissions relating to the application at the national level of EU legislation. Thus, a case was taken in November 2000 by Monsanto Agricoltura Italia SpA and others to the Italian courts, challenging the validity of an Italian decree that temporarily suspended trade in and use of certain novel foods within Italy,[94] and seeking compensation for loss claimed to result from the decree. Similarly, a number of consumers' associations contested another Italian decree implementing a Commission Directive on infants' food, the effect of which was that the presence of GMOs in a proportion not exceeding 1 per cent of the ingredients making up baby foods and follow-on formulae, caused by adventitious contamination, did not need to be indicated in the labelling of those foods and formulae. The case was also referred to the Court for a preliminary ruling on the basis of Article 234 TEC.[95]

5. CONCLUSIONS

The above analysis shows that, in Europe, the development and use of biotechnology always take place consistently with respect for a number of fundamental rights and freedoms. By taking a proactive role, the EU regulator aims at developing policies that allow the exploitation of biotechnology in a responsible manner, consistent with European values and standards.[96] Through its detailed regulation on approval procedures, risk assessment, post-marketing monitoring, traceability and labelling, the EU tries to reconcile the protection of the fundamental rights to the environment, health and consumers' choice and protection with other fundamental freedoms such as scientific research and conduct of a business. In Europe, the existence of a highly regulated sector ensures the primacy of the rule of law, of the *stato di diritto* over an auto-regulated market, and thus the accountability of all actors involved—public institutions included.

In the only area in which the Community has refrained from legislating—coexistence—the result is a regulatory jungle of 'GMO free' zones that, outside all constitutional balance, may ultimately prevent the effective exercise of basic rights of research, freedom of economic activity and even of consumers' freedom of choice.

Thus, in conclusion, contrary to the view that human rights correspond to the absence of state intervention, the opposite approach of responsible

---

[94] The decree enacted a safeguard measure pursuant to Art 12 of Regulation 258/97, n 42 above.

[95] See Judgment of 26 May 2005 in Case C–132/03, *Ministero della Salute v Codacons and Others*, not yet reported.

[96] See *Life Science and Biotechnology—A Strategy for Europe*, n 29 above.

and democratic regulation appears to be more in line with international human rights provisions, such as Article 28 of the Universal Declaration of Human Rights.[97] That Article, as Professor Francioni clearly explains,[98] by establishing a right to 'a social and international order in which the rights and freedoms set forth in this Declaration can be fully realised', calls upon governments to be responsible and to act in order to ensure social (and thus also scientific and economic) conditions in which human rights can be safeguarded. This provision is particularly meaningful in the field of GMOs, because of the impact human rights have on biotechnologies.

---

[97] Adopted and proclaimed by the UN with GA Resolution 217A(III) on 10 Dec 1948. See www.un.org/Overview/rights.html.
[98] F Francioni in this volume.

9

Biogenetic Resources and Indigenous Peoples' Rights

FEDERICO LENZERINI*

1. BIOPROSPECTING OF INDIGENOUS PEOPLES' BIOGENETIC RESOURCES

FOR MANY CENTURIES, since Europeans first came into contact with indigenous peoples, rites performed by medicine men for treating diseases, characterised by the use of plants and other natural ingredients, have been considered by Europeans mainly as manifestations of superstitious beliefs typical of savage and primitive groups. Blinded by stereotypes, racial European scientists were often prevented from improving their medical skills by sharing indigenous knowledge relating to the curative properties of natural elements, knowledge which had been developed throughout the centuries in the context of an intimate and respectful relationship with nature as the core element of the holistic vision of life of indigenous peoples.

This approach has drastically changed over recent decades. The Western world has become aware of the high value of indigenous knowledge concerning nature and its enormous potential to improve human life, not just in the field of medical treatment. Lands inhabited by indigenous peoples have thus become the most potentially valuable areas for bioprospecting, not only on account of their virtual 'virginity' (from the perspective of Western culture) and richness in biodiversity, but also of the extremely precious knowledge, developed by autochthonous peoples throughout the centuries, concerning medical or other commercially exploitable properties of local resources. Nevertheless, such new awareness has not led to the development

* Juris Dr at the University of Siena, where he is a researcher in international law. He is also a consultant to UNESCO. Email: lenzerini@unisi.it. The author wishes to thank Professor Francesco Francioni of the European University Institute for the precious advice given during the writing of this work, Dr Ana Filipa Vrdoljak of the European University Institute for her help and kindness, and Ms Veronica Pulcini (Juris Dr, University of Siena), for her valuable help in the research.

of adequate legal norms prescribing respect for indigenous knowledge concerning the sustainable use of natural resources for the enhancement of the quality of life. On the contrary, it has spurred the appropriation of such knowledge, making it a source of huge income through its commercial use. This appropriation persists today, usually without recognition of the legitimate owners' rights, thus profiting from their unfamiliarity with typically Western legal-economic concepts, such as 'ownership' and 'intellectual property', which have recently been incorporated into the body of international law (particularly in the TRIPS Agreement).[1]

A few examples will serve to illustrate the point. Over a century ago the Guajajara tribe of Brazil discovered that one of the autochthonous plants on its land, the *Pilocarpus jaborandi*, could be used as a treatment for glaucoma. For years the Guajajara applied their tribal medical knowledge of this plant, developing and sharpening its medical uses. But, at present it is no longer available to the Guajajara, while the Brazilian government exports it worldwide as a glaucoma treatment obtaining revenue of US $25 million per year. This amount pales into insignificance when compared to the income of the pharmaceutical company which patented the *Pilocarpus jaborandi*.[2]

A similar case concerns the plant *Banisteriopsis caapi*, located in the Western Amazon basin. For centuries, shamans of local indigenous peoples used the plant to produce a ceremonial drink called *ayahuasca* (literally meaning 'vine of the soul'), which is used in religious and medicinal ceremonies to diagnose and treat diseases, meet spirits and predict the future. Only selected and especially skilled individuals are able to get *ayahuasca* from the *Banisteriopsis caapi*, using special processes which differ among the various tribes living in Western Amazonia and are transmitted orally from generation to generation. Of course, none of these processes has ever been patented or made the subject of any other kind of legal protection. On 17 June 1986 the American citizen Loren Miller obtained for the International Plant Medicine Corporation of US patent number US5751P for *Banisteriopsis caapi (cv) 'Da Vine'*, described as '[a] new and distinct Banisteriopsis caapi plant named "Da Vine" which is particularly characterized by the rose colour of its flower petals which fade with age to near white, and its medical properties'.[3] Miller thus claimed he had discovered a previously unknown variety of the plant, characterised

---

[1] See n 95 below.

[2] See DA Posey and G Dutfield, *Beyond Intellectual Property: Toward Traditional Resource Rights for Indigenous Peoples and Local Communities* (Ottawa, International Development Research Centre, 1996), 53; C Hamilton, 'The Human Genome Diversity Project and the New Biological Imperialism' (2001) 41 *Santa Clara Law Review* 619, at 621 f.

[3] See http://12.espacenet.com/espacenet/viewer?PN=US5751P&CY=ch&LG=en&DB= EPD (last visited 30 Sept 2004).

by the particular flower colour, in a domestic garden in the Amazonian rainforest. In 1994 the Center for International Environmental Law (CIEL), on behalf of the Coordinating Body of Indigenous Organisations of the Amazon Basin (COICA), filed a request to re-examine the patent, based on the claims that *Da Vine* was neither new nor distinct and on the sacred nature of the *Banisteriopsis caapi* for the indigenous tribes of the Amazon Region, which made the patent contrary to the public morality aspects of the US Patent Act.[4] In 1999 the United States Patent and Trademark Office (USPTO), persuaded by the arguments raised by CIEL, revoked the patent. Nevertheless, in 2001, on the basis of new arguments presented by Loren Miller, the USPTO reversed its decision and declared that the patent for *Da Vine* had to stand.[5]

Patenting of indigenous biodiversity-related knowledge has been very common in recent years with regard to natural resources whose properties were discovered and developed throughout the centuries by indigenous peoples of India. For example, on 4 May 1999, the Indian plant *karela* (bitter gourd), along with *jamun*, *brinjal* and *gurmar*, the previous utilisation of which, through the centuries, by local indigenous groups for the treatment of diabetes was widely documented, became the object of US patent number 5,900,240 as a newly invented herbal anti-diabetic treatment. Of course, the patent absolutely ignores the fact that such plants have been used by the autochthonous communities of India for thousands of years as a treatment for diabetes.[6] This is not the only example of 'biopiracy' concerning Indian resources, a phenomenon which has been defined as 'epidemic'.[7]

Finally, the situation in Africa is not dissimilar. A landmark case is Madagascar's Rosy Periwinkle, which was traditionally used by local communities as an anti-diabetic. In 1954 technicians of the American firm Eli Lilly obtained two alkaloids from this plant, *Vinblastine* and *Vincristine*, which were deemed to have cancer treating properties. Although the interest of Eli Lilly in the Rosy Periwinkle had been motivated by its traditional use for treating diabetes, and only after subsequent studies

---

[4] Patent Act, 35 USCS Sects 1–376.

[5] All the information concerning the case of the *Banisteriopsis caapi* has been taken from www.amazonlink.org/biopiracy/ayahuasca.htm (last visited 30 Sept 2004). For other examples regarding unauthorised and unregulated taking of biological samples from the lands belonging to the indigenous peoples of Amazonia see www.amazonlink.org/biopiracy (last visited 30 Sept 2004).

[6] See K Reddy, 'Will India lose its Ayurvedic Heritage?', in (1999) 29 *Sword of Truth*, 19 July 1999, available at http://ilovehyderabad.com/columns/columns-will-india-lose-its-ayurvedic-heritage-html (last visited 18 aug 2006) V Shiva, 'The US Patent System Legalizes Theft and Biopiracy', *The Hindu*, 28 July 1999, available at www.organicconsumers.org/Patent/uspatsys.cfm (last visited 30 Sept 2004). See also, generally, D Sharma, 'Selling Biodiversity. Benefit Sharing is a Dead Concept', 3 May 2004, available at www.mindfully.org/WTO/2004/Selling-Biodiversity-Sharma3may04.htm (last visited 30 Sept 2004).

[7] See Shiva, n 6 above.

was it found to have anti-cancer properties as well, the US company maintained that the link with traditional knowledge had been extinguished by the fact that Madagascar's local communities had never used the Rosy Periwinkle for treating cancer. Eli Lilly thus obtained a patent for the two alkaloids isolated from the plant, and from the drugs produced by using such alkaloids it earned about $100 million per year during the whole course of the patent, without paying any compensation either to the indigenous peoples who had developed knowledge about the plant or to the government of Madagascar.[8]

2. THE LAST FRONTIER OF BIOIMPERIALISM: HARVESTING GENETIC SAMPLES OF INDIGENOUS PEOPLES

With the advancement of research on the human genome, the members of indigenous communities have become the target of a new form of bio-prospecting. In the context of the efforts made by the scientific community to discover all the secrets of the genetic heritage of humankind, the most important manifestation of which is certainly the Human Genome Diversity Project (HGDP),[9] a particularly relevant role is played by indigenous peoples' genetic material. The reason for this is that such peoples have generally evolved in a state of relative isolation; consequently, since their primitive genetic features have not been 'contaminated' by any cross–breeding with other ethnic groups, their genomes retain their original character. The information contained in their genetic material may thus tell scientists much more about the evolution of human genetic history than that collected from peoples which have irremediably merged with other ethnic groups. Such information may eventually lead to the identification of genes which confer resistance or vulnerability to diseases and to the development of appropriate medical tests and treatments.

---

[8] See Case Western Reserve University, 'Case Study: Rosy Periwinkle (Madagascar)', available at http://home.cwru.edu/~ijd3/autorship/rosy.html (last visited 30 Sept 2004). For other examples of biopiracy see G Stenton, 'Biopiracy within the Pharmaceutical Industry: A Stark Illustration of just how Abusive, Manipulative and Perverse the Patenting Process can be towards Countries of the South' (2003) 1 *Hertfordshire Law Journal* 30.

[9] The HGDP is an international consortium of governments, universities and scientists, conceived in 1991 by the eminent geneticist Luca Cavalli-Sforza of Stanford University, the purpose of which is the collection of biological samples from different population groups throughout the world for understanding human genetic diversity. Since the beginning of the implementation of the project, Cavalli-Sforza and other geneticists and anthropologists have been visiting several ethnic groups around the world collecting genetic samples from their members. See Hamilton, n 2, above passim; 'The Human Genome Diversity Project', *GenEthics News*, issue 10, available at www.hgalert.org/topics/personalInfo/hgdp.htm (last visited 30 Sept 2004). See also K H Ching, 'Indigenous Self-determination in an Age of Genetic Patenting: Recognizing an Emerging Human Rights Norm' (1997) 66 *Fordham Law Review* 687, at 692 ff.

The HGDP was launched in 1991, following a practice, already developed among Western scientists, which consisted in collecting genetic samples from indigenous groups. For example, in 1989 some American scientists had taken genetic materials from 24 members of the Hagahai tribe located in the Madang Province of Papua New Guinea, discovering that a cell-line derivation was potentially valuable for diagnosing adult leukaemia and chronic degenerative neurological diseases. A patent application was submitted by the US National Institutes of Health (NIH) for this cell-line, which was actually patented.[10] The patent became the object of intense international debate, including an official inquiry by the Government of Papua New Guinea asking whether the patent violated its sovereignty.[11] Following growing international pressure, on 24 October 1996 NHI surrendered its 'rights' to the patent at the USPTO.[12]

As can be imagined, the HGDP generated a powerful reaction on the part of most of the targeted indigenous groups. Apart from the (serious) difficulty, *ex se*, of obtaining informed consent from individuals who are totally unfamiliar with Western science,[13] the practice of collecting genetic samples from indigenous peoples is absolutely incompatible with their dignity and, thus, intolerable for two reasons in particular. The first is spiritual in character: certain indigenous peoples, such as the Maori of New Zealand, consider their genetic heritage as their 'life spirit', a distinctive individual spiritual realm composed of the experiences inherited from the ancestors that will also constitute the essence of the identity of future generations.[14] It is very close to the idea of the soul for Christians. Consequently, the collection of samples of such genetic heritage without the prior, informed and fully mindful individual and collective consent of the individuals and groups concerned would constitute an intolerable intrusion into their intimacy and a desecration of their dignity as human beings and/or peoples.

---

[10] See A O Wu, 'Surpassing the Material: the Human Rights Implications of Informed Consent in Bioprospecting Cells Derived from Indigenous Peoples Groups' (2000) 78 *Washington University Law Quarterly* 979, at 983; Ching, n 9, above at 701 ff; V Tejera, 'Tripping Over Property Rights: It is Possible to Reconcile the Convention on Biological Diversity with Article 27 of the TRIPs Agreement?' (1999) 33 *New England Law Review* 967, at 967 ff.

[11] See Ching, n 9 above, at 702.

[12] Ibid, at 702.

[13] On this problem see Hamilton, n 2 above, at 625 ff.

[14] See ibid, at 626. See also Ching, n 9 above, at 688, stating that '[s]ome [indigenous] peoples' religious or philosophic beliefs do not permit the patenting of life'. See also D Harry and F Dukepo, *Indians, Genes and Genetics: What Indians Should Know About the New Biotechnology.* (Nixou, Nev, indigenous peoples'coliation against biopiracy, 1998), quoted in UN doc E/CN.4/Sub.2/2004/38 of 23 July 2004 ('Human Rights and the Human Genome—Preliminary Report submitted by the Special Rapporteur, Iulia-Antonella Motoc'), para 30: '[s]cientists say it's just DNA. For an Indian, it is not just DNA, it's part of a person, it is sacred, with deep religious significance. It is part of the essence of the person. To us, any part of ourselves is sacred'.

Secondly, it is the very philosophy underlying the Project which has correctly appeared to be absolutely unacceptable from the perspective of indigenous peoples. Before the launch of the project, the scientists who conceived it wrote that:

> [I]solated populations are being rapidly merged with their neighbours ... destroying irrevocably the information needed to reconstruct our evolutionary history ... It would be tragically ironic if, during the same decade that biological tools for understanding our species were created, major opportunities for applying them were squandered.[15]

The letter, as well as other early HGDP documents, does not express any concern about the possible disappearance of many ethnic groups, only emphasising the need to collect genetic samples from members of them before such disappearance. In other words, it appeared that the identity of indigenous peoples was considered of importance only for their genetic peculiarities; once samples of their genetic heritage had been collected, it was not important if they, as distinct ethnic groups, disappeared forever. This perception is confirmed by the terminology used in the context of the HGDP: the ethnic groups considered of interest were defined as 'isolates of historical interest', abbreviated to IHIs.[16] This led a representative of indigenous peoples to comment bitterly:

> After being subjected to ethnocide and genocide for 500 years, which is why we are endangered, the alternative is for our DNA to be stored and collected ... why don't they address the causes of our being endangered, instead of spending $20 million for five years to collect and store us in cold laboratories?[17]

On 19 February 1995 the indigenous peoples of the Western hemisphere, representing 17 populations located throughout the American continent, issued a Declaration on the HGDP at a meeting in Phoenix, Arizona.[18] Emphasising that the 'principle of harmony requires that we do not violate the principles of Creation by manipulating and changing the natural order', they rejected 'all programs involving genetic technology', particularly the HGDP, and 'the patenting of all natural genetic materials'. Furthermore, they strongly reaffirmed that 'indigenous peoples have the fundamental rights to deny access to, refuse to participate in, or to allow removal or appropriation by external scientific projects of any genetic

---

[15] See the letter written in 1991 by Luca Cavalli-Sforza and other scientists to the scientific journal *Genomics*, emphasising the need for a systematic study of human genetic diversity, quoted in 'The Human Genome Diversity project', n 9 above.

[16] See ibid.

[17] See ibid. (declaration by Victoria Tauli-Corpuz, Cordillera People's Alliance, Philippines).

[18] See *Declaration of Indigenous Peoples of the Western Hemisphere Regarding the Human Genome Diversity Project*, 19 Feb 1995, available at www.indians.org/welker/genome.htm (last visited 30 Sept 2004).

materials'. They thus demanded that 'the Human Genome Diversity Project and any other such scientific project cease any attempts to seduce or coerce participation in their projects through promises of benefits and financial gain in order to obtain consent and participation of indigenous peoples', and asked for 'an immediate moratorium on collections and/or patenting of genetic materials from indigenous persons and communities by any scientific project, health organization, governments, independent agencies, or individual researchers'.[19]

In the same year, a similar position was embraced by a group of Asian indigenous peoples at a meeting in Sabah, Malaysia, which also emphasised, in more general terms, that sampling and patenting of indigenous genetic materials amounts to neo-colonialism, and that indigenous peoples' control over their own genetic heritage is an essential part of their fight for self-determination.[20]

---

[19] The HGDP has also been rejected by other organisations representing various indigenous groups. Eg, the Canada-based World Council of Indigenous People declared that its members 'categorically reject and condemn the Human Genome Diversity Project' (see CJ Hanley, 'Indigenous Peoples Resist Worldwide Gene Study', *Los Angeles Times*, 7 July 1996, A8); similarly, the 1995 Beijing Declaration of Indigenous Women, issued during the United Nations Fourth World Conference on Women, states that '[b]ioprospecting, which is nothing but the alienation of our invaluable intellectual and cultural heritage through scientific collection missions and ethnobotanical research, is another feature of recolonization. After colonizing our lands and appropriating our natural resources, they are now appropriating our human genetic resources, through the Human Genome Diversity Project. Their bid for the patenting of life forms is the ultimate colonization and commodification of everything we hold sacred. It won't matter anymore that we will disappear because we will be 'immortalized' as 'isolates of historic interest' by the Human Genome Diversity Project' (see para 8); the Declaration thus includes the request 'that the Human Genome Diversity Project be condemned and stopped' (see para 41); the full text of the Declaration is available at www.ipcb.org/resolutions/htmls/dec_beijing.html (last visited 30 Sept 2004). See Ching, n 9 above, at 693, n 42.

[20] See Sabah Declaration of the Regional Meeting of Indigenous Peoples' Representatives on the Conservation and Protection of Indigenous Peoples' Knowledge Systems, TVRC Tambunan, Sabah, Malaysia, 24–27 Feb 1995, quoted by Ching, n 9 above, at 721, n 42. In this sense see UN doc E/CN.4/Sub.2/2004/38, n 14 above, para 31; according to the Special Rapporteur, the problem 'is the usurpation of group identity in favour of an individualistic one that may not be reflected in other cultures . . . By overriding the wishes of the group and conferring only with an individual, researchers diminish the authority of the group to make compulsory decisions concerning its members. This violates the draft United Nations declaration on the rights of indigenous peoples by compromising the right to self-determination and cultural independence.' For further statements concerning the position of indigenous peoples with regard to the HGDP see UN Working Group on Indigenous Populations, 'Standard-Setting Activities: Evolution of Standards Concerning the Rights of Indigenous Peoples—Human Genome Diversity Research and Indigenous Peoples—Note by the Secretariat', UN doc E/CN.4/Sub.2/AC.4/1998/4 of 4 June 1998; in particular, para 4 notes that the American Indian Law Alliance 'and other opponents call the HGDP the "Vampire Project", referring to the taking of blood and skin samples from living humans'; also, according to para 12, '[s]ome indigenous peoples see the Project as a new form of colonialism, with sinister overtones'. The document also accurately deals with the legal framework concerning its subject matter (see paras 26–35), and concludes by emphasising that '[i]t is arguable that there is a developing awareness of and sensitivity to the ethical and legal issues surrounding the collection of human genome', and that '[s]ome concerns of

The position of indigenous peoples on the issue in point is clear. At the international level their expectations are defended by Article 10 of the UNESCO Universal Declaration on the Human Genome and Human Rights,[21] which implicitly refers to indigenous peoples by stating that '[n]o research or research applications concerning the human genome, in particular in the fields of biology, genetics and medicine, should prevail over respect for the human rights, fundamental freedoms and human dignity of individuals or, where applicable, of *groups of people*'.[22] This provision clearly implies that bioprospecting of indigenous genetic heritage is not permissible if it may result in the infringement of the dignity, *as self-perceived*, of such peoples. This topic is thus strictly related to the safeguarding of fundamental human rights of the members of indigenous communities, in the sense that bioprospecting and patenting of indigenous human genetic materials (within the limits in which such materials may be considered as patentable[23]) may amount to the actual infringement of such rights.

At first sight it could appear that the situation is rather different with regard to bioprospecting and patenting of non-human indigenous biogenetic resources and related traditional knowledge. It could in fact be simply maintained that such issue only translates into a problem of intellectual property rights (IPRs) over the concerned plant varieties and/or related knowledge. Nevertheless, this perspective changes if one takes into account that knowledge concerning non-human biogenetic resources and practices related to such knowledge often represent an essential element of the *collective* identity of an indigenous group, and that, consequently, bioprospecting and patenting of biogenetic resources traditionally belonging to indigenous peoples, as *well as of traditional knowledge related to them*,[24] may actually result in a serious threat to the integrity of the idiosyncratic identity of the peoples concerned.

Collective rights are no less important than individual rights according to the holistic vision of life of indigenous peoples. Thus, in the same way that such peoples fight strongly against the collection of their human genetic samples for the development of scientific programmes carried

---

indigenous peoples … cannot be adequately addressed without a complete ban on projects such as the HGDP, and the patenting of human genome materials' (see para 39).

[21] The full text of the Declaration is available at www.unesco.org (last visited 3 Oct 2004).

[22] Emphasis added. See, on this point, N Lenoir, 'Universal Declaration on the Human Genome and Human Rights: The First Legal and Ethical Framework at the Global Level' (1999) 30 *Columbia Human Rights Law Review* 537, at 556 ff.

[23] On this problem see sect 4 below.

[24] In this sense it is necessary to emphasise that protection must be granted to *both* biogenetic resources as such (when they are located in indigenous ancestral lands) and traditional knowledge related to such resources. They are in fact inextricably interrelated and represent two different but inseparable elements of the same concept, that is the *special holistic relationship of indigenous peoples with their biogenetic resources*. The simultaneous protection of both

out in a world dominated by the Western vision of life, they also firmly oppose the appropriation and patenting by commercial enterprises of their traditional and other natural elements and related knowledge.

Now, the question is whether and to what extent the international community, in the framework of international law as created and developed in the context of the Western world, supports this fight by indigenous peoples for the preservation by their identity and for preventing their disappearance.

3. INDIGENOUS PEOPLES IN THE INTERNATIONAL LEGAL ARENA AND THEIR RIGHTS TO GOVERN ACCESS TO THEIR BIOGENETIC MATERIAL

In recent years international action for the rights of indigenous peoples has made significant steps that have led to the wide recognition of certain rights in their favour, which are considered essential by indigenous peoples themselves for the preservation of their very cultural identity and anthropological distinctiveness. The most noteworthy conquests in this respect have been made at the level of state practice rather than in the context of international treaty law. Indigenous rights over their ancestral lands have been recognised, at both the judicial and legislative levels, by most countries in which indigenous groups are particularly concentrated, especially hunting and fishing rights[25] and, in more general terms, native title to land. Following the historic decision of the High Court of Australia in the *Mabo II case*,[26] a growing number of domestic judicial bodies has recognised such title in very extensive terms. On 14 October 2003, the Constitutional Court of South Africa declared that the indigenous *Richtersveld* Community, a group of descendants of the *San* (so-called Bushmen, the most ancient people of Africa), 'is entitled ... to restitution of the right to ownership ... (including its minerals and precious stones) and to the exclusive use and occupation' of their traditional land,[27] of which they had been dispossessed in the first half of the twentieth century to allow the exploitation of diamond resources situated therein. The decision was taken pursuant to section 2(1) of the Restitution of Land Rights

these elements is thus essential for granting indigenous peoples the concrete enjoyment of their relationship with nature, essential to the survival of their distinctive identity.

[25] On this topic see F Lenzerini, 'The Interplay between Environmental Protection and Human and Peoples' Rights in International Law' (2002) 10 *African Yearbook of International Law* 63, at 89 ff.

[26] See *Mabo and Others v Queensland (No. 2)* (1992) 175 Commonwealth Law Reports 1.

[27] See *Alexkor Limited and the Government of the Republic of South Africa v The Richtersveld Community and Others*, 14 Oct 2003, available at www.constitutionallaw.co.za/alert/cases/alexkor.pdf (last visited 15 Oct 2004), para 103.

Act of 1994, since the appropriation of the subject land was considered by the Court to be a result of past racially discriminatory laws or practices.

Such state practice has developed in the context of an international legal framework increasingly sympathetic to indigenous peoples' rights, where the 'assimilationist' policy pursued by 1957 ILO Convention number 107[28] has been replaced by the open-minded approach followed by the 1989 Convention number 169,[29] which recognises and defends the right of indigenous peoples to preserve their own identity and to maintain and transmit to future generations their traditional way of life and beliefs, as well as a significant range of rights, including land rights. In such a global context, certain legal principles are emerging, if not already crystallised, in the context of general international law which, by means of the extensive or analogical application of the relevant international provisions or as a result of the development of an unambiguous practice, are clearly supportive of the right of indigenous peoples to preserve and manage their biogenetic resources.

3.1 Binding International Law Provisions

The first binding principle of relevance for the purpose of this chapter is solemnly proclaimed by the so-called 'International Bill of Human Rights'. Articles 1 of both the International Covenant on Economic, Social and Cultural Rights (ICESCR)[30] and the International Covenant on Civil and Political Rights (ICCPR)[31] proclaim the right of all peoples (including indigenous ones) to self-determination, previously expressed also by Articles 1(2) and 55 of the United Nations Charter, which encompasses the right 'freely [to] dispose of their natural health and resources, without prejudice to any obligations arising out of international economic co-operation, *based upon the principle of mutual benefit'.*[32] Although the right of peoples (which, in light of an evolutionary interpretation of the term, include non-national peoples) freely to dispose of their own natural resources is subordinated to the application of international economic treaties, the reference to the principle of mutual benefit implies that the 'obligations arising out of international economic co-operation' contemplated by the provision in point must be implemented so as to guarantee that all the relevant actors involved receive adequate benefits from the

---

[28] The full text of the Convention is available at www.ilo.org/ilolex/english/convdisp1. htm (last visited 30 Sept 2004).
[29] The full text of the Convention is available at www.ilo.org/ilolex/english/convdisp1. htm (last visited 30 Sept 2004).
[30] 993 UNTS 3.
[31] 999 UNTS 171.
[32] Emphasis added.

operation of such co-operation. This implies, inter alia, that indigenous knowledge may not be lawfully exploited without the profits obtained from such exploitation being shared with the autochthonous communities concerned. In addition, the right 'of all peoples to enjoy and utilize fully and freely their natural wealth and resources' is considered by both Article 25 of the ICESCR and Article 47 of the ICCPR as *inherent*, thus implying that it is absolutely non-derogable.

The principle of self-determination is to be considered part of customary international law, at least with regard to its *internal* dimension (while it has developed as a 'right to independence' solely in the context of decolonisation). Whereas the concept of 'internal self-determination' amounts to a limited degree of legal and administrative autonomy (without challenging the territorial integrity of the state in which a people is located) and to the right of any distinct community to participate in taking decisions regarding itself, it certainly includes the right of any people to retain control over its traditional knowledge on biodiversity. This conclusion is particularly evident with regard to indigenous peoples, for at least two reasons: first, the relationship of indigenous peoples with their local environment ('Mother Earth') is particularly intimate and profound, as a result of their holistic vision of life; their environment constitutes a crucial element of their own identity, and the appropriation of their environmental knowledge may amount to jeopardising their very existence as distinct communities. The meaning of the special relationship of indigenous peoples with their lands and natural resources has recently been emphasised by the World Bank:

> [f]or indigenous peoples, secure, effective, collective ownership rights over the lands, territories, and resources they have traditionally owned or otherwise occupied and used *are fundamental to economic and social development, to physical and cultural integrity*, to livelihoods and sustenance. Secure rights to own and control lands, territories, and resources are also essential for the maintenance of the worldviews and spirituality of indigenous peoples—in short, to their very survival as viable territorial communities. *Without secure and enforceable property rights, indigenous peoples' means of subsistence are permanently threatened.* Loss or degradation of land and resources results in deprivation of the basics required to sustain life and to maintain an adequate standard of living. Failure to recognize and respect these rights undermines efforts to alleviate indigenous peoples' poverty and to achieve sustainable development.[33]

Since, as highlighted by the World Bank, 'secure, effective, collective ownership rights over the lands, territories, and resources they have traditionally owned or otherwise occupied and used are fundamental ... to

---

[33] See *Striking a Better Balance*, Final Report of the World Bank Independent Extractive Industries Review, 15 Jan 2004, available at www.eireview.org/html/EIRFinalReport.html (last visited 30 Sept 2004), i; at 40 (emphasis added).

physical and cultural integrity' of indigenous peoples, the lack of adequate protection of these rights could eventually lead to the disappearance (that is to say 'elimination') of such peoples in their distinctive cultural identity. It is certainly pleonastic to explain why cultural 'elimination' of a people is not compatible with the principle of self-determination.

Secondly, indigenous peoples are a preferential target of bioprospecting, both for their knowledge regarding the curative properties of plants and the 'originality' of their genetic heritage. They are thus the object of particularly invasive external pressure which, taking into account their special idiosyncratic vulnerability, may constitute an insuperable obstacle to the realisation of their right to self-determination. As emphasised by the Human Rights Committee in its General Comment on Article 1 of the ICCPR:

> [t]he right of self determination is of particular importance because its realization is an essential condition for the effective guarantee and observance of individual human rights and for the promotion and strengthening of those rights.[34]

This implies, a fortiori, that biopiracy against indigenous peoples, by endangering the realisation of their right to self-determination, constitutes a serious threat to the enjoyment of their individual and collective fundamental human rights, that is to say a potential violation of a number of basic principles of general international law and *jus cogens*.

In addition to the right of self-determination, other collective rights have also developed in the context of international human rights law, which encompass the recognition of a considerable degree of sovereignty of indigenous peoples over their natural (including biogenetic) resources. In particular, Article 27 of the ICCPR states that:

> [i]n those States in which ethnic, religious or linguistic minorities exist, persons belonging to such minorities shall not be denied the right, in community with the other members of their group, to enjoy their own culture, to profess and practise their own religion, or to use their own language.

In its General Comment on this provision, the Human Rights Committee held that:

> culture manifests itself in many forms, including a particular way of life associated with the use of land resources, especially in the case of indigenous peoples. That right may include such traditional activities as fishing or hunting and the right to live in reserves protected by law. The enjoyment of these rights may require positive legal measures to ensure the effective participation of members of minority communities in decisions which affect them … The protection of these rights is directed towards ensuring the survival and continued development of the cultural, religious and social identity of the minorities concerned, thus enriching

---

[34] See *General Comment No 12: The Right to Self-determination of Peoples (Art 1)*, 13 Mar 1984, available at www.ohchr.org/english/bodies/hrc/comments.htm (last visited 30 Sept 2004).

the fabric of society as a whole. Accordingly, the Committee observes that these rights must be protected as such and should not be confused with other personal rights conferred on one and all under the Covenant. *States parties, therefore, have an obligation to ensure that the exercise of these rights is fully protected.*[35]

Also, Article 21 of the African Charter on Human and Peoples' Rights[36] expressly states that '[a]ll peoples shall freely dispose of their wealth and natural resources. This right shall be exercised in the exclusive interest of the people. In no case shall a people be deprived of it'. In the famous case of the Ogoni people, decided in 2001, the African Commission on Human and Peoples' Rights found that the term 'peoples' referred to in Article 21 includes a distinct indigenous people within a state and does not refer just to the national people as a whole.[37]

International case law is unanimously orientated towards an approach according to which both States and private actors must respect indigenous title over their lands and natural resources. For example, in *Ominayak and the Lubicon Lake Band v Canada*,[38] the Human Rights Committee found that state concessions for oil and gas exploitation violated the rights of the Lubicon Lake people provided for by Article 27 of the ICCPR, on account of the fact that they were capable of destroying its traditional hunting and trapping territory and thus its means of subsistence and traditional way of life. In *Awas Tingni Mayagna (Sumo) Indigenous Community v Nicaragua*,[39] the Inter-American Court of Human Rights held that the right to property affirmed by Article 21 of the American Convention on Human Rights[40] encompasses the collective rights of indigenous peoples over their ancestral land and the natural resources they have traditionally used,[41] by virtue of the fact that the 'right to property' has a meaning in international law which transcends its connotation according to domestic law.[42] The Court also held that the indigenous right of ownership of their lands and resources is founded on indigenous customary law rather than on domestic law of the territorial state.[43] These findings have recently

---

[35] See Human Rights Committee General Comment No 23 of 6 Apr 1994, 'The Rights of Minorities (Art. 27)', available at www.ohchr.ch/english/bodies/hrc/comments.htm (last visited 30 Sept 2004), para 7 ff (emphasis added).

[36] (1982) 21 ILM 58.

[37] See *Communication 155/96, The Social and Economic Rights Action Center and the Center for Economic and Social Rights / Nigeria*, OAU doc ACHPR/COMM/A044/1 of 27 May 2002, para 55 ff.

[38] See Communication No 167/1984, 26 Mar 1990, available at http://heiwww.unige.ch/humanrts/undocs/session45/167-1984.htm (last visited 30 Sept 2004).

[39] Judgment of 31 Aug 2001, available at www.indianlaw.org/IACHR_Judgement_Official_English.pdf (last visited on 30 Sept 2004).

[40] OAS Treaty Series n 36.

[41] See para 153.

[42] See para 146.

[43] See para 146 ff.

been reiterated and elaborated upon by the Inter-American Commission on Human Rights in *Mary and Carrie Dann v United States*[44] and *Maya Indigenous Communities of the Toledo District v Belize*.[45]

Finally, even the Committee on the Elimination of Racial Discrimination has urged states parties to the 1966 Convention on the Elimination of All Forms of Racial Discrimination.[46]

> to recognize and protect the rights of indigenous peoples to own, develop, control and use their communal lands, territories and resources and, where they have been deprived of their lands and territories traditionally owned or otherwise inhabited or used without their free and informed consent, to take steps to return those lands and territories. Only when this is for factual reasons not possible, the right to restitution should be substituted by the right to just, fair and prompt compensation. Such compensation should as far as possible take the form of lands and territories.[47]

In the context of such practice, 'natural resources' evidently means any kind of such resources, and no reason exists for asserting that a distinction should be drawn between sub-surface and surface resources (including biogenetic ones).[48]

With regard to international law specifically dealing with indigenous peoples, ILO Convention number 169,[49] after emphasising, in the Preamble, 'the distinctive contributions of indigenous and tribal peoples to the cultural diversity and social and ecological harmony of humankind',[50] contemplates a series of rights which are implicitly valuable in view of the protection of such peoples against invasive bioprospecting. Articles 4 and 5 lay down the obligation to recognise and protect, inter alia, the cultures and environment of the peoples concerned, as well as their social, cultural, religious and spiritual values and practices. Article 13 adds that, in applying the provisions of the Convention, states parties must respect 'the special importance for the cultures and spiritual values of the peoples concerned of their relationship with [their] lands or territories, … and

---

[44] Report of 15 Oct 2001, available at http://heiwww.unige.ch/humanrts/cases/113-01.html (last visited 30 Sept 2004).

[45] Report of 24 Oct 2003, available at www.indianlaw.org/200310PrelimRpt.pdf (last visited 30 Sept 2004).

[46] 660 UNTS 195.

[47] See *General Recommendation XXIII (51) concerning Indigenous Peoples*, 18 Aug 1997, available at www.austlii.edu.au/journals.OLD/AILR/1998/6.html (last visited 30 Sept 2004), para 5.

[48] See, for a similar view, UN Sub-Commission on the Promotion and Protection of Human Rights, 'Indigenous Peoples' Permanent Sovereignty over Natural Resources', Final Report of the Special Rapporteur, Erica-Irene A Daes, Addendum, doc E/CN.4/Sub.2/2004/30/Add.1 of 12 July 2004, para 11.

[49] See n 29 above.

[50] See seventh sentence.

in particular the collective aspects of this relationship'. In any event, the most significant provision for the purpose of this chapter is undoubtedly Article 15, which states that '[t]he rights of the peoples concerned to the natural resources pertaining to their lands shall be specially safeguarded. These rights include the right of these peoples to participate in the use, management and conservation of these resources'. This provision, although granting a simple right of participation instead of a right of ownership as envisaged by the other norms concerning land rights (Part II of the Convention),[51] contemplates a general principle which is *directly* applicable to biogenetic resources. Since this principle concerns the management of natural resources located in the ancestral lands belonging to indigenous peoples, it in fact encompasses the obligation of states to prevent illicit appropriation of such resources. In addition, paragraph (2) of the Article in point provides for a more specific provision, affirming that:

> [i]n cases in which the State retains the ownership of mineral or sub-surface resources or rights to other resources pertaining to lands, governments shall establish or maintain procedures through which they shall consult these peoples, with a view to ascertaining whether and to what degree their interests would be prejudiced, before undertaking or permitting any programmes for the exploration or exploitation of such resources pertaining to their lands. The peoples concerned shall wherever possible participate in the benefits of such activities, and shall receive fair compensation for any damages which they may sustain as a result of such activities.

Although this provision was specifically conceived with regard to mineral and similar sub-surface resources, it may be considered applicable by analogy, in the light of the principle of evolutionary interpretation implied by Article 31 of the Vienna Convention on the Law of Treaties (considered in its entirety), also to biogenetic resources. The fact that these resources are not expressly contemplated by the provision in point is almost certainly due to the fact that, at the time of the drafting of the Convention, the economic potentialities of the possible exploitation of indigenous knowledge concerning nature had not yet been fully realised. If this assumption corresponds to reality, then Article 15(2) of ILO Convention number 169 presupposes that in the case of exploitation of indigenous biogenetic resources and related knowledge, they should *at least* receive fair compensation for such exploitation (since the 'damage' contemplated by the provision is *in re ipsa*).

---

[51] On this point see G Citroni, 'Pueblos indigenas y biotecnologías: aspectos jurídicos', paper presented at the Conference on 'Biotechnology and International Law', Siena, 8–9 Oct 2004 (on file with the author), para 1.2.

In the context of international environmental law, the most pertinent norm with respect to the issue in point is Article 8(j) of the 1992 Convention on Biological Diversity (CBD).[52] This provision, in the context of the *in situ* conservation of biodiversity, establishes an obligation to:

> respect, preserve and maintain knowledge, innovations and practices of indigenous and local communities embodying traditional lifestyles relevant for the conservation and sustainable use of biological diversity and promote their wider application with the approval and involvement of the holders of such knowledge, innovations and practices and encourage the equitable sharing of the benefits arising from the utilization of such knowledge, innovations and practices.

The above provision is weakened by the fact that not only are the duties it lists barely defined but the provision is to be implemented 'as far as possible', 'as appropriate' and 'subject to [the] national legislation' of the state party concerned. However, Article 8(j) CBD has *at least* the merit of recognising the special role of indigenous peoples for the preservation of biodiversity and that the exploitation of indigenous biogenetic resources and/or related traditional knowledge may not be freely carried out, but is conditional on the obligation of ensuring adequate involvement of the communities concerned as well as adequate sharing of benefits.[53]

Finally, the relevance of cultural rights of indigenous communities is worth emphasising. In this perspective, the Convention for the Safeguarding of Intangible Cultural Heritage, adopted by the UNESCO General Conference on 17 October 2003, has special significance. The Convention, after recognising the particularly relevant role of indigenous communities 'in the production, safeguarding, maintenance and re-creation of immaterial cultural heritage',[54] points out that access to such heritage (which includes indigenous traditional knowledge, in particular 'knowledge and practices concerning nature and the universe')[55] must be ensured while 'respecting customary practices governing access to specific aspects of [it]'. Among these practices, those governing access to biogenetic resources traditionally owned by indigenous peoples are certainly to be included. In any event, at least in principle, the most significant provision of this Convention for the purposes of this chapter is undoubtedly Article 15, which calls upon states parties 'to endeavour to ensure the widest possible participation

---

[52] The full text of the Convention is available at www.biodiv.org/doc/legal/cbd-en.pdf (last visited 30 Sept 2004).

[53] In my opinion the aptitude of benefit sharing adequately satisfying indigenous needs is somewhat disputable. On this point see section 4 below.

[54] See Preamble, sixth sentence.

[55] See the definition of 'intangible cultural heritage' provided for by Art 2, particularly para (2)(d).

of communities … that create, maintain and transmit' intangible cultural heritage within the framework of national safeguarding activities of such heritage, and to involve those communities 'actively in its management'. In principle, the relevance of this provision would consist in the fact that, by requiring states to involve local communities in the management of all activities aimed at safeguarding their traditional intangible heritage (including knowledge on biogenetic resources), it would preclude states parties from acting in such a way as to deprive—or allow private operators to deprive—them of such heritage. The use of the conditional is necessary because the possible significance of the provision in point for the purposes of this work is made void by Article 3(b)'s express reservation to the effect that nothing in the Convention may be interpreted as 'affecting the rights and obligations of States parties deriving from any international instruments relating to intellectual property rights or to the use of biological and ecological resources of which they are parties'.

3.2 Soft Law

Besides the binding instruments referred to in the previous paragraph, a myriad of *soft law* international provisions directly or indirectly recognise the 'sovereign' right of indigenous peoples over their natural resources and related knowledge.

In the context of the protection of cultural heritage, Annex II to the 2001 UNESCO Universal Declaration on Cultural Diversity[56] considers respect and protection for 'traditional knowledge, in particular that of indigenous peoples [and recognition of] the contribution of traditional knowledge, particularly to environmental protection and the management of natural resources',[57] as one of the objectives to be achieved for ensuring the application of the Declaration.

As regards international environmental law, the fundamental contribution of indigenous communities in preserving the integrity of the global environment has been recognised since 1992, when Principle 22 of the 1992 Rio Declaration on Environment and Development[58] emphasised the vital role of such communities 'in environmental management and development *because of their knowledge and traditional practices*',[59] also stressing that 'states should recognize and duly support their identity, culture and interests and enable their effective participation in the achievement of sustainable

---

[56] The full text of the Declaration (annexes included) is available at www.unesco.org (last visited 3 Oct 2004).
[57] See para 14.
[58] See UN doc A/CONF.151/26 (Vol. I) of 12 Aug 1992, Annex I.
[59] Emphasis added.

development'. In addition, Chapter 26 of Agenda 21,[60] also adopted by the 1992 Rio Conference on Environment and Development (UNCED), recognised the special relationship of indigenous peoples with their ancestral lands, characterised by the development 'over many generations [of] a holistic traditional scientific knowledge of their lands, natural resources and environment'.[61] Among the measures to be adopted for accommodating, promoting and strengthening the role of such peoples in implementing 'environmentally sound and sustainable development'[62], States are called, upon inter alia, to recognise 'that the lands of indigenous people and their communities should be protected from activities that are environmentally unsound or that the indigenous people concerned consider to be socially and culturally inappropriate'[63] and to '[a]dopt or strengthen appropriate policies and/or legal instruments that will protect indigenous intellectual and cultural property and the right to preserve customary and administrative systems and practices'.[64]

These commitments were reiterated by the World Summit on Sustainable Development held in Johannesburg in 2002. Paragraph 25 of the Johannesburg Declaration on Sustainable Development,[65] adopted on 4 September of that year, reaffirms 'the vital role of the indigenous peoples in sustainable development'. The means of safeguarding and fostering such role are highlighted in the Plan of Implementation of the Summit.[66] In particular, the Plan emphasises the need to recognise, subject to national legislation:

> the rights of local and indigenous communities who are holders of traditional knowledge, innovations and practices, and, with the approval and involvement of the holders of such knowledge, innovations and practices, develop and implement benefit-sharing mechanisms on mutually agreed terms for the use of such knowledge, innovations and practices.[67]

In addition, the Plan stresses the necessity of promoting 'the effective participation of indigenous and local communities in decision and policy-making concerning the use of their traditional knowledge',[68] and 'the preservation, development and use of effective traditional medicine

---

[60] The full text of Agenda 21 is available at www.unep.org/Documents/Default. asp?DocumentID=52 (last visited 3 Oct 2004).

[61] See para 26.1.

[62] Ibid.

[63] See para 26.3(a)(ii).

[64] See para 26.4(b).

[65] See www.joburg.org.za/clean_city/johannesburgdeclaration.pdf (last visited 3 Oct 2004).

[66] The full text of the Plan of Implementation is available at www.un.org/esa/sustdev/documents/WSSD_POI_PD/English/WSSD_PlanImpl.pdf (last visited 3 Oct 2004).

[67] See para 42(j).

[68] See para 42(l).

knowledge and practices, where appropriate, in combination with modern medicine, recognizing indigenous and local communities as custodians of traditional knowledge and practices, while promoting effective protection of traditional knowledge, as appropriate, consistent with international law'.[69] Finally, the opportunity of increasing, consistently with national law, 'the use of scientific knowledge and technology, and increase[ing] the beneficial use of local and indigenous knowledge in a manner respectful of the holders of that knowledge' is also stressed.[70]

In the framework of the United Nations the 'indigenous question' is, at the time of writing, the object of an intense debate. 2004 was the final year of the International Decade of the World's Indigenous Peoples, proclaimed by the General Assembly in 1993 and officially launched on 10 December 1994.[71] One of the unaccomplished aims that should had been achieved by the end of the Decade was the adoption of the 1994 *Draft United Nations Declaration on the Rights of Indigenous Peoples*[72] which, however, was adopted by the new Human Rights Council on 29 June 2006, during the second Decade of the worlds Indigenous Peoples, launched at the end of 2004.[73] Even within the limits of a soft law instrument, the Declaration will represent (after its final adoption lay the UN General Assembly) a major step in the affirmation of indigenous peoples' rights in the framework of international law. Its text entails a comprehensive catalogue of rights which fully encompass, inter alia, the idea of indigenous sovereignty over biogenetic resources (including knowledge related to them). In particular, having proclaimed the right of such peoples to self-determination,[74] it declares the collective and individual right of indigeuous people not to be 'subjected to forced assimilation or destruction of their culture', which includes the prevention of and redress for any action having 'the aim or effect of depriving them of their integrity as distinct peoples, or of their cultural values or ethnic identities [or] … of dispossessing them of their lands, territories or resources'.[75] Article 11 proclaims the right 'to practise and revitalize their cultural traditions and customs', which may include the right to 'restitution of cultural, intellectual, religious and spiritual property taken without their free and informed consent or in violation of their laws, traditions and customs'. More specifically, Article 24 sets out the right 'to their traditional medicines and to maintain their health practices, including the right to the *conservation of their vital medicinal plants*,

---

[69] See para 47(h).
[70] See para 103(a).
[71] See Res 48/163 of 21 Dec 1993.
[72] See UN doc A/HRC/1/L 10 of 30 June 2006, 58 ff.
[73] The second International Decade of the World's Indigenous People, starting from 1 Jan 2005, has been proclaimed by the UN GA by Res 59/174 of 20 Dec 2004.
[74] See Art 3.
[75] See Art 8.

animals and minerals'.[76] The rights of indigenous peoples, inter alia, 'to maintain and strengthen their distinctive spiritual relationship with their traditionally or otherwise occupied and used lands, territories, waters and coastal seas and other resources they have traditionally owned or otherwise occupied or used, and to uphold their responsibilities to future generations in this regard' (Article 25), 'to own, use, develop and control [their] lands territories and resources' (encompassing the right to 'legal recognition and protection to these lands, territories and resources ... conducted with due respect to the customs, traditions and land tequre systems of the indigenous peoples concerned[77]), and 'to determine and develop priorities and strategies for the development or use of their lands, territories and other resources' (article 32) are also proclaimed. However, the most significant provision for the purposes of the present chapter is probably Article 31 par. 1, according to which '[i]ndigenous peoples have the right to maintain, control, protect and develop their cultural heritage, traditional knowledge and traditional cultural expressions, as well as the manifestations of their sciences, technologies and cultures, including human and genetic resources, seeds, medicines, knowledge of the properties of fauna and flora, oral traditions, literatures, designs, sports and traditional games and visual and performing arts. They also have the right to maintain, control, protect and develop their intellectual property over such cultural heritage, traditional knowledge, and traditional cultural expressions'. In sum, a number of rights enshrined lay the Declaration would be infringed in the event that biogenetic resources and/or related traditional knowledge, which are usually part of the very cultural identity of indigenous peoples, were appropriated and patented by non-indigenous actors.

In 2004, the Special Rapporteur of the UN Sub-Commission on the Promotion and Protection on Human Rights, Erica-Irene A Daes, completed her final report on 'Indigenous peoples' permanent sovereignty over natural resources'.[78] The Report emphasises, inter alia, '[t]he unfairness and adverse impacts of the misappropriation of indigenous peoples' genetic and other biological resources, sometimes termed "biopiracy"',[79] and the fact that:

> [t]he inadequacy and unfairness in present legal regimes regarding bio-prospecting, patents, and other intellectual property laws have deprived indig-enous peoples of valuable economic resources and have resulted in damage to indigenous cultures as well.[80]

---

[76] Emphasis added.

[77] See Art 26.

[78] See UN docs E/CN.4/Sub.2/2004/30 of 13 July 2004 and E/CN.4/Sub.2/2004/30/Add.1 of 12 July 2004 (see n 48 above and corresponding text).

[79] See E/CN.4/Sub.2/2004/30, para 36.

[80] Ibid, para 36.

Sovereignty of indigenous peoples over their traditional lands and resources is seen as 'permanent' because 'it is intended to refer to an inalienable human right of indigenous peoples ... [which] arises out of the right of self-determination, the right to own property, the right to exist as a people, and the right to be free from discrimination, among other rights, all of which are inalienable'.[81]

At the regional level, the question of indigenous biogenetic resources is specifically addressed by the Proposed American Declaration on the Rights of Indigenous Peoples, adopted in 1997 by the Inter-American Commission on Human Rights.[82] Article XX, entitled 'Intellectual Property Rights', reads as follows:

> 1. Indigenous peoples have the right to the recognition and the full owner-ship, control and protection of their cultural, artistic, spiritual, technological and scientific heritage, and legal protection for their intellectual property through trademarks, patents, copyright and other such procedures as established under domestic law; as well as to special measures to ensure them legal status and institutional capacity to develop, use, share, market and bequeath that heritage to future generations. 2. Indigenous peoples have the right to control, develop and protect their sciences and technologies, including their human and genetic resources in general, seed, medicine, knowledge of plant and animal life, original designs and procedure. 3. The states shall take appropriate measures to ensure participation of the indigenous peoples in the determination of the conditions for the utilization, both public and private, of the rights listed in the previous paragraphs 1 and 2.

3.3 International Practice

The inference that the rise of international legal instruments (both of bind-ing and soft-law character), supporting the fight of indigenous peoples against misappropriation of their biogenetic resources and related knowl-edge, is gradually generating a general conscience sharing the idea of the need to prevent and repress such phenomenon (which could soon crystal-lise into a principle of general international law), is supported by consis-tent practice that has evolved in recent times. In particular, the number of patents revoked after it was ascertained that the 'inventions' patented consisted in nothing but the appropriation of pre-existing indigenous traditional knowledge is indeed increasing.

The case of the Neem Tree, a tropical evergreen native to South Asia, represents a landmark example of such practice.[83] Various components

---

[81] Ibid, para 47.

[82] The full text of the Proposed Declaration is available at www.cidh.org/Indigenous.htm (last visited 3 Oct 2004).

[83] See Case Western Reserve University, 'Neem Seed (India)', available at http://home. cwru.edu/~ijd3/autorship/neem.html (last visited 30 Sept 2004).

of this plant (eg, bark, leaves and seeds) have in India been traditionally used for centuries for their anti-bacterial qualities in a number of ways, particularly as a pesticide, but also for the production of soap, toothpaste and spermicides. In 1994 the US company WR Grace, which had previously 'discovered' the pesticidal properties of the Neem seed, obtained a patent from the European Patent Office on account of the claimed innovative character of its process for extracting the seed emulsion used as pesticide, considered as an improvement on the traditional means of extracting such emulsion which had been used in India for centuries. After the patent had been released, WR Grace started suing Indian companies that produced the emulsion, depriving a large number of Indian farmers of their primary means of subsistence and leading to a dramatic increase in the price of the pesticide, which consequently became unsustainable for most Indians. This course of action triggered the violent reactions of Indian farmers as well as of local and international NGOs. Finally, in 2000, after a manager of an Indian agricultural company had proved that he (like many others Indians) had been using the patented extraction process of the Neem seed emulsion several years before the claim was filed, the European Patent Office revoked the patent.[84]

A similar case is that of *Cupuaçu*, a tree located in the Amazon rainforest which belongs to the Cocoa family, the fruit of which has been used for centuries by local indigenous communities as a primary food source as well as for treating abdominal pains and, after being blessed by a shaman, for facilitating difficult births. Since the *Cupuaçu* tree is similar to the cocoa tree, its main economic potential is embodied in its possible use for the production of chocolate-like foodstuffs. A series of patents were granted for the production of *Cupuaçu* chocolate, almost all of them registered by the Japanese ASAHI Foods Company.[85] For years a number of NGOs have conducted a campaign for the cancellation of such patents based on the fact that the Japanese company was exploiting traditional knowledge owned by some Amazonian indigenous communities. Finally, on 1 March 2004, the Japanese Patent Office, persuaded by the arguments put forward by such NGOs, decided to cancel the *Cupuaçu* trade mark.[86]

A different category of cases is represented by those disputes which have been settled by means of agreements reached between the patenting companies and the traditional owners of the patented knowledge. These cases may also be considered as representing, at least to a limited extent, victories obtained by the communities concerned against those who try unfairly to take their resources and knowledge, since the companies

---

[84] See K Hoggan, 'Neem Tree Patent Revoked', BBC News, 11 May 2000, available at http://news.bbc.co.uk/1/hi/sci/tech/745028.stm (last visited 30 Sept 2004).

[85] See 'The Cupuaçu Case', available at www.amazonlink.org/biopiracy/cupuacu.htm (last visited 30 Sept 2004).

[86] See 'Trademark Cupuaçu cancelled in Japan', available at www.amazonlink.org/biopiracy/2004_03_01.htm (last visited 30 Sept 2004)

involved have been bound actually to surrender all or part of their claims over such knowledge. By way of example, one may cite the case of the Hoodia,[87] a cactus native to the Kalahari desert which has been used for centuries by the local San speaking tribes (commonly and inappropriately referred to by the term 'Bushmen') to stave off hunger and thirst during extensive hunting expeditions in the desert.[88] In 1997, having extracted from the cactus the molecule which curbs appetite, called 'P57', the South African Council for Scientific and Industrial Research (CSIR) obtained a patent and sold the licensing rights to an English pharmaceutical company, Phytopharm, which then sold such rights to the American firm Pfizer for $25 million. This sale was the object of intense media coverage, and only at that point did the San become aware of what was happening, making strong protests. The Chief Executive Officer of Phytopharm defended the company by stating that CISR had led him to believe that the tribes which used the Hoodia cactus were extinct. In 2001 the San tribes threatened the CISR with litigation; to avoid scrutiny by the media and the international community CISR agreed to enter into negotiations with the San, and on 9 April 2002 the conclusion of a Memorandum of Understanding was announced, in which the right of the tribes concerned to share the benefits obtained by the commercial exploitation of the Hoodia cactus was recognised.[89]

---

[87] See, on this case, Case Western Reserve University, 'Case Study: Hoodia Cactus (South Africa)', available at http://home.cwru.edu/~ijd3/autorship/hoodia.html (last visited 30 Sept 2004); J Limson, 'Focus on biopiracy in Africa', in *Science in Africa*, Sept 2002, available at www.scienceinafrica.co.za/2002/september/biopiracy.htm (last visited 30 Sept 2004).

[88] On the traditional San hunting technique see F Lenzerini, 'Intangible Cultural Heritage in Danger: A Part of the Human Memory that Is Disappearing', in *Symposium: The Transmission and Present State of Cultural Heritage*, Kyoto, 2002, at 72 ff.

[89] Although not concerning biogenetic resources, the famous American case of *Crazy Horse* is also worth mentioning as an example of illegitimate appropriation of the spiritual heritage of indigenous peoples (see Case Western Reserve University, 'Case Study: Crazy Horse', available at http://home.cwru.edu/~ijd3/autorship/crazyhorse.html – (last visited 30 Sept 2004)). Crazy Horse was a famous leader of the Lakota Sioux people, known as a spiritual warrior and one of the bravest defenders of his people, who lived from 1849 to 1877 and was, inter alia, one of the main protagonists of the landmark victory of Natives in Little Bighorn in 1876. Having said this, it is clearly unnecessary to explain why his memory represents a particularly important icon for Native Americans. On 17 Mar 1992 Hornell Brewing Company introduced The Original Crazy Horse Malt Liquor into the American market, distributed nationwide by another firm, using images and symbols which are sacred to Native peoples. Following the complaints of the Oglala people of Pine Ridge and other Native Tribes, in Oct 1992 President Bush signed a law banning the sale in the national territory of any alcoholic beverage bearing the Crazy Horse label (See Pub. L No 102-393, para 633, 106 Stat 1729). Nevertheless, in Apr 1993 a New York federal court overturned the statute, finding that it infringed Hornell's right to commercial free speech and that it was thus contrary to the First Amendment of the Constitution, since '[t]he Crazy Horse Malt Liquor label is indisputably commercial speech' (See *Hornell Brewing Co. v Brady*, 819 F Supp 1227, 1233 (EDNY 1993)). The controversy continued for several years before the Rosebud Tribal jurisdiction (District of South Dakota) until, on 25 Apr 2001, an agreement was concluded between Strohs Brewing Company (which in 1996 had taken over the firm G Heileman Co, the bottler for Hornell Brewing Company, which had filed for bankruptcy relief) and the

Finally, in some instances claims concerning patents over indigenous biogenetic resources and/or traditional knowledge have been voluntarily surrendered by those who originally applied for the patents, as a result of the strong international pressure they were subjected to. In addition to the *Hagahai* case, described above,[90] one may cite that of the Guaymi Indians from Panama concerning human genetic resources. In 1993 a researcher from the NIH extracted a blood sample from a Guaymi woman, leading to the discovery that the blood of this tribe contained a gene which conferred natural resistance against leukaemia. The US Department of Commerce filed an application for both national and international patents on the woman's cell-line; the application was casually found by a NGO (Rural Advancement Foundation International—RAFI) while one of its employees was exploring a database of patent applications. RAFI immediately contacted the Guaymi people, discovering that neither the tribe nor the woman concerned knew anything about the research developed from her blood sample and the patent application. This triggered a strong reaction by the representatives of the Guaymi, who lamented the invasion of their genetic privacy and violation of their integrity of life and of their deepest sense of morality. As a result of such protests, the patent application was later withdrawn.[91]

The case of the Mayan communities of Chiapas is also of significance. Such communities, represented by the *Consejo Estatal de Parteras y Médicos Indígenas Tradicionales de Chiapas* (*Consejo de Chiapas*), have for years strongly opposed the Drug discovery and biodiversity among the Maya of Mexico project, begun in 1988 by the US Government-financed International Co-operative Biodiversity Groups (ICBG) with the purpose of using indigenous knowledge for discovering, isolating and evaluating pharmacological compounds from plants and micro-organisms traditionally used in Mayan medicine.[92] The project was opposed by the *Consejo de Chiapas* as a typical example of biopiracy, since they considered it theft of the traditional knowledge developed and collectively owned by them for centuries. ICBG claimed, on its part, that the project had huge educational purposes, that the knowledge developed as a result of its research was of an extraordinary character (that is to say *beyond* or *outside* the ordinary knowledge of local communities) and that, according to Mexican law, the natural resources of Chiapas, as well as of the entire Mexican territory, were the exclusive property of Mexico, thus excluding the validity of any claim made by the Mayan or other indigenous

Estate of Crazy Horse and the Rosebud Sioux Tribe (which defended the name of Crazy Horse). The agreement included, inter alia, a public apology for the misuse of the name 'Crazy Horse', assignment to the Estate of all intellectual property rights in the name of Crazy Horse, and agreement to never use such name in any commercial project.

[90] See nn 10, 11 and 12 above and corresponding text.
[91] See Wu, n 10 above, at 985 ff; Ching, n 9 above, at 700.
[92] See *Primal Seeds News*, 1 Dec 1999, available at www.primalseeds.org/news1dec.htm (last visited 30 Sept 2004).

communities.[93] In any event, at the beginning of 2002 the project was terminated due to the widespread opposition it had caused.[94]

4. THE TENSION BETWEEN DIFFERENT INTERNATIONAL LEGAL REGIMES: POSSIBLE SOLUTIONS

Sometimes the 'decentralised', even 'anarchic', character of international law leads to paradoxical situations characterised by the simultaneous existence of provisions which protect a given value and norms safeguarding other values incompatible with the former. With regard to the object of this work, whereas, as seen in the previous sections, international law provides for a comprehensive *corpus* of norms (also supported by significant international practice) which are potentially enforceable for the safeguarding of indigenous peoples against misappropriation of their biogenetic resources and related knowledge, at the same time it also offers formidable instruments for supporting the activities of private investors in the bio-genetic sector by protecting their taking of such resources and/or knowledge through existing treaties on patentability and IPRs, particularly the TRIPS Agreement.[95] Article 27 of the TRIPS Agreement, after having established that 'patents shall be available for any inventions, whether products or processes, in all fields of technology, provided that they are new, involve an inventive step and are capable of industrial application',[96] provides for a particular regime of protection of IPRs with regard to plant varieties. According to paragraph (3)(b) plants may be excluded from patentability, but the protection of plant varieties is to be regulated 'either by patents or by an effective *sui generis* system or by any combination thereof'. In requiring states parties to provide some kind of means for the protection of IPRs on plants, Article 27 clears the way to allow the rights of indigenous communities relating to their biogenetic resources and related traditional knowledge to be usurped.

In this sense, Article 27(3)(b) of the TRIPS Agreement is manifestly at odds with Article 8(j) CBD. One should *sic et sempliciter* resolve such incompatibility by applying the principle of *lex posterior derogat priori* as set out in Article 30 of the Vienna Convention on the Law of Treaties,[97]

---

[93] See EA Berlin and B Berlin, 'Knowledge? Whose Pro perty? Whose Benefits? The Case of OMIECH, RAFI and the Maya ICBG', 13 Dec 1999, available at http://guallart.dac.uga.edu/ICBGreply.html (last visited 30 Sept 2004).

[94] See 'Chiapas Update February 2002', available at www.indymedia.org.uk/en/2002/02/22180.html (last visited 30 Sept 2004).

[95] The full text of the TRIPS Agreement is available at www.wto.org/english/docs_e/legal_e/27-trips.pdf (last visited 3 Oct 2004).

[96] See para 1.

[97] Art 30(3) of the 1969 Vienna Convention on the Law of Treaties (1155 UNTS 331) states that '[w]hen all the parties to the earlier treaty are parties also to the later treaty but the earlier treaty is not terminated or suspended in operation under article 59, the earlier treaty applies only to the extent that its provisions are compatible with those of the latter treaty'.

concluding that the TRIPS Agreement prevails over the CBD by virtue of the 'later in time' rule. This conclusion would be also reinforced by the compatibility clause included in Article 22 of the CBD, which, in making reference to 'any *existing* international agreement',[98] excludes *tout court* that the Convention may be invoked in order to affect the parties' rights and obligations deriving from successive treaties, even in the event that, as provided for in existing agreements, 'the exercise of those rights and obligations would cause a serious damage or threat to biological diversity'. Nevertheless, this is a conclusion that cannot be agreed with on account of *at least* two reasons: (a) it does not take into account the fact that bioprospecting of indigenous biogenetic resources (when it actually amounts to biopiracy) is usually capable of leading to the infringement of basic human rights of the indigenous communities concerned (at both the collective and individual levels), causing the infringement of norms which, being included in the scope of Article 103 of the Charter of the United Nations, supersede states' obligations existing under 'any other international agreement';[99] b) it does not correspond to the official position shared by virtually all states which, in the context of both the CBD and WTO, agree that there is a need to harmonise the two provisions in point (when they do not explicitly assert that the CBD should prevail over the TRIPS Agreement).[100]

Article 27(3)(b) of the TRIPS Agreement itself, demonstrating its provisional character, expressly provides that its content 'shall be

---

[98] Emphasis added.

[99] On this point see section 5 below.

[100] The vision that the CBD should prevail over the TRIPS Agreement has been supported, eg, by the 'Environmental for Europe' Ministerial Conference held in Aarhus on 23–25 June 1998 (which included more than 50 European, North American and Central Asian ministers of environment who declared that environmental agreements should prevail over other international agreements, particularly those concerning trade), by the EU–ACP Joint Assembly held in Brussels on 21–24 Sept 1998, by the European Parliament in Oct 1998 (see Resolution A4-0347/98 on the European Community Biodiversity Strategy (COM(98)0042)) and by the Southern and Eastern African Trade Information and Negotiations Initiative, held in Kampala on 4–9 Mar 1999; see GRAIN, 'TRIPS versus Biodiversity: Options for the 1999 Review of Article 27.3(b) in the context of CBD', Annex 1, 'Political appeals for the Primacy of CBD over TRIPS', available at www.acts.or.ke/prog/biodiversity/trips/grain. doc (last visited 18 Aug 2006). As for the WTO framework and the different positions expressed by states parties with regard to the issue at stake, see 'TRIPS: Reviews, Article 27.3(b) and Related Issues. Background and the Current Situation', available at www.wto. org/english/tratop_e/trips_e/art27_3b_background_e.htm (last visited 3 Oct 2004), and the documents listed therein, particularly docs. IP/C/W/368 of 8 Aug 2002 ('The Relationship between the TRIPS Agreement and the Convention of Biological Diversity. Summary of Issues Raised and Points Made'), IP/C/W/369 of 8 Aug 2002 ('Review of the Provisions of Article 27.3(b). Summary of Issues Raised and Points Made'), and IP/C/W/370 of 8 Aug 2002 ('The Protection of Traditional Knowledge and Folklore. Summary of Issues Raised and Points Made'); see also, on the points raised by parties, docs. IP/C/W/257 (United States), IP/C/W/383 (EU), IP/C/W/400/Rev.1 (Switzerland), IP/C/W/403 (Bolivia, Brazil, Cuba, Ecuador, India, Peru, Thailand, and Venezuela) and IP/C/W/404 (The African Group).

reviewed four years after the date of entry into force of the WTO Agreement'. Since the Agreement entered into force in 1995, the revision of the provision in point was due to take place in 1999. At the time of writing debate on this revision is still taking place, and a timely solution of the question is unlikely, since states are still profoundly divided on the subject. The problem of traditional knowledge is central to the entire debate, as demonstrated by the fact that Article 19 of the Doha Ministerial Declaration[101] instructs the Council for TRIPS, 'in pursuing its work programme [regarding] ... the review of Article 27(3)(b) ... to examine, *inter alia*, the relationship between the TRIPS Agreement and the Convention on Biological Diversity [and] the *protection of traditional knowledge* and folklore'.[102]

The core problem to be resolved in order to ensure adequate protection of indigenous rights against illicit bioprospecting of their biogenetic resources is represented by the need to reconcile the operation of IPRs (as protected by Article 27(3)(b) of TRIPS) with the exigency of preserving indigenous title over such resources and related knowledge.

In attempting to outline the possible solutions to this problem, a clear distinction is to be drawn between human genetic materials and other natural materials and related traditional knowledge. As regards the former, the principle of non-patentability of human genetic material, and particularly of the human genome (at least in its 'natural state'), may be considered accepted by the international community as a whole, as demonstrated by the recognition of the human genome as the 'heritage of humankind'[103] and by relevant international practice both at the public and at the private level.[104] Nevertheless, a tendency to draw a distinction between the elements of the human body as such (which are not patentable at all) and the results of inventive, scientific or technical work associated with human biological data is emerging in various contexts, especially at the European level.[105] According to the European Court of Justice, the patentability of such kind of work implies that 'biological data existing in their natural state in human beings ... where necessary for the achievement and exploitation of a particular industrial application', are also to be considered patentable.[106]

---

[101] The text of the Declaration is available at www.wto.org/english/thewto_e/minist_e/min01_e/mindecl_e.pdf (last visited 3 Oct 2004).

[102] Emphasis added.

[103] See, in particular, Art 1 of the Universal Declaration on the Human Genome and Human Rights, n 21 above.

[104] On this topic see F Lenzerini, 'Biotechnology, Human Dignity and the Human Genome' in F Francioni and T Scovazzi (eds), *Biotechnology and International Law* (Oxford, Hart Publishing, 2006), 285 at 302 ff.

[105] See ibid, 289 ff.

[106] See Case C–377/98 *Netherlands v European Parliament and Council* [2001] ECR I–7079., para 75.

This means that anyone could obtain a patent concerning a particular process by means of which, for example (as in the case of the Guaymi Indians), a gene having special curative properties could be obtained by way of an indigenous blood sample, and such a patent could lawfully include the relevant cell-line, since it would actually be necessary for the exploitation of the technical process concerned. In any case, according to relevant international and domestic law, bioprospecting on human genetic materials is subject to the respect of particularly strict rules, especially the principle of prior, free and truly informed consent,[107] which prevent researchers from taking such materials without the approval of the person(s) concerned. The proper application of such rules (which, with regard to indigenous peoples, imply the necessity of obtaining consent not only from the individual concerned, but *also from the group of which such person is a member*), besides being a prerequisite for the safeguarding of fundamental individual rights of the members of indigenous communities, is capable of preventing, a priori, the unlawful appropriation of indigenous human genetic materials which, as previously emphasised,[108] represent a basic feature of indigenous individual and collective identity. In this case, the operation of IPRs is thus conditioned on the respect for human rights of the individuals (and groups) concerned, to the effect that the protection of such rights may inescapably preclude the application of the rules on IPRs.

Concerning non-human natural materials and related knowledge, we start out from the assumption that, as stated by the UN Sub-Commission on Prevention of Discrimination and Protection of Minorities in its resolution 1992/35:

> there is a relationship, in the laws or philosophies of indigenous peoples, between cultural property and intellectual property, and that the protection of both is essential to the indigenous peoples' cultural and economic survival and development.

This is to say that the protection of indigenous 'intellectual property' over their biological resources and related traditional knowledge is essential for preserving their cultural integrity and, a fortiori, their very identity as peoples. I therefore think that it is essential that such property is *effectively* safeguarded by international law. A model of protection is effective when it actually and concretely ensures the preservation and maintenance of the value which is the object of the protection itself, irrespective of the modalities used for fulfilling such task (on the condition that, of course, such modalities are lawful). Thus, protection is not effective when it

---

[107] On this point see Lenzerini, n 104 above, 306 ff.
[108] See section 2 above.

is managed by recourse to safeguarding models which, despite being appropriate for the prevailing cultural archetype, do not work when applied to social contexts characterised by a drastically different cultural background. As a result the above safeguarding models require structural and methodological changes in order to grant the same degree of protection they usually bestow in their 'natural environment'. In other words, the very concept of 'intellectual property' may be extraneous as such to the indigenous vision of life, and needs thus to be re-adapted to their holistic conception of the world.

Having said this, in my opinion four models of protection exist which could *in abstracto* be used for granting protection to indigenous intellectual property over their biogenetic resources and related traditional knowledge: (a) use of existing mechanisms of intellectual property protection; (b) benefit sharing; (c) non-recognition of IPRs on indigenous biogenetic resources and related traditional knowledge; (d) use of *sui generis* systems of protection of intellectual property.

4.1 Use of Existing Mechanisms of Intellectual Property Protection

The first option which could be used for safeguarding indigenous biogenetic resources and related traditional knowledge lies in the utilisation of the existing models of intellectual property protection, particularly of existing schemes for patenting inventions. With regard to this option, as already emphasised in a very efficacious fashion, the 'existing forms of legal protection of cultural and intellectual property ... are not only inadequate for the protection of indigenous peoples' heritage, but [are also] *inherently unsuitable*'.[109] The very notion of patent is extraneous to the culture of most indigenous peoples, being based on concepts like 'commercial exploitability' and 'economic potential', which are not compatible with their holistic vision of life. Not to mention the financial and technical barriers which should be overcome in order to make a patent application by an indigenous person concretely practicable.[110] In

---

[109] See UN Sub-Commission on Prevention of Discrimination and Protection of Minorities, 'Study on the Protection of the Cultural and Intellectual Property of Indigenous Peoples by Erica-Irene Daes, Special Rapporteur of the Sub-Commission on Prevention of Discrimination and Protection of Minorities and Chairperson of the Working Group on Indigenous Populations', UN doc. E/CN.4/Sub.2/1993/28 of 28 July 1993, para 32.

[110] On this point see M Hanning, 'An Examination of the Possibility to Secure Intellectual Property Rights for Plant Genetic Resources Developed by Indigenous Peoples of the Nafta States: Domestic Legislation Under the International Convention for Protection of New Plant Varieties' (1996) 13 *Arizona Journal of International and Comparative Law* 175, at 202.

addition, even leaving these arguments aside, there are some practical obstacles which make recourse to existing schemes of patent protection absolutely inappropriate for the protection of indigenous biogenetic resources and related traditional knowledge. First, patents are only available to new inventions which 'involve an innovative step';[111] to refer to *traditional* knowledge as new would simply be a contradiction in terms. Secondly, patents are generally granted to individuals (either physical or legal), while indigenous biogenetic resources and related traditional knowledge, although such knowledge is usually kept by a few charismatic individuals, represent a common heritage of the community, and are thus of a collective character.[112] Finally, patents grant over the invention exclusive rights which are of limited duration, and after the expiry of such limited period it comes into the public domain, while indigenous traditional knowledge is transmitted from generation to generation and it should thus be protected without being subject to temporal restrictions.

4.2 Benefit-sharing

The second option could be that of ensuring in favour of the communities that have taken care of biogenetic resources or developed the related traditional knowledge a share in the financial income deriving from the commercial or industrial exploitation of such resources or knowledge carried out by people or enterprises external to the communities concerned. Although this option, which is the one privileged by Decision VI/24 of the Conference of the Parties to the CBD adopting the Bonn Guidelines on Access to Genetic Resources and Fair and Equitable Sharing of Benefits Arising out of their Utilization,[113] could actually be satisfactory in certain circumstances, in general terms there are two major objections which can make it inadequate, the first being of cultural in character, the second of a practical nature. The cultural obstacle is again to be ascribed to the idiosyncratic indigenous vision of life, in the context of which biological resources and related traditional knowledge are often part of the very cultural identity of a community. While in Western culture there is a strong tendency to consider that everything may be bought with money, indigenous peoples usually follow the opposite approach, considering the intrinsic and intimate features of the individual and the community as absolutely inalienable,

---

[111] See Art 27(1) TRIPS Agreement.

[112] On this point see, inter alia, Tejera, n 10 above, at 974.

[113] The text of the decision is available at www.biodiv.org/decisions (last visited 3 Oct 2004).

and are not willing to accept that something which belongs to their quintessence may be sold for money or other material commodities.

In addition, some notable obstacles of a practical character would also make the option of benefit-sharing hardly practicable with regard to indigenous biogenetic resources and related traditional knowledge. First, as pointed out by a scholar,[114] the concrete operation of benefit sharing would require the informed consent of the communities concerned prior to accessing the resources and/or the related knowledge; this would mean, in practical terms, that members of such communities should be fully aware of the use to be made of their resources and/or knowledge and of the meaning and implications of their industrial applications, again according to Western technical concepts which are generally extraneous and hardly comprehensible for indigenous culture (this *in no way* implies any sort of cultural 'inferiority': it is simply diversity). Thus, the question, even in the event that such consent would be obtained, is could it be really *informed*, thus effectively capable of preventing any use of indigenous biogenetic resources and/or related knowledge which would be against the will of the communities concerned?

Also, the problem of the duration of the protection, previously raised with regard to the option considered at 4.1 above, would extend to benefit sharing. It would indeed be a form of protection limited to the temporal period in which the individuals or enterprises using the resources and/or knowledge concerned would retain the exclusive right to exploit them, after which the indigenous communities concerned would be forever deprived of their rights to their own resources and/or knowledge, as well as of any reward deriving from them.

4.3 Non-recognition of IPRs on Indigenous Biogenetic Resources and Related Traditional Knowledge

In light of the above considerations, the best solution for adequately preserving indigenous biogenetic resources and related traditional knowledge could appear to be that of excluding *tout court the possibility of obtaining the recognition of IPRs on such resources and knowledge*. In particular, this solution could appear appropriate on account of the essential relevance of such resources and knowledge for the very collective identity of indigenous peoples. As it would be inconceivable for the Christian culture to make the soul the object of economic exploitation, it may equally be

---

[114] See H Ullrich, 'Biotechnology Related Traditional Knowledge and International Patent Law: Romantics v. Economics?' in Francioni and Scovazzi (eds.), n 104 above, para 2(b)(*ii*).

inconceivable to accept such a course of action in relation to indigenous knowledge about nature and the related resources, which are often part of the intimate relationship between the individual and Mother Earth, thus recalling the idea of the soul for Christians.

Nevertheless, even with regard to this third option there is a major objection which shows how it may in fact be, inappropriate. Culture is not a static entity. It continuously reinvents itself and develops. To pretend that indigenous peoples crystallise their existing cultural model forever would simply be a blatant application of the paternalistic approach which characterises the attitude of most Western practitioners concerning such peoples. It is thus necessary that those indigenous peoples that wish to exploit the economic potentialities of their biogenetic resources and related traditional knowledge, with a view to improving their life conditions and wellbeing, are free to do this in accordance with their philosophies and beliefs. It is just a matter of self-determination and freedom of choice. Nobody should be able to force indigenous peoples to give a commercial or industrial dimension to their biogenetic resources and related traditional knowledge if they are not willing to do so, but, on the contrary, those indigenous communities that wish to use their intellectual resources for improving their social and economic conditions should not be precluded from having the chance of doing that. The example of the Brazilian Sateré-Mawé community (represented by the *Conselho Geral de Tribo Sateré-Mawé*), which has developed a system for commercialising worldwide the *guaraná* (derived from the seeds of the Amazonian fast-growing perennial vine *Paullinia capania* variety *sorbilis*), cultivated according to its traditional knowledge (thus making possible the real improvement of the level of well-being of the people and the preservation of its natural environment),[115] clearly demonstrates that excluding a priori any form of protection of indigenous intellectual rights would be inappropriate and unjust in many concrete cases. On the contrary, protection of indigenous IPRs is essential when the communities concerned are willing commercially or industrially to exploit their resources and/or related knowledge (or to grant external enterprises access to such resources and/or knowledge), since it is self-evident that, without an adequate system for the safeguarding of such rights, anyone could freely take and exploit such resources and related knowledge, thus unjustly depriving its legitimate owners of their inherent title over them. At the same time, it is necessary that such protection is put into practice by adopting models shaped on the basis of the specific character and

---

[115] See 'Guaranà nativo degli indios dei Sateré-Mawé. Dall'Amazzonia l'energia della solidarietà', available at www.commercioequo.org/news/PCN_giugno04/prodotti%20produttori.htm (last visited 3 Oct 2004).

nature of the kind of knowledge to be safeguarded and taking into full account the cultural idiosyncrasy and wishes of the peoples concerned.

4.4 Use of *sui generis* Systems of Protection of Intellectual Property

In sum, in line with the idea supported by a number of scholars[116] and states, particularly developing countries,[117] in my opinion the best option for ensuring adequate protection to indigenous biogenetic resources and related traditional knowledge is represented by *sui generis* systems for the safeguarding of such resources and knowledge, consistently with the rule laid down by Article 27(3)(b) of the TRIPS Agreement with regard to plant varieties.[118] An essential prerequisite for the adequacy of such systems is that they are specifically shaped on the basis of the concrete nature of and the holistic cultural framework characterising the knowledge to be safeguarded. It is thus essential fully to take into account the unique characteristics of traditional knowledge, and, a fortiori, to consider the following elements:[119] (a) indigenous traditional knowledge is not of an individual, but of collective character, and extends to the community as a whole;[120] (b) the practical and spiritual elements of such knowledge are inseparably linked (this is particularly relevant for the identification of the holders of such knowledge); (c) protection should be granted by taking into account the specific peculiarities of the communities concerned and their knowledge; the specific modalities of protection should thus be decided by the members of such communities themselves, so as to ensure that it is actually suitable to preserve their idiosyncratic cultural characters (for this reason, the contractual models typical of Western society are generally inappropriate); (d) in principle, the kind of protection to be granted should not be limited in time, so as to allow the communities concerned to continue to transmit their knowledge from generation to generation; (e) indigenous knowledge is constantly subject to evolution and change, and protection should thus be granted in a flexible way, so as to have the potential of being continuously re-adapted as a result of the

---

[116] See, inter alia, T Cottier and M Panizzon, 'Legal Perspectives on Traditional Knowledge. The Case for Intellectual Property Protection' (2004) 7 *Journal of International Economic Law* 371, at 381 ff.

[117] See WTO doc IP/C/W/370 of 8 Aug 2002, n 100 above.

[118] See text following n 95 above.

[119] See, on this point, T Zamudio, 'Traditional Biodiversity-Related Knowledge and Practices in Latin America', paper presented at the Conference on 'Biotechnology and International Law', Siena, 8–9 Oct 2004 (on file with the author), at 24; SAS Kishi and L Machado, 'Access to Genetic Patrimony and to Traditional Associated Knowledge', a paper given to the same conference, para 8. For an analysis of some proposed alternative models see T Simpson, *Indigenous Heritage and Self-Determination* (Copenhagen, 1997), 132 ff.

[120] On this point see, inter alia, ibid, at 18 f.

developments of such knowledge; (f) the members of the communities concerned should participate in the access process and sharing of benefits resulting from their resources and/or traditional knowledge consistently with the customary law of such communities; (g) any external access to such resources and/or knowledge should be previously authorised, on the basis of principle of the prior, free and informed consent, by the representatives of the communities concerned; (h) where traditional knowledge is of a secret character and is thus known only to the members of the communities concerned, they should not be forced to disclose it.

One of the best qualities of the approach based on the adoption of sui generis systems of protection of biogenetic resources and related knowledge, which has already been translated into concrete action in the context of both domestic and supranational practice,[121] is that it is in line with the provisions of Article 27(3)(b) of the TRIPS Agreement in its present state (although it does not expressly refer to 'traditional knowledge'), and it may thus be applied irrespectively of whether or not the revision of such provision, as planned by its own text, is fulfilled.

---

[121] One remarkable example of such practice is represented, at the domestic level, by the Indian Model Biodiversity Related Community Intellectual Rights Act (available at www. prodiversitas.bioetica.org/draftindia.htm last visited 3 Oct 2004), which states, inter alia, that communities 'will have rights on germplasm covering the whole range of biological diversity of all genera and species, including micro-organisms in the Indian territory under sovereign right' (see cl 7(i)). In short, the Act also recognises that the 'communities of India have unencumbered access to their biological wealth for their survival needs and traditional uses', and that the 'local communities shall at all times and in perpetuity be the lawful and sole owners, custodians and stewards of biological resources, knowledge and innovation related to them'; such ownership can in no way be impaired (see cl 7(ii)), but it may be shared with both other communities and/or the state (see cl 8). Any community may at any time grant free access to its resources and knowledge to other communities 'wherever situated without any payment or reward provided always that such resources and innovation is not acquired for commercial exploitation' (see cl 9). If the communities concerned wish commercially to exploit their knowledge and related resources they will determine, in partnership with the *ad hoc* National Biodiversity Authority, terms for access of enterprises to such resources and terms of intellectual property rights (see cl 10). No commercial acts regarding biological resources and related knowledge owned by local communities (eg production, offering for sale, marketing, export or import) may be performed without the joint authorisation of the rights' holder and the state (see cl 12(a)(iii)), except, of course, with regard to local communities themselves, which may use their traditional knowledge and resources for their traditional practices, including traditional commercial exploitation (see cl 12(a)(iv)). Researchers retain the right to have free and complete access to protected resources for research purposes, but any subsequent commercial utilisation is subject to the rule established by cl 10 (see cl 12(b)). All acts of sale, disposal, use for commercial purposes of protected resources, without the express written consent of the communities concerned and the national authority, or non-payment of adequate compensation (as decided by the communities and the authority) are considered violations of 'biodiversity related community intellectual rights' (see cl 12(c)). At the international level, the African Model Legislation for the Protection of the Rights of Local Communities, Farmers and Breeders, and for the Regulation of Access to Biological Resources, developed by the Organisation of African Unity (available at www.grain.org/brl_files/oau-model-law-en.pdf, last visited 3 Oct 2004), is worth mentioning. It recognises and protects, inter alia, community rights over biological resources (see cl 16) 'as they are enshrined and protected under the norms, practices and

On the other hand, the main difficulty attached to such an approach lies in the identification of the holders of indigenous biogenetic resources and related traditional knowledge. This type of knowledge has an intrinsic 'cultural character', in the sense that, as already emphasised, it is strictly intertwined with the cultural and anthropocentric identity of a given people and it is transmitted from generation to generation. Although indigenous peoples have a special link with their ancestral lands, history shows plenty of cases of cultural interchange between different communities for various reasons (eg, movement of traditionally hunted herds of animals or transhumance), and consequently traditional knowledge has also been geographically spread outside the original physical boundaries of the community which traditionally held such knowledge. When this happens, knowledge that was originally owned by a single community becomes part of the cultural heritage of other groups, which, although they are generally members of the same ethnic stock, are distinct and often live in different lands. As a consequence, the extension of a given knowledge proper of a particular ethnic group may not be easily determinable on the basis of a geographical approach, since it may be shared by communities which belong to the same ethnic stock but live in different parts (even on different continents) of the world. Thus, the model of the 'denomination of origin' based on the geographical location of a given product or method of production is not appropriate for the safeguarding of indigenous traditional knowledge. It is thus desirable that an archetype of 'cultural denomination' is developed at both the national and international levels, based on the classification of any precisely identified knowledge on account of specific common cultural elements which, irrespective of the geographical proximity or remoteness of its different holders, could permit their recognition as the concurrent owners of such knowledge.

5. CONCLUSION

The analysis carried out in the previous sections has led to the conclusion that the best means of ensuring adequate protection to indigenous

customary law found in, and recognized by, the concerned local and indigenous communities, whether such law is written or not' (see cl 17). The right of local communities to refuse access to their biological resources and knowledge, 'where such access will be detrimental to the integrity of their natural and cultural heritage', is also recognised (see cl 19). In addition, local communities have an inalienable right 'to access, use, exchange or share their biological resources in sustaining their livelihood systems as regulated by their customary practices and laws' (see cl 21), and their 'Community Intellectual Rights' shall 'at all times remain inalienable, and shall be further protected' by means of ad hoc mechanisms (see cl 23). In the event that the communities concerned allow external access to their biological resources, they retain the right 'to withdraw consent or place restrictions to activities relating to access where such activities are likely to be detrimental to their socio-economic life, or their natural or cultural heritage' (see cl 20).

biogenetic resources and related traditional knowledge against biopiracy and, at the same time, of resolving the tension between the CBD and the TRIPS Agreement, is the development of sui generis systems aimed at safeguarding indigenous IPRs which should be shaped on the basis of the indigenous holistic vision of life. In particular, the cornerstone of such systems should rest on the freedom and exclusive right of the peoples concerned to decide *whether or not* they wish to utilise their biological resources and/or related traditional knowledge for commercial or industrial purposes and, in the event that they wish to do so, to determine the terms for this commercial exploitation on account of their specific peculiarities so as to preserve their integrity as peoples and retain the opportunity of transmitting the relevant resources and knowledge to future generations according to their own traditions. It is only by respecting these conditions that international law on IPRs would, with regard to indigenous biogenetic resources and related traditional knowledge, be fully consistent with international human rights law since, as highlighted in the previous sections, any other course of action could result in the violation of fundamental rights, at both the collective and individual levels, of indigenous peoples. In this sense, the pre-eminence of human rights over IPRs is undisputable, by reason of the fact that the protection of human dignity is included among the purposes of the United Nations that, by virtue of Article 103 of the UN Charter, override any obligation deriving from any other international agreement. The assumption that IPRs must surrender in the event that their operation would necessarily imply a serious infringement of fundamental rights is also confirmed by recent WTO practice, particularly the General Council decision of 30 August 2003, which authorises a waiver from certain obligations set out in the TRIPS Agreement with respect to pharmaceutical products aimed at treating pandemic diseases (with the purpose of safeguarding the human right to health by means of allowing wider access to such products in developing countries), and which was finally transformed into a permanent amendment to the TRIPS Agreement on 6 December 2005.[122]

As a consequence, it is essential that IPRs be implemented in such a way as to ensure the full respect for fundamental rights, including collective rights of indigenous peoples. This is also important for ensuring the legitimacy of the system of IPRs itself, since, in the event that it conflicts with international principles on human rights, states could (*rectius*, would be bound to), pursuant to Article 103 of the UN Charter, lawfully derogate from it, with the ultimate result of destabilising the very system of IPRs.

---

[122] See 'Implementation of paragraph 6 of the Doha Declaration on the TRIPS Agreement and Public Health', Decision of the General Council of 30 Aug 2003, WTO doc WT/L/540 of 1 Sept 2003; 'Amendment of the TRIPS Agreement', Decision of 6 Dec 2005, WTO doc WT/L/641 of 8 Dec 2005.

Part IV

Intellectual Property Rights and Trade Issues

10

Biotechnology, Human Rights and International Economic Law

ERNST-ULRICH PETERSMANN*

1. INTRODUCTION: REGULATION OF BIOTECHNOLOGY AS A CHALLENGE FOR THE UNITY AND LEGITIMACY OF INTERNATIONAL LAW

BIOTECHNOLOGY AND ITS legal regulation are driven by diverse economic, scientific, moral, religious, environmental, political and legal interests of producers, consumers, governments and other social groups (like researchers, churches). Assessments of the advantages and risks of agricultural, food, industrial, environmental, medical or military uses of biotechnologies often vary among countries (eg, depending on their respective genetic resources, industries, democratic preferences). They remain fraught with numerous uncertainties about potential abuses of biotechnologies, of related intellectual property rights, or about harm caused by involuntary cross-breeding between genetically modified organisms (GMOs) and natural organisms. The application of biotechnologies to parts of the human body, animals, food and agriculture raises ethical and human rights concerns and new legal questions (eg, regarding the legal status of human stem cells, the embryo, the foetus) that may be answered differently depending upon one's normative value premises.[1] Hence, national and international regulation of biotechnologies—for

* Professor of International and European Law at the European University Institute and joint chair at the Robert Schumann Centre for Advanced Studies, Florence. He was previously Professor at the University of Geneva and the Graduate Institute of International Studies, and Legal adviser to the GATT and WTO. Email: ulrich.petersmann@iue.it.

[1] Cf. the distinction between three ethical constituencies (ie, utilitarians, human rights advocates and 'the dignitarian alliance') by R Brownsword in this volume. The European Court of Human Rights, in its recent judgment of 8 July 2004 in *Vo v France* (App. no 53924/00), explicitly left open the controversial question whether Art 2 of the European Convention on Human Rights (ECHR) (right to life) also protects the human embryo and the foetus' right to life: 'it is neither desirable, nor even possible as matters stand, to answer in the abstract the question whether the unborn child is a person for the purposes of Article 2 of the Convention' (para 85).

instance in Europe, the United States, in less-developed countries (LDCs) and specialised international organisations regulating specific biotech products and processes from particular agricultural, commercial, environmental, medical and/or political perspectives—remains fragmented and contested.[2] The numerous conflicts of interests—for instance between the highly concentrated biotech industries in developed countries (eg, private interests in patent protection for genetic engineering) and public interests in LDCs (eg, interests in more equitable sharing of the benefits derived from their genetic resources and from their 'traditional knowledge')—render a worldwide harmonisation of the diverse regulation of biotechnologies difficult. The justification of some biotech rules by human rights (eg, in the 1997 United Nations Educational, Scientific and Cultural Organisation (UNESCO) Declaration on the Human Genome and Human Rights, the 1997 Council of Europe Convention on Human Rights and Biomedicine) or by indigenous peoples' rights may call for new legal and judicial approaches to the 'balancing' of private and public interests.[3]

How should governments, courts and international organisations respond to the regulatory differences, uncertainties and governance problems, for instance if state-centred international treaties and their science-based risk assessment requirements conflict with the democratic preferences inside states (eg, for 'GMO free zones')? Is the claim by international lawyers[4] true that general international law (eg, on the interpretation of treaties) can resolve conflicts among the diverse international treaty regimes—such as the 1992 UN Convention on Biodiversity, its 2000 Cartagena Protocol on Biosafety (CPB), the 2001 Food and Agricultural Organisation (FAO) Treaty on Plant Genetic Resources, the 1997 UNESCO Declaration on the Human Genome and Human Rights, the 1994 World Trade Organisation (WTO) Agreements on Sanitary and Phytosanitary Measures (SPS) and Trade-Related Intellectual Property Rights (TRIPS), the biotechnology rules adopted by regional organisations like the European Union (EU), the Council of Europe and the Andean Community? Or is this normative postulate of the legal unity of international law inconsistent with the diversity of national and international regulations and their sometimes conflicting value premises

---

[2] On the fragmented regulation of biotechnologies since the 1992 UN Convention on Biodiversity and the emergence of only a few general principles defining 'common concerns of humanity', equitable benefit-sharing, precautionary measures and mutual supportiveness of trade, environmental and other biotech rules, see R Pavoni in this volume; S Maljean-Dubois, 'Bioéthique et Droit International' [2000] *Annuaire Français de Droit International* 82.

[3] See the contributions by F Francioni and F Lenzerini respectively in this volume.

[4] Cf J Pauwelyn, *Conflict of Norms in Public International Law: How WTO Law Relates to Other Rules of International Law* (Cambridge, Cambridge University Press, 2003), 37–8. According to Pauwelyn, 'it is for the party claiming that a treaty has 'contracted out' of general international law to prove it' (at 213).

(eg, state sovereignty vs 'inalienable' human rights)? Do conflicting legal regimes justify a presumption in favour of exhaustive regulation within the special regime?[5] Does the regulatory competition promote progressive reforms of international law or endanger its legal unity and consistency?

All international treaties, notwithstanding their common legal foundations (eg, in the *ius cogens* principles of *pacta sunt servanda* and good faith), remain incomplete and confronted with the task of reconciling the special treaty rules with other fields of national and international law. WTO law, for instance, promotes welfare-increasing market access regulations subject to numerous 'exceptions' and 'public interest provisions' protecting the sovereign rights of each WTO Member to determine its national level of protection of health, the environment and of other non-economic policy goals. Also other 'specialised agencies' like the World Intellectual Property Organisation (WIPO) and multilateral environmental agreements (MEAs), focus on particular regulatory problems. The general international law rules for the promotion of mutually coherent treaty interpretations (such as freedom of contract, *lex specialis*, *lex posterior*, *lex superior*, *ius cogens*) often offer no 'mathematical' or mechanical solutions for resolving normative conflicts among international rules and their underlying values (eg, state sovereignty, human rights, science-based risk assessments, economic cost-benefit analyses). For instance, the scientific logic of risk assessment procedures (as required by the CPB and the SPS Agreement) and the legal logic of intergovernmental adjudication (eg, in the WTO) may not overcome divergent democratic preferences and different national human rights traditions underlying conflicting regulatory regimes. Yet, even antagonistic interactions between science, law and democratic procedures may enhance the quality of public discourse and the democratic acceptability of rules by taking into account the legitimate interests of affected persons.[6]

Sections 2 and 3 of this contribution discuss regulatory challenges of biotechnologies for the legitimacy and methodology of international economic law. Sections 4 and 5 argue that—in addition to the formal rules of international law (eg, on treaty interpretation and state responsibility)

---

[5] Many WTO Members and WTO lawyers perceive WTO law as a separate trade regime; according to this view, the limitation of the jurisdiction of WTO dispute settlement bodies to 'the covered agreements' confirms that '[i]n international law, every tribunal is a self-contained system (unless otherwise provided)', as stated by the International Criminal Tribunal for Yugoslavia in its *Tadic* judgment (1996) 35 ILM 32, at para 11. For an explanation of this narrow perception of the WTO as—in principle—a self-contained legal regime see, eg, J Neumann, *Die Koordination des WTO-Rechts mit anderen völkerrechtlichen Ordnungen: Konflikte des materiellen Rechts und Konkurrenzen der Streibeilegung* (Berlin, Duncker & Humblot, 2002).

[6] On public discourse as a source of democratic legitimacy of rules see J Habermas, *Between Facts and Norms—Contribution to a Discourse Theory of Law and Democracy* (Oxford, Polity Press, 1996), 315–28.

for resolving conflicts of international law rules[7]—human rights offer a coherent, comprehensive and flexible legal framework for balancing the diverse private and public interests among each other—not only on national and regional levels of governance and biotechnology regulation, but also in international economic regulation. Just as the European Court of Human Rights (ECtHR) rightly emphasises the need for an 'evolutive interpretation' of the European Convention on Human Rights (ECHR) as a 'living instrument which must be interpreted in the light of present-day conditions',[8] so must international economic law be construed with due regard to human rights and to the common concerns reflected in the emerging principles and rules of international biotechnology law.

2. BIOTECHNOLOGY AND HUMAN RIGHTS: CHALLENGES TO THE LEGITIMACY AND METHODOLOGY OF ECONOMIC LAW

Biotechnology industries and their biotech products, services and production processes are penetrating ever more sectors of the economy. The opportunities of industrial and environmental biotechnology (eg, for industrial enzymes, bioplastics, biofuels, bioremediation, biofertilisers, biopesticides), of agricultural and food biotechnology (eg, for enhanced productivity of crops, genetically modified food and feed) as well as of health-related biotechnology (eg, for transplantation of human tissues, biomedical treatment of genetic diseases) depend not only on science, but also on market opportunities, public attitudes and national and international regulation.[9] Some of the regulatory regimes were designed in order to deal with specific biotechnological problems and the conservation and sustainable use of biological diversity, such as the 2000 CPB which was negotiated under the auspices of the Convention on Biodiversity (CBD) and entered into force in September 2003. Other legal regimes set out general rules for international trade and domestic regulation with few or no specific references to the new bio-economy, such as the more than 20 agreements constituting the law of the WTO. This fragmented regulation of the economic, environmental, technological and human rights dimensions of biotechnology in most worldwide treaties, and the resultant inconsistencies (eg, between the precautionary rules, risk assessment requirements, trade and labelling rules in the CPB and in WTO law), challenge the consistency, legitimacy and effectiveness of international

---

[7] Pauwelyn, n 4 above, at 170 ff, defines 'conflict' as mutually exclusive obligations and the incompatibility of the exercise of a right with an obligation.

[8] See *Tyrer v UK*, judgment of 25 Apr 1978, Series A no. 26, 15–16, para 31, and subsequent case law (eg, above n 1, at para 82).

[9] Cf *The Biotechnology Promise. Capacity-Building for Participation of Developing Countries in the Bioeconomy*, UN 2004.

economic law.[10] European integration law, by contrast, and the Council of Europe Convention on Human Rights and Biomedicine (1997), with its Additional Protocols on the prohibition of cloning human beings (1998), the transplantation of organs and tissues of human origin (2002) and biomedical research (2004), regulate economic, environmental and human rights dimensions of biotechnologies in a more integrated manner.[11]

Biotechnology enables new kinds of production processes (eg, genetic manipulation of plants, therapeutic cloning of animal stem cells), products (like GMOs) and services (like treatment of genetic diseases) the legal regulation of which is increasingly influenced by national and regional human rights concepts, democratic preferences and risk approaches that may legitimately differ among countries (eg, the 'equivalence principle' being more widely accepted in North American than in European regulation of GM food and feed). Divergent biotechnology regulations may then give rise to border adjustment measures (such as EC import prohibitions, labelling, monitoring and traceability requirements) which may be challenged in dispute settlement proceedings as being inconsistent with the rules of the General Agreement on Tariffs and Trade (GATT) and the WTO Agreements on Technical Barriers to Trade (TBT), SPS, the Agreement on Agriculture (AA), the General Agreement on Trade in Services (GATS) and/or the TRIPS Agreement. In response to such legal challenges, international dispute settlement bodies (eg, in the EC and the WTO) may have to respond to requests to interpret trade rules in conformity with human rights, for instance if restrictions on trade in human stem cells, gene therapies or trade-related patents in isolated body parts are justified on grounds of human rights. International trade thus creates pressures:

— to reduce the fragmentation of international treaty regimes, the regulatory diversity among national jurisdictions, and the investment and transaction costs needed for the development of biotechnology; and

— to promote more 'integrated governance regimes' and legal harmonisation aimed at maximising biotechnological opportunities and minimising related risks.

Yet, whenever these economic pressures run counter to democratic preferences for the domestic regulation of biotechnology and of the modern

---

[10] For a detailed comparison of WTO and CPB rules, and certain inconsistencies of their respective provisions for risk assessment procedures and precautionary measures, see M Böckenförde, *Grüne Gentechnik und Welthandel. Das Biosafety-Protokoll und seine Auswirkungen auf das Regime der WTO* (Berlin, Springer, 2004).

[11] Cf R Pavoni, *Biodiversita e biotecnologie nel diritto internazionale e communitario* (Milan, Giuffrè, 2004); T Christoforou, 'The Regulation of Genetically Modified Organisms in the EU: The Interplay of Science, Law and Politics' (2004) 41 *Common Market Law Review* 637.

bio-economy, the narrow focus of WTO rules is increasingly challenged. The more people perceive international trade and international economic law only as *instruments* for realising non-economic objectives, the stronger are calls for a 'human rights approach'—as requested by the UN High Commissioner for Human Rights (UNHCHR) and UNESCO and practised inside the EU and the Council of Europe—in order to ensure that international trade regulation remains consistent with national constitutional laws, democratic preferences, human rights and basic human needs (see section 3 below).

Intergovernmental agreements—such as the WTO Agreement, the Biosafety Protocol, the 2001 FAO Treaty on Plant Genetic Resources for Food and Agriculture and human rights conventions—fail to regulate their mutual interrelationships in a precise manner. It is often only in dispute settlement proceedings that courts and other dispute settlement bodies have to clarify the legal relationships concerned, for example to what extent GATT, TBT and SPS rules can be applied cumulatively to biotechnology regulations, and whether departures from WTO rules may be justifiable by MEAs (like the CBP) or human rights. The WTO panel proceeding by Argentina, Canada and the US against the EC approval procedures for GMOs and GM products, as well as against national import bans and internal restrictions on GMOs and GM products in some EC Member States, illustrates the complexity and limits of WTO law and WTO dispute settlement proceedings over regulatory differences if they are due to divergent democratic preferences and risk approaches rather than to economic protectionism. Whereas trade and environmental diplomats tend to emphasise the special character of their treaty regimes, judges may be required to interpret treaty rules as parts of broader legal systems (cf section 4).

Such disputes also raise questions about what the Preamble to the WTO Agreement calls 'the basic principles and ... objectives underlying this multilateral trading system'. Under GATT 1947, these objectives and principles tended to be defined by economists and politicians without regard to international law. Since 1995, WTO case law has increasingly identified the general international *legal principles* underlying the WTO legal system (like legal security, good faith, transparency, prohibition of abuse of rights, rule of law, non-discrimination, interpretation of WTO law as part of the international legal system).[12] As every WTO Member

---

[12] On the jurisdiction of WTO dispute settlement bodies and their 'judicial function' to define these 'basic legal principles' see EU Petersmann, 'From 'Member-Driven Governance' to Constitutionally Limited 'Multilevel Trade Governance' in the WTO' in G Sacerdoti *et al* (eds), *The WTO at 10: The Role of the Dispute Settlement System* (Cambridge, Cambridge University Press, 2006), chap 7.

has ratified one or more UN human rights conventions, the human rights obligations of all WTO Members (eg, under the UN Charter) should induce WTO dispute settlement bodies to interpret WTO rules, and the explicit WTO commitment to 'sustainable development', in conformity with the universal human rights obligations of WTO Members.[13] The judicial balancing of rights and obligations of WTO Members by the WTO Appellate Body could lead to different conclusions depending on whether the balancing focuses only on state-centred rules or also on *erga omnes* obligations under human rights law (cf section 5).

The (quasi-) judicial clarification of WTO rules by the Appellate Body—similar to the judicial clarification of the 'basic principles' underlying European integration law by the EC Court of Justice (ECJ) and the ECtHR—illustrates the legitimate differences between *intergovernmental* and *judicial approaches* to international economic regulation. Most states adopt a 'realist approach' to international relations, ie, they perceive international law as a system based on 'sovereign equality of states' and the pursuit of state interests rather than as a system committed to the protection of human rights. Intergovernmental organisations are viewed as mere frameworks for intergovernmental bargaining driven by national interests and by the relative power of governments. As human rights are nowhere mentioned in WTO law, trade diplomats often emphasise that disregard for human rights should not prevent WTO Members from enhancing economic welfare and open markets through trade liberalisation. Hence, governments often define the scope of jurisdiction of WTO bodies more narrowly than international judges required to interpret WTO law with due regard to 'any relevant rules of international law applicable in the relations between the parties'.[14] The WTO Appellate Body, for instance, has refused to interpret WTO law 'in clinical isolation' from other fields of international law and has insisted on its autonomous powers to adopt 'evolutionary interpretations' of WTO rules, for instance by taking into account MEAs in the interpretation of WTO rules and by admitting *amicus curiae* briefs, notwithstanding the declared political opposition by almost all WTO Members to the admissibility of *amicus curiae* briefs.[15] It is noteworthy that all the *amicus curiae* submissions in the WTO dispute settlement proceedings challenging the EC approval

---

[13] See EU Petersmann, 'Human Rights and International Trade Law—Defining and Connecting the Two Fields' in T Cottier, J Pauwelyn and L Bürgi (eds), *Human Rights and International Trade* (Oxford, Oxford University Press, 2005).

[14] Cf Art 31(3)(c) Vienna Convention on the Law of Treaties (VCLT).

[15] On the methods of treaty interpretation applied by WTO dispute settlement bodies see EU Petersmann, 'Future challenges for the WTO Dispute Settlement System', in Y Taniguchi et al (eds), *The WTO at Ten; Dispute Settlement, Multilateral Negotiations, Regional Integration* (Cambridge University Press, forthcoming).

procedures for GMOs argued in favour of the environmental and health protection concerns raised by the EC rather than in support of the export interests of the biotech industries in the complaining countries.[16]

3. THE 'HUMAN RIGHTS APPROACH' TO TRADE AND BIOTECHNOLOGY ADVOCATED BY THE UN HIGH COMMISSIONER FOR HUMAN RIGHTS

The fulfillment of many human rights (eg, to food, health, education) depends on access to scarce goods and services (eg, drinking water, cheap medicines, health and educational services). Also enjoyment of civil and political human rights (eg, personal freedom, rule of law, access to justice, democratic self-government) requires economic resources (eg, for financing democratic and law-enforcement institutions). The widespread, yet unnecessary, poverty, health problems and legal insecurity (eg, among the more than 1 billion people living on one dollar a day or less) bear witness to the fact that UN member states and UN law have so far failed to realise the UN objective of 'universal respect for, and observance of, human rights and fundamental freedoms for all' and 'creation of conditions of stability and well-being which are necessary for peaceful and friendly relations among nations' (Article 55 of the UN Charter). Even though international trade is essential for increasing the availability and quality of scarce resources, UN human rights bodies, until recently, tended to ignore international trade law or, as in a report for the UN Commission on Human Rights of 2001, discredited the WTO as 'a veritable nightmare' for developing countries and women.[17] The diversity of national human rights traditions, the longstanding neglect by UN human rights law—in contrast to EU fundamental rights and national constitutions in many countries—of economic liberties (eg, freedom of profession) and welfare-creation through division of labour among free citizens, and one-sided claims to legal superiority of UN human rights conventions over international economic law, are among the reasons why most trade diplomats prefer leaving human rights and labour standards to specialised UN and ILO bodies and oppose discussions on human rights and labour rights in the WTO.

---

[16] Cf EU Petersmann, 'The WTO Dispute over Genetically Modified Organisms: Interface Problems of International Trade Law, Environmental and Biotechnology Law' in F Francioni and T Scovazzi (eds), *Biotechnology and International Law* (Oxford, Hart Publishing, 2006).

[17] *Globalization and its impact on the full enjoyment of human rights*, ECOSOC document E/CN.4/Sub.2/2000/12 of 15 June 2000, at para 15. Apart from a reference to patents and their possibly adverse effects on pharmaceutical prices (depending on the competition, patent and social laws of the countries concerned), the report nowhere identifies conflicts between WTO rules and human rights.

3.1. Human Rights Dimensions of Trade and Biotechnology Law

In response to the widespread criticism of the anti-market bias of such 'nightmare reports', the UNHCHR has recently published more differenti- ated reports analysing human rights dimensions of the WTO Agreements on TRIPS, [18] the AA,[19] GATS,[20] international investment agreements,[21] non-discrimination in the context of globalisation,[22] the impact of trade rules on the right of everyone to the enjoyment of the highest attainable standard of physical and mental health,[23] and on the relationship of the right to development and the WTO.[24] The reports call for a 'human rights approach to trade' which:

(i)   sets the promotion and protection of human rights as objectives of trade liberalization, not exceptions;

(ii) examines the effect of trade liberalization on individuals and seeks to devise trade law and policy to take into account the rights of all individu- als, in particular vulnerable individuals and groups;

(iii) emphasises the role of the State in the process of liberalization — not only as negotiators of trade law and setters of trade policy, but also as the primary duty bearer of human rights;

(iv) seeks consistency between the progressive liberalization of trade and the progressive realization of human rights;

(v)   requires a constant examination of the impact of trade liberalization on the enjoyment of human rights;

(vi) promotes international cooperation for the realization of human rights and freedoms in the context of trade liberalization.[25]

The UNHCHR emphasises that, because every WTO Member has ratified one or more UN human rights conventions and has human rights obligations also under general international law, human rights may be

---

[18] *The Impact of the Agreement on Trade-Related Aspects of Intellectual Property Rights on Human Rights*, E/CN.4/Sub.2/2001/13 (27 June 2001).

[19] *Globalization and its Impact on the Full Enjoyment of Human Rights*, E/CN.4/2002/54 (15 Jan 2002).

[20] *Liberalization of Trade in Services and Human Rights*, E/CN.4/Sub.2/2002/9 (18 June 2002).

[21] *Human Rights, Trade and Investment*, E/CN.4/Sub.2/2003/9 (2 July 2003).

[22] *Analytical Study of the High Commissioner for Human Rights on the Fundamental Principle of Non-discrimination in the Context of Globalization*, E/CN.4/2004/40 (15 Jan 2004).

[23] *The Right of Everyone to the Enjoyment of the Highest Attainable Standard of Physical and Mental Health*. Report by the Special Rapporteur Paul Hunt on his Mission to the WTO, E/CN.4/2004/49/Add.1 (1 Mar 2004).

[24] *Mainstreaming the Right to Development into International Trade Law and Policy at the WTO*, E.CN.4/Sub.2/2004/17 (9 June 2004).

[25] Document E/CN.4/Sub.2/2002/9, 2. For a discussion of these reports see EU Petersmann, 'The Human Rights Approach Advocated by the UN High Commissioner for Human Rights and by the ILO: Is it Relevant for WTO Law and Policy?' [2004] *Journal of International Economic Law* 605.

'relevant context' for the interpretation and application of WTO rules.[26] This 'interpretative approach' focusing on the human rights obligations of every WTO Member avoids hierarchical claims of legal primacy of human rights over trade rules. According to the UNHCHR, the needed human rights approach to international trade must recognise as 'entitlements the basic needs necessary to lead a life in dignity and ensures their protection in the processes of economic liberalization'; these entitlements cannot be 'left subject to the whims of the market'.[27] The UNHCHR differentiates between obligations to respect human rights (eg, by refraining from interfering in the enjoyment of such rights), to protect human rights (eg, by preventing violations of such rights by third parties), and to fulfill human rights (eg, by taking appropriate legislative, administrative, budgetary, judicial and other measures towards the full realisation of such rights); even if obligations to fulfil human rights may depend on the 'available resources' (cf Art 2 of the ICESCR), obligations to respect and protect human rights are legally binding. In contrast to the often one-sided focus in American human rights literature on the use of trade sanctions for promoting respect for human rights *abroad*, the UNHCHR reports analyse the human rights dimensions of trade liberalisation, trade restrictions and other trade regulations in a broader perspective, emphasising both potential synergies and potential conflicts between human rights and trade rules. A more recent report rightly emphasises that 'outwardly-directed' trade sanctions for promoting human rights abroad in foreign jurisdictions demand far greater justification and respect for the 'sovereign equality' of states than 'inwardly-directed' trade measures for protecting the human rights of domestic citisens inside the national jurisdiction.[28]

As enjoyment of human rights depends on availability, accessibility, acceptability and quality of traded goods and services, the UN reports acknowledge and discuss the relevance of WTO rules for access to low-priced goods and services, for limitations of 'market failures' (eg, inadequate supply of public goods like essential medicines for poor people), and for protection and fulfilment of human rights. The reports underline that what are referred to—in numerous WTO provisions—as *rights* of WTO Members to regulate may be *duties* to regulate under human rights law (eg, so as to protect and fulfill human rights of access to

---

[26] 'Relevant context' is used here in a broader sense than in Art 31(2) VCLT, ie, including international rules that have to be 'taken into account, together with the context', pursuant to Art 31(3) VCLT.

[27] E/CN.4/Sub.2/2002/9, n 20 above, para 6; E/CN.4/2002/54, n 19 above, para 9.

[28] *Raising Human Rights Concerns in the WTO: The Human Rights Implications of General Exception Clauses in WTO Agreements*, E/CN.4/2005.

water, food, essential medicines, basic health care and education services at affordable prices). The UNHCHR suggests, inter alia:

— to recognise the promotion of human rights as an objective of the WTO;
— to encourage interpretations of WTO rules that are compatiblewith international human rights as progressively clarified in the 'General Comments' adopted by UN human rights bodies;
— to carry out 'human rights assessments' of WTO rules, and
— to develop intergovernmental protection of human rights so as to ensure that trade rules and policies promote the human rights and basic needs of all.

In the area of human rights and biotechnology, another UN report identified benefit-sharing and patenting of genetic material (eg, by promoting rights to property, health, to benefit from scientific progress), discrimination (eg, in employment and insurance), gender and sex (eg, discrimination and participation of women), and reproductive and therapeutic human cloning as priority issues (eg, use of biotechnology to determine traits of children, prevent genetic diseases). In analysing biotechnology issues in the light of all human rights and obligations of all actors involved, and in seeking to balance the various rights in order to maximise respect for all rights and right-holders, a human rights approach is said to:

— place emphasis on participation of individuals in decision-making;
— introduce accountability for actions and decisions, which can allow individuals to complain about decisions affecting them adversely;
— seek non-discrimination among individuals through the equal enjoyment of rights and obligations by all individuals;
— empower individuals by allowing them to use rights as leverage for action and legitimizing their 'voice' in decision-making; and
— link decision-making at every level to the agreed human rights norms at the international level as set out in the various human rights covenants and treaties.[29]

3.2. Liberal Trade Rules and Human Rights: Synergies and Conflicts

The reports by the UNHCHR identify potential synergies between trade rules and human rights: 'trade liberalization is generally a positive contributor to poverty alleviation—it allows people to exploit their productive potential, assists economic growth, curtails arbitrary policy

---

[29] *Human Rights and Bioethics*, E/CN.4/2003/98 (10 Feb 2003), 8.

interventions and helps to insulate against shocks'.[30] Also intellectual property rights may act as incentives for innovation in the pharmaceutical industry and for the transfer of technology to LDCs.[31] Yet, the UNHCHR also emphasises the potential conflicts between 'existential' human rights and 'instrumental' WTO rules (eg, on protection of intellectual property rights and investor rights), for example if trade rules lead to higher prices (eg, of food, seeds, pharmaceutical products), cause unemployment, or entail other 'market failures' (eg, in the supply of essential medicines for tropical diseases).[32] According to the High Commissioner, the needed 'human rights assessments' of trade rules and trade policies must focus on the rights and basic needs of vulnerable individuals and of the most disadvantaged communities, whose human rights risk being adversely affected most in the process of trade liberalisation.[33] This contrasts with the focus of WTO rules on future welfare effects of liberal trade for consumers, and on the adjustment costs for import-competing producers, rather than on individual rights of other, adversely affected citizens. The numerous 'public interest clauses' and 'exceptions' in WTO law perceive the relationship between trade and non-trade rules as a matter of national sovereignty and of intergovernmental co-ordination, rather than in terms of subordinating trade law to a particular set of UN human rights standards the interpretation of which, like that of many UN human rights covenants, remains contested among states.

The macroeconomic objectives and state-centered rules of WTO law (eg, on most-favoured-nation treatment of imported goods, dispute settlement among WTO Members) differ fundamentally from the individualist objectives and individual entitlements of human rights. Even where WTO rules require governments to protect individual rights (such as trade-related intellectual property rights, individual access to courts), the rationale of WTO rules remains instrumental (utilitarian) rather than based on human rights. The High Commissioner emphasises the need for using WTO rules (eg, on special and differential treatment of developing countries), WTO safeguard clauses and WTO 'exceptions' for actively promoting mutually coherent interpretations of WTO law and human rights. The UNHCHR criticises the lack of guidance and of monitoring mechanisms in WTO law for ensuring the taking into account

---

[30] E/CN.4/2002/54, n 19 above, para 33.

[31] See the report on TRIPS, n 19 above.

[32] Eg, the report on TRIPS, ibid, notes that 'the commercial motivation of IPRs means that research is directed, first and foremost, towards 'profitable' disease. Diseases that predominantly affect people in poorer countries—in particular tuberculosis and malaria—still remain relatively under-researched' . . . 'questions remain as to whether the patent system will ensure investment for medicines needed by the poor. Of the 1,223 new chemical entities developed between 1975 and 1996, only 11 were for the treatment of tropical disease' (para 38).

[33] E/CN.4/2002/54, n 18 above, para 34.

of human rights. The reports do, however, not identify concrete conflicts between human rights and WTO law. In view of the 'constitutional function' of WTO guarantees of freedom, non-discrimination, rule of law and of other 'public interests' for limiting welfare-reducing government policies (eg, discriminatory market restrictions), for enhancing the use of non-discriminatory policy instruments,[34] and for protecting the priority of non-economic values (as reflected in the numerous 'public interest clauses' in WTO law), conflicts between the often flexible WTO rules and human rights appear unlikely at the level of international principles.[35]

Yet, even though the WTO objective of 'sustainable development' and the numerous WTO 'exceptions' appear to offer enough policy space for taking into account universal and other human rights obligations of WTO Members, WTO law in no way ensures that human rights obligations are actually taken into account in the legislative and administrative implementation of WTO rules and in their judicial protection (or, usually, disregard) by domestic courts. The fact that none of the UN human rights conventions has been ratified by all WTO Members (see the non-ratification by more than 30 WTO Members, including the USA, of the UN Covenant on Economic, Social and Cultural Rights) entails that WTO law must respect the diversity of human rights traditions in WTO member states. Whereas the WTO guarantees of freedom, non-discrimination and rule of law frequently go far beyond those in national laws, UN human rights conventions often set out only minimum standards that remain far below those in the domestic constitutional laws of many countries (eg, in the EU). Hence, just as human rights activists often ignore the contribution of trade rules to the fulfilment of human rights, so do WTO diplomats and WTO judges prefer to avoid human rights discourse in WTO bodies in view of the disagreement over the human rights dimensions of WTO rules. In view of the narrow limitation of the 'terms of reference' of WTO dispute settlement bodies to the 'covered WTO agreements', it remains controversial whether—in WTO dispute settlement proceedings—the parties to WTO disputes may invoke human rights law not only as relevant context for the interpretation of WTO rules, but also directly for justifying departures from their WTO obligations (eg, in the case of US trade sanctions in response to human rights violations in Myanmar).[36]

---

[34] National policy autonomy is safeguarded, eg, by GATT Art III, XVI and XXVIII, GATS Arts VI and XIX, Arts 8, 30, 31, 40 of the TRIPS Agreement and by the numerous WTO 'exceptions' and safeguard clauses.

[35] Cf EU Petersmann, 'Human Rights and the Law of the WTO' (2003) 37 *Journal of World Trade* 241.

[36] Cf, eg, J Pauwelyn, 'Winning a WTO Dispute Based on Non-WTO Law: Questions of Jurisdiction and Merits' (2003) 37 *Journal of World Trade* 997.

4. INTERNATIONAL TRADE LAW, GMOS AND GM PRODUCTS

The subject of this contribution—the impact of biotechnology and of human rights on international economic law and international trade—lies at the interface of four of the most dynamic developments of modern international law: human rights, international trade and economic law, intellectual property law and international environmental law. WTO law was drafted as a worldwide framework agreement with only few, specific references to biotechnology concerns, notably in the TRIPS Agreement (eg, Article 27(3) regarding the patentability of microbiological processes). International environmental law (such as the 2000 Biosafety Protocol, the 2001 FAO Treaty on Plant Genetic Resources for Food and Agriculture), human rights instruments (such as UNESCO's 1997 Universal Declaration on the Human Genome and Human Rights, the 1997 Council of Europe Convention on Human Rights and Biomedicine) and EU law, by contrast, include an increasing number of specific provisions regulating the human rights dimensions, environmental and trade problems of biotechnologies. As biotechnologies, trade in genetically modified organisms (GMOs),[37] GM products, related services and intellectual property rights often have transnational effects on many countries, there are good reasons to believe that 'genetic commerce will change the world',[38] including international economic law.

Just as the completion of the EC's common market required comprehensive harmonisation measures in the field of biotechnology (such as the 1998 EC Directive on the legal protection of biotechnological inventions, the 2001 EC Directive on the deliberate release of GMOs into the environment, the 2003 EC Regulation on genetically modified food and feed) so as to ensure the free movement of goods and services (eg, GM food), so do the WTO Agreements on SPS, TBT and TRIPS reflect the worldwide need for harmonising product standards, production and process methods, risk assessment and risk management procedures, and trade-related intellectual property rights so as better to protect human needs and democratic preferences, including the need to reduce unnecessary trade restrictions, trade discrimination and transaction costs.[39] The 1994 and 2001 Ministerial Declarations had mandated

[37] A GMO can be defined as an organism in which the genetic material has been altered–usually by insertion of foreign genes into the cell of the receiving organism—in a way that does not occur naturally by mating and/or natural recombination. Genes are composed of DNA (Deoxyribonucleic Acid) molecules that contain genetic information determining the proteins which the cell will produce. The first GMO crop was produced only in 1983 and became ready for commercialisation only in the early 1990s.

[38] Cf J Rifkin, *The Biotech Century: How Genetic Commerce Will Change the World* (London, Victor Gollancz,1998).

[39] For a detailed analysis of the revision of the GMO and biotech rules of the EC since 1998 see Christoforou, n 11 above.

the WTO Committee on Trade and Environment to study, and make recommendations for reconciling, international trade and environmental rules. So far, WTO Members have not yet identified concrete conflicts between WTO rules and environmental rules that could not be prevented by interpreting trade and environmental rules in a mutually coherent manner. As worldwide human rights and environmental agreements do not provide for compulsory international adjudication, it may be the WTO dispute settlement bodies that, once again, may have to clarify the legal interrelationships between trade rules, environmental and biotechnology rules, and human rights.

4. 1. International Trade Disputes Resulting from Divergent Attitudes towards GMOs

The attitudes towards GMOs, GM products, related services and intellectual property rights continue to vary considerably (eg, among farmers, consumers, scientists, biotech firms) depending on the biotechnological capacities of countries, their risk perceptions and benefits of biotechnology, the level of risk that individuals are willing to accept, scientific risk assessments, administrative risk management capacities, the dependence of countries on food imports, and their production and exports of GM products. Scientific risk assessments in Europe and North America have emphasised the potential advantages of GM products (eg, in terms of increased agricultural output, higher quality food like 'vitamin A rice', protection of health and of the environment by lesser use of pesticides) and the lack of evidence of biosafety risks.[40] The GMO regulations and 'conventional' risk assessments of some countries (eg, in North America) proceed from the 'equivalence principle' relying on existing scientific evidence and substantial equivalence of GM foods with natural products, including their equivalent safety. Other countries (notably in Europe) apply a 'no risk' and 'precautionary approach' and consider GM products not only as inherently different from conventional products, but also as requiring further research and precautionary controls in view of the scientific uncertainty regarding their potential

---

[40] In the WTO dispute brought by Argentina, Canada and the USA against the alleged EC moratorium on the approval of biotech products, as well as the national marketing and import bans on biotech products maintained by EC Member States, the EC referred to certain adverse environmental effects determined in the UK Field Trials from 1999 to 2003. Yet, there does not appear to exist any definite scientific evidence of harm by agricultural, food and pharmaceutical GM products to human health (eg, in terms of altering antibiotic resistance and fostering allergic reactions).

health and environmental risks.[41] Some states (like Switzerland) apply temporary *moratoria* prohibiting the commercial cultivation of all GMOs and all GMO imports.

Since EC Directive 90/220/EEC of 23 April 1990 on the deliberate release of genetically modified organisms, the EC has enacted a large number of Directives and Regulations introducing approval procedures, labelling and traceability requirements for the release and marketing of GMOs, medicinal products containing or produced from GMOs, novel foods and novel food ingredients, as well as quality and safety standards for the donation, procurement, testing, processing, preservation, storage and distribution of human tissues and cells.[42] Following the adoption of the CBD (1992) and its CPB (2000),[43] more than 100 countries are reviewing or developing their biosafety rules, often with the help of the capacity-building initiative by the Global Environment Facility of the United Nations Environment Programme. The international policy differences entail that exports from major producers of GM crops (such as Argentina, Brazil, China, Canada, the USA) are confronted with non-tariff market access barriers in countries that have adopted a 'no risk approach' (eg, in the EC) or insist on remaining 'GM free zones' (eg, in Upper Austria, Tuscany, some African countries) so as to avoid contamination of their conventional agricultural products.

In 2003, Argentina, Canada and the US started a WTO dispute settlement proceeding against the alleged 'moratorium' in the EC approval procedures for GMOs and GM products, as well as against the import bans and internal restrictions on GMOs and GM products in some EC Member States.[44] The EC had approved the commercial release of 18 GM products up to 1998. Between October 1998 and 2004, however, GM products were approved only under the simplified procedure (based on 'substantial equivalence') provided for in the Novel Foods Regulation 258/97 but no longer under EC Directive 90/220

---

[41] For a comparative analysis of domestic legislation on agro-biotechnology in selected developed and developing countries, including the complex EC approval and risk assessment procedures for GMOs, novel foods and novel food ingredients, see S Zarilli, *International Trade in GMOs and GM Products: National and Multilateral Legal Frameworks* (Geneva, UN, 2005), chap II.

[42] For detailed references to, and analyses of, the large number of EC directives and regulations on GMOs, see ibid, and L Stökl, *Der Welthandelsrechtliche Gentechnikkonflikt* (Berlin, Duncker, 2003); Christoforou, n 11 above.

[43] For the texts see, eg, D Hunter, J Salzman and D Zaelke, *International Environmental Law and Policy, Treaty Supplement* (New York, Foundation Press, 2002), 345 ff. The CBD has been signed by 188 contracting parties (including Argentina, Canada and the EC), but not by the US. The CPB was ratified by more than 110 countries (including Argentina, Canada and the EC) as of Dec 2004 and entered into force on 11 Sept 2003; its legal relationships with WTO law are analysed by Zarilli, n 41 above, chap III, and M Böckenförde, n 10 above. The issue of GMOs is also addressed in the Preamble to and Art 6 of the 1998 Aarhus Convention on Access to Information, Public Participation in Decision-making and Access to Justice on Environmental Matters which entered into force in Oct 2001.

[44] See Zarilli, n 41 above, chapter IV, and EU Petersmann, n 16 above.

and its successor Directive 2001/18, even if the competent EC Scientific Committee and the European Food Safety Authority had confirmed the absence of health and biosafety risks.[45] Based on the various safeguard clauses in the EC Directives and Regulations, six EC Member States adopted marketing or import bans on biotech products that had previously been approved by the EC. Several EC and Member State officials had spoken of a 'moratorium' pending the adoption and entry into force of new, stricter EC Directives and Regulations for the approval, labelling and traceability of GMOs and GM products. Following the entry into force of these new EC rules in April 2004, several GM products have been approved by the EC since May 2004. Yet, the WTO complainants continued to argue that the *de facto* moratorium practised by the EC from 1998 to April 2004, and the national import prohibitions for GMO products imposed by several EC Member States notwithstanding prior positive risk assessments by the competent EC authorities for the products concerned, violated the WTO obligations of the EC and EC Member States (eg, under GATT 1994, the SPS and TBT Agreements). The EC defended its measures as being consistent not only with WTO law, but also with the Biosafety Protocol (see section 4.3 below).

The biotech regulations of the EC and of individual EC Member States have also been challenged in an increasing number of dispute settlement proceedings in the European Court of Justice (ECJ), for example regarding the competences of the EC to conclude the CPB,[46] the interpretation and validity of the EC regulations for novel foods,[47] the consistency of the EC rules for biotechnological inventions with human rights,[48] and national provisions on banning the use of GMOs.[49]

4.2. Trade-related GMO Rules in the 2000 Cartagena Protocol on Biosafety

The safe transfer, handling and use of 'living modified organisms' (LMOs), including GMOs, and 'advance informed agreement' procedures for imports of LMOs, are specifically regulated in the 2000 CPB, which entered into force in September 2003, having been ratified by

---

[45] On the European Food Safety Authority and the EC legal framework for GMOs, see, eg, G Majone (ed), *Risk Regulation in the EU: Between Enlargement and Internationalization* (Florence, EUI, 2004).

[46] Opinion 2/00 [2001] ECR I–9713.

[47] Case C–236/01 *Monsanto Agricoltura Italia SpA v Presideuza del Consigliodei Ministry* [2003] ECR I–8105, regarding EC Regulation 258/97 on Novel Foods.

[48] Case C–377/98 *Netherlands v European Parliament and Council* [2001] ECR I–7079, regarding EC Directive 98/44 on legal protection of biotechnological inventions.

[49] In a judgment of 5 Oct 2005 (Joined Cases T–366/03, T–235/04, *Land Oberösterreich and Republic of Austria v EC Commission*), not yet reported, the EC Court of First Instance confirmed that the ban on the use of GMOs in Upper Austria violated EC law.

50 countries.[50] As a Protocol to the CBD, it is primarily concerned with the conservation and sustainable use of biodiversity, including risks to human health. Overlaps and potential conflicts with WTO rules appear to exist with regard to their respective precautionary approaches, risk assessment procedures, trade and labelling rules. For example:

(1) The right to adopt precautionary import restrictions 'in cases where relevant scientific evidence is insufficient' is more limited under Article 5(7) of the SPS Agreement (eg, by obligations to seek additional information and review the measures within a reasonable period of time) than under the CPB (cf Articles 10, 11: imports may be restricted without time limits and without seeking additional information to reach scientific certainty). According to WTO case law, Article 5(7) of the SPS Agreement permits provisional SPS measures (i) 'in cases where relevant scientific evidence is insufficient' and (ii) 'on the basis of available pertinent information' provided Members (iii) 'seek to obtain the additional information necessary for a more objective assessment of risk' and (iv) 'review the sanitary or phytosanitary measure within a reasonable period of time'.[51]

(2) The risk assessment and risk management requirements in Article 5 of the SPS Agreement are in some respects more stringent (eg, concerning 'the objective of minimizing negative trade effects' and avoiding 'arbitrary or unjustifiable distinctions in the levels' of protection) than those in Article 15 of the CPB (which allows the importing country to request the exporter to carry out the risk assessment).

(3) Article 26 of the CPB appears to permit broader justifications of trade-restrictive measures by 'socio-economic considerations arising from the impact of LMOs on the conservation and sustainable use of biological diversity' compared with Article 5(3) of the SPS Agreement.

(4) The documentation and labelling requirements in Article 18 of the CPB for the handling, safe transport, packaging, segregation and identification of LMOs may run counter to those of the TBT Agreement (eg, its requirement of non-discriminatory treatment of 'like products').

At a meeting in February 2004, the parties to the CPB agreed on specific trade rules for LMOs relating, inter alia, to documentation and handling

---

[50] See n 43 above. The definition of LMOs in the Protocol ('any living organism that possesses a novel combination of genetic material obtained through the use of modern biotechnology') differs from the definition of GMOs in EC Directive 2001/18 ('an organism, with the exception of human beings, in which the genetic material has been altered in a way that does not occur naturally by mating and/or natural recombination'). The Protocol covers LMOs for voluntary introduction into the environment (eg, seeds for planting), for contained use and for direct use as food, feed or for processing (eg, GM crops like soyabean, maize, tomato, cotton), but not consumer products derived from LMOs (such as corn flakes, tomato ketchup, seed-oil).

[51] On these four cumulative requirements of Art 5(7) see the Appellate Body report on *Japan–Measures Affecting Agricultural Products* (WT/DS76/AB/R of Oct 1998). In *Japan–Measures Affecting the Importation of Apples* (WT/DS245/AB/R of Nov 2003), the Appellate

requirements, risk assessment, risk management, the socio-economic factors that may be taken into account in decision-making processes, liability and redress in the event of accidents or incidents where LMOs cause damage (eg, to human health or biodiversity).[52] The potential impact of the Biosafety Protocol on the interpretation of WTO rules in WTO dispute settlement proceedings remains to be clarified.[53] The Preamble to the Biosafety Protocol addresses the relationship between the Biosafety Protocol and other international agreements, including trade agreements, in an ambiguous manner by stating that 'this Protocol shall not be interpreted as implying a change in the rights and obligations of a Party under any existing international agreements', subject to the 'understanding that the above recital is not intended to subordinate this Protocol to other international agreements'.

4.3. WTO Agreements Relevant for Trade in GMOs and GM Products

In the above-mentioned WTO dispute settlement proceeding, Argentina, Canada and the US challenge the consistency of the EC's adoption of a moratorium on the approval of biotech products, as well as the national marketing and import bans on biotech products maintained by EC Member States, with the EC's obligations under the WTO Agreements on SPS, the Agreement on Agriculture, the TBT Agreement and GATT 1994. The legal relationships among these various WTO agreements are only in part explicitly regulated. The TBT Agreement states that its provisions do not apply to SPS measures (Article 1(5)). The General Interpretative Note to Annex 1A of the WTO Agreement makes clear that, in the event of a conflict between GATT 1994 and the SPS/TBT Agreements, the more specific provisions in the SPS/TBT Agreements shall prevail. To the extent that there is no such conflict, it is now well established—according to the WTO Appellate Body—'that the WTO Agreement is a 'Single Undertaking' and therefore all WTO obligations are generally cumulative and Members must comply with all of them simultaneously'.[54] While measures conforming to the SPS Agreement are presumed to be in accordance with GATT 1994 (cf Article 2(4) of the SPS), there is no equivalent legal presumption for measures conforming to the

---

Body clarified that 'relevant scientific evidence will be 'insufficient' within the meaning of Article 5.7 if the body of available scientific evidence does not allow, in quantitative or qualitative terms, the performance of an adequate assessment of risks as required under Article 5.1 and as defined in Annex A to the SPS Agreement' (para 179).

[52] Cf Zarilli, n 41 above, at 26.
[53] Cf Böckenförde, n 10 above, at 417 ff.
[54] Appellate Body Report, *Korea-Dairy Safeguards*, WT/DS98/AB/R, adopted 12 Jan 2000, paras 23, 24 and 594.

TBT Agreement. It remains to be clarified to what extent inconsistencies of TBT or SPS measures with GATT provisions may be justifiable under the exception clauses of GATT 1994 (such as Article XX); and whether, as claimed in the GMO dispute pending in the WTO, the different components of measures pursuing health as well as environmental protection objectives can be examined cumulatively under the SPS as well as TBT legal disciplines.

The SPS Agreement establishes 'a multilateral framework of rules and disciplines to guide the development, adoption and enforcement of sanitary and phytosanitary measures in order to minimize their negative effects on trade' (Preamble). The Agreement recognises that 'Members have the right to take sanitary and phytosanitary measures necessary for the protection of human, animal or plant life or health, provided that such measures are not inconsistent with this Agreement' (Article 2(1)). Members shall ensure that any SPS measure 'is applied only to the extent necessary to protect human, animal or plant life or health, is based on scientific principles and is not maintained without sufficient scientific evidence' (Article 2(2)), except if a Member provisionally adopts an SPS measure 'in cases where relevant scientific evidence is insufficient' (Article 5(7)). SPS measures in conformity with the SPS Agreement 'shall be presumed to be in accordance with the obligations … under GATT 1994 which relate to the use of sanitary or phytosanitary measures' (Article 2(4)).

The TBT Agreement aims 'to ensure that technical regulations and standards, including packaging, marking and labelling requirements, and procedures for assessment of conformity with technical regulations and standards do not create unnecessary obstacles to international trade' (Preamble). It recognises 'that no country should be prevented from taking measures necessary to ensure the quality of its exports, or for the protection of human, animal or plant life or health, of the environment, or for the prevention of deceptive practices, at the levels it considers appropriate, subject to the requirement that they are not applied in a manner which would constitute a means of arbitrary or unjustifiable discrimination between countries where the same conditions prevail or a disguised restriction on international trade, and are otherwise in accordance with the provisions of this Agreement' (Preamble).

In the above-mentioned WTO dispute, the US claims that the EC practised a general *de facto* moratorium, without scientific and legal justification, which delayed the 27 pending applications for placing biotech products on the EC market at various stages of the approval process, in violation of Articles 2(2), 2(3), 5(1), 5(5), 7 and 8 of the SPS Agreement. The EC's failure to consider for approval each of the 27 pending applications for biotech products also constituted, according to the US, separate 'SPS measures' (= product specific *moratoria*) that:

— (1) imposed 'undue delay' on the completion of the approval procedures, and applied approval procedures in a non-transparent manner in violation of Article 8 of the SPS Agreement;
— (2) failed to publish promptly the product specific moratoria in violation of Article 7;
— (3) failed to base the product-specific moratoria on risk assessments and scientific principles in violation of Articles 2(2) and 5(1); and
— (4) applied arbitrary or unjustifiable distinctions in its level of protection which resulted in discrimination or a disguised restriction on trade in violation of Articles 2(3) and 5(5) of the SPS Agreement.

As regards the national EC Member State bans, the US claims that they are 'based on' neither risk assessments as required under Article 5(1), nor on scientific principles or 'sufficient scientific evidence' as required under Article 2(2) of the SPS Agreement. According to the US, none of the EC Member State bans bore a 'rational relationship' to the only relevant risk assessments on the record in this dispute, ie, the EC Scientific Committees positive opinions. In the face of positive risk assessments with respect to each product subject to a Member State ban, the Member State bans did not meet any of the four criteria necessary to bring them within the scope of Article 5(7). While the US has not made any claim under the TBT Agreement (although it reserved the right to do so), Argentina and Canada also claimed violations of the TBT Agreement (notably Articles 2(1), (2), (9) and 5) in case the Panel should find that the product-specific marketing bans are not SPS measures, or that those parts of the measures which do not pursue SPS objectives are covered by the TBT Agreement in addition to the SPS Agreement. According to the US, the Greek import ban violates Article XI(1) GATT. Argentina and Canada claim also a violation of the national treatment requirement in Article III(4) of GATT 1994 because GM products were treated less favourably than 'like' conventional products.

5. GENETIC ENGINEERING, BIOMEDICINE, HUMAN RIGHTS AND TRADE

As explained above (sections 2 and 3), the universal recognition of 'inalienable' human rights by all 191 UN member states justifies examining the interrelationships between trade, economic, biotechnology rules and human rights not only from intergovernmental perspectives, but also in the light of the *erga omnes* obligations of governments *vis-à-vis* their citisens relying on the rule of law. It is individuals, not states, who are the subjects of human rights and the sources of democratic legitimacy, and who produce, trade in and consume goods and services (like biomedicine). From a human rights perspective, governments and intergovernmental rules and policies are *instruments* for the protection of the human rights and basic needs of citisens. Human rights offer a coherent framework for promoting, protecting and balancing individual and public interests, not only in the field of health and biotechnology policies, but also in international economic law regulating the international division of labour among producers, investors, traders and consumers. UN human rights instruments and WTO law leave every WTO Member broad national policy discretion for reconciling the utilitarian WTO objectives of promoting economic welfare and 'sustainable development' with due respect for human rights. The competing utilitarian, state-centred and human rights philosophies can and should be reconciled in a mutually coherent manner.

5.1. Human Rights as Constitutional Principles and Restraints of International Economic Law?

Modern international law is based on 'universal respect for, and observance of, human rights and fundamental freedoms for all' (Articles 1, 55 of the UN Charter). All UN human rights covenants (such as the 1966 Covenants on civil, political, economic, social and cultural rights) emphasise that human rights 'derive from the inherent dignity of the human person'; 'recognition of the inherent dignity and of the equal and inalienable rights of all members of the human family is the foundation of freedom, justice and peace in the world'. This universal legal obligation to respect human dignity as a source of 'inalienable' human rights can be seen as a *ius cogens* 'core obligation' of all UN member states limiting all national and intergovernmental powers for the benefit of human rights.[55] Beyond this core obligation, however, the legal status and role of human rights in international biotechnology and trade regulation remains contested among governments in many respects. For example:

— To what extent have the relevant guarantees in the Universal Declaration of Human Rights (UDHR) evolved into general legal obligations to protect, eg, rights to liberty (Article 3), property (Article 17), freedom of thought (Article 18), rights to food and medical care (Article 25), and

---

[55] Cf EU Petersmann, 'Taking Human Dignity, Poverty and Empowerment of Individuals More Seriously: Rejoinder to Alston' [2002] *European Journal of International Law* 845; *idem*, n 13 above, at 25–6. On the impact of universal human rights on the sources and structure of international law see O. de Frouville, *L'Intangibilité des droits de l'homme en droit international* (Paris, Pedone, 2004), at 25–6, 266.

rights to protection of moral and material interests resulting from scientific production (Article 27)?

— Are intellectual property rights (such as copyrights and other authors' rights in the integrity of the work and of its author) to some extent protected by human rights (eg, in case of 'takings' and expropriation of private property)? How must intellectual property rights be legally limited in order to remain compatible with human rights (eg, of access to essential medicines)?

— Are UN human rights conventions of a higher legal rank than international trade, environmental and intellectual property agreements? Or do the latter also promote basic human needs the legal regulation of which must be co-ordinated and 'balanced' with human rights depending upon the particular context?

— Are international trade rules sufficiently flexible for protecting and promoting human rights and divergent democratic preferences, including national and EU guarantees (such as freedom of profession and other economic liberties) that are not recognised in UN human rights conventions?

In the inevitable 'balancing' and co-ordination of human rights with other international rules, it is important to distinguish the 'empowering functions' from the 'limiting functions' of human rights, and to take the 'inalienable', 'indivisible' core of human rights and the legitimate diversity of national human rights traditions into account as relevant 'context' for the interpretation of international trade and biotechnology rules.

5.1.1. Human Dignity and Human Rights as Empowerment of Individuals

According to Article 1 of the UDHR, '[a]ll human beings are born free and equal in dignity and rights' (paragraph (1)); '[t]hey are endowed with reason and conscience and should act towards one another in a spirit of brotherhood' (paragraph (2)). 'Reason and conscience' are thus defining elements of humanity and dignity.[56] The recognition of the need for moral conduct 'in a spirit of brotherhood', of rights to democratic self-government (Article 21 of the UDHR) and 'social security' (Articles 22, 23 of the UDHR), and of everyone's 'duties to the community in which alone the free and full development of his personality is possible' (Article 29 of the UDHR), further acknowledges the social responsibility and vulnerability of individuals in their pursuit of an 'existence worthy of human dignity' (Article 23 of the UDHR). The legally and institutionally fragmented UN human rights conventions—through their recognition and legal protection of civil, political, economic, social and cultural rights—reflect a broad conception of human dignity as requiring

---

[56] Cf K Dicke, 'The Founding Function of Human Dignity in the Universal Declaration of Human Rights' in D Kretzmer and E Klein (eds), *The Concept of Human Dignity* (The Hague, Kluwer, 2002), 111,117.

'inalienable', national and international legal safeguards in all areas of personal self-development at home and abroad.

Human dignity is likewise recognised as a constitutive principle and fundamental human right in Article 1 of the EU Charter of Fundamental Rights proclaimed by the European Parliament, the EU Commission and the EU Council in December 2000 and incorporated into Part II of the 2004 Treaty establishing a Constitution for Europe (TCE).[57] According to the ECJ, 'it is for the Court of Justice, in its review of the compatibility of acts of the institutions with the general principles of Community law, to ensure that the fundamental right to human dignity and integrity is observed'.[58] Like UN law, EU law does not offer a precise legal definition of the concept of human dignity. Yet, by defining and legally protecting specific dignity rights (Articles II–61–65), freedoms (Articles II–66–79), equality rights (Articles II–80–86), solidarity rights (Articles II–87–98), EU citizen rights (Articles II–99–106) as well as rights of access to justice (Articles II–107–110) that are all 'founded on the indivisible, universal values of human dignity, freedom, equality and solidarity' (Preamble to the EU Charter of Fundamental Rights), the EU Treaty Constitution protects human dignity and the 'indivisibility' of human rights in a comprehensive manner complementing other national and international human rights guarantees (like the ECHR, cf Article II–112 TCE).

In national human rights laws, only a few national constitutional systems (eg, in Germany, India, Israel, South Africa) explicitly recognise human dignity as a constitutional value or human right. Political and legal conceptions of human rights continue to differ considerably, for example regarding the legal status of human dignity as a constitutional principle or human right. Even where human dignity is not explicitly mentioned in human rights conventions (like the ECHR) and national constitutions (like the US and French Constitutions), the ECtHR and national courts have often referred to human dignity as one of the values underpinning human rights (eg, Article 3 of the ECHR concerning inhuman or degrading treatment): 'the very essence of the Convention is respect for human dignity and human freedom'.[59] Arguably, the link between human dignity and human rights implies that 'dignitarian arguments'

---

[57] The text of the Charter of Fundamental Rights is published at [2000] OJ C/364/1, and is also included in the Treaty establishing a Constitution for Europe [2004] OJ C/310/1. On the controversy over whether Art 1 recognises a fundamental right to human dignity or merely an objective constitutional principle, see the commentary on Art 1 of the Charter by M Borowsky in J Meyer (ed), *Die Grundrechtscharta der EU* (Baden-Baden, Nomos, 2003), 45–66.

[58] Case C–377/98, n 48 above, at para 70.

[59] European Court of Human Rights, *Pretty v United Kingdom, no 2346/02, ECHR. 2002-III, para 65; SW v UK and CR v UK*, judgment of 22 Nov 1995, Series A no 335–B, 363, paras 42, 44.

should not operate as absolute constraints ('conversation stoppers') but must be balanced with other human rights (as 'conversation guarantees') requiring respect for individual autonomy (eg, informed consent as a potential justification of biomedical treatments).[60]

5.1.2. Respect for Human Dignity Requires Additional Constitutional Rights

Regardless of whether human dignity is recognised as the most funda-mental human right from which all other rights flow (as, eg, in German and Israeli constitutional law), or whether human dignity is viewed only as an objective constitutional principle, human dignity as empowerment requires respect for the legal and democratic autonomy of citizens to define their respective human rights and other 'fundamental freedoms' through constitutional contracts. Constitutionalism rests on the premise that—in order to limit abuses of liberty ('paradox of liberty') and avoid conflicts between rational long-term interests (eg, in equal freedoms) and emotional short-term inclinations (eg, to hand over government powers to a dictator, as in Germany in 1933)—liberty must be constitutionally protected and restrained by self-imposed rules of a higher legal rank. German constitutional law (eg, Articles 2, 12, 14 of the Basic Law) and EU law, for example, protect 'fundamental freedoms and non-discrimination' (Article I–4 of the TCE) in the economy no less than in the polity. The German constitutional concept of a general right to liberty (in addition to specific liberty rights and human rights[61]), like the ECJ judicial protection of the EC Treaty's intergovernmental 'market freedoms' as 'fundamental rights' of EU citizens,[62] can be morally justified by the Kantian view of human dignity as requiring individual freedom to be exercised in a 'just' manner respecting maximum equal freedom of all others.[63]

According to Kant, respect for human dignity and justice requires human beings to be treated as ends in themselves, respecting their individual

---

[60] On the competing bioethical philosophies see Brownsword in this volume.

[61] See R Alexy, *A Theory of Constitutional Rights* (Oxford, Oxford University Press, 2002), chap 7.

[62] On the relevance of Kant's legal theory for explaining the legitimacy and liberties of EU constitutional law see EJ Mestmäcker, *Wirtschaft und Verfassung in der EU* (Baden-Baden, Nomos, 2003), 78, 133 ff.

[63] According to Kant, 'freedom constitutes man's worth', and 'freedom (independence from the constraints of another's will) insofar as it is compatible with the freedom of every-one else in accordance with a universal law, is the sole and original right that belongs to each human being by virtue of his humanity' (cf I Kant, 'The Metaphysics of Morals' in H Reiss (ed), *Kant—Political Writings* (Cambridge, Cambirdge University Press, 1991), 136; AD Rosen, *Kant's Theory of Justice* (Ithaca, NY, Cornell University Press, 1993), 9,42). For Kant, equality and all other human rights derive from this universal birthright of freedom. Justice requires, according to Kant, exercising one's freedom in accordance with the moral 'categori-cal imperative' that individuals and a just constitution must allow the 'greatest freedom'

choices, and protecting maximum equal freedoms of individuals through ever more precise, national, international and cosmopolitan constitutional rules.[64] As everybody 'may seek his happiness in whatever way he sees fit, so long as he does not infringe upon the freedom of others to pursue a similar end which can be reconciled with the freedom of everyone else within a workable general law',[65] treating a person as an end-in-herself and respecting her freely chosen objectives may require legal protection of maximum equal freedom in all areas of personal self-development, as recognised in Article 2 of the German Basic Law and in the comprehensive guarantees of 'fundamental freedoms' in the EU Charter of Fundamental Rights and the 2004 EU Treaty Constitution.[66] For only the individual herself can know, and decide on, her own ends, preferences and personal self-development. Hence, human dignity and human liberty are indivisible, as recognised in the 1993 Vienna Declaration adopted by the World Conference on Human Rights ('[a]ll human rights are universal, indivisible and interdependent and interrelated'[67]) as well as in the foundation of the EU on 'the indivisible, universal values of human dignity, freedom, equality and solidarity' (Preamble to the EU Charter of Fundamental Rights). According to Kant, law should be defined and protected as 'the sum total of those conditions within which the will of one person can be reconciled with the will of another in accordance with a universal law of freedom'; it follows from the moral 'categorical imperative' that 'every action which by itself or by its maxim enables the

---

for each individual along with 'the most precise specification and preservation of the limits of this freedom so that it can coexist with the freedom of others'. According to Kant, only such national, international and cosmopolitan 'constitutional rules of justice'–based on ever more precisely defined equal freedoms and democratic self-government–offer the prospect of 'perpetual peace'.

[64] On Kant's moral 'categorical imperatives' for acting in accordance with universal laws ('[a]ct only in accordance with that maxim through which you can at the same time will that it become a universal law'), for respecting human dignity by treating individuals and humanity as ends in themselves ('[s]o act that you use humanity, whether in your own person or that of another, always at the same time as an end, never merely as a means'), and for respecting individual autonomy ('the idea of the will of every rational being as a will giving universal law') and individual right ('[a]ny action is right if it can coexist with everyone's freedom according to a universal law'), see also AW Wood, *Kant's Ethical Thought* (Cambridge, Cambridge University Press, 1999). On Kant's theory of the antagonistic human nature promoting market competition and constitutional 'rules of justice', and on the Kantian requirement of transforming the 'wild freedom' in the state of nature into lawful freedom through ever more precise national, international and cosmopolitan constitutional rules see also EU Petersmann, 'How to Constitutionalize International Law and Foreign Policy for the Benefit of Civil Society?' (1999) 20 *Michigan Journal of International Law* 1.

[65] I Kant, 'On the Common Saying: This May be True in Theory, but it does not Apply in Practice' in Reiss, n 53 above, at 74.

[66] On the constitutional concept of a general right to liberty see Alexy, n 61 above, chap 7.

[67] Para 5 of the Declaration, reproduced in *The United Nations and Human Rights 1945-1995* (New York, UN, 1995) 450.

freedom of each individual's will to co-exist with the freedom of everyone else in accordance with a universal law is *right*.[68] Such constitutional protection of 'fundamental freedoms' (eg, in EU law) may be no less important for personal self-development and for mutually beneficial co-operation among free citizens (including division of labour across discriminatory border barriers) than the classical civil and political human rights, even if such economic freedoms (such as freedom of profession, property rights) are not mentioned in UN human rights conventions. The historical reasons for the incomplete protection of individual liberties in UN human rights instruments must not lead to the wrong conclusion that other, non-enumerated personal liberties do not deserve constitutional protection.

5.1.3. Human Dignity and Human Rights as Constitutional Constraints

The universal *ius cogens* obligations to respect human dignity and inalienable human rights require not only empowerment of individuals through constitutional rights, but also constitutional restraints on all abuses of public and private powers.[69] Human dignity reveals itself in moral and rational human choice which merits respect as the source of all values[70] and justifies a prima facie right to maximum equal freedom in all areas of personal self-development, subject to democratic legislation and other human rights as 'constitutional restraints'. From this perspective, personal self-development by means of exercising *economic* freedoms (eg, freedom of profession and economic liberty as protected in Articles 2, 12 of the German Basic Law, Articles 15, 16 of the EU Charter of Fundamental Rights) may be no less rooted in human dignity than the exercise of personal freedom in non-economic areas.

The 1997 UNESCO Declaration on the Human Genome and Human Rights, the 1997 Council of Europe Convention on Human Rights and Biomedicine, as well as other modern human rights instruments define the legal limits of modern science and medicine in terms of human dignity and human rights empowering and restraining individual freedom.[71] The necessary legal and judicial balancing of

---

[68] I Kant, 'The Metaphysics of Morals' in Reiss, n 53 above, 133.

[69] On the 'constitutional approach' to international law see EU Petersmann, 'Constitutional Primacy and Indivisibility of Human Rights in International Law?' in S Griller (ed), *International Economic Governance and Non-Economic Concerns* (Vienna, Springer, 2003), 211–66. On 'human dignity as empowerment' and 'human dignity as constraint' see D Beyleveld and R Brownsword, *Human Dignity in Bioethics and Biolaw* (Oxford, Oxford University Press, 2001).

[70] Cf AD Rosen, n 53 above, at 62. On the distinction, eg, in German law and in the case law of the ECtHR, between human dignity as an absolute core limit to state interference, and secondary rights deriving from the concept of human dignity and subject to limitation, see JA Frowein, 'Human Dignity in International Law' in Kretzmer and Klein, n 56 above, at 123.

[71] Cf Beyleveld and Brownsword, n 69 above.

conflicting human rights claims—for example, in the famous French dwarf-throwing dispute,[72] of the autonomy rights invoked by the dwarf (such as respect for freedom of profession and for self-respect gained by being employed) and by the municipalities prohibiting 'dwarf-throwing' as contrary to *ordre public* (in the sense of a duty not to compromise one's own human dignity)—may legitimately differ among democratic communities, depending on their individual or communitarian value preferences. The idea of dignity as a virtue might justify higher standards of individual responsibility than in situations characterised by inadequate moral and rational autonomy (eg, of children, sick patients, poor and unemployed individuals). Economists—from Adam Smith via Friedrich Hayek up to Nobel Prize laureate Amartya Sen—emphasise the importance of division of labour and consumer-driven market economies for enabling and promoting individual freedom as the ultimate goal of economic life and the most efficient means of realising general welfare.[73] Yet, there are far-reaching differences between the liberal Smithian conception of freedom (eg, as non-interference with individual liberty), the constitutional Hayekian conception (eg, of liberty as constitutional, legislative and judicial guarantees against arbitrary domination of the individual), Sen's social empowerment concept of positive individual freedom, and the related conceptions of the individual (eg, as an atomistic, autonomous being or as individuals embedded in social relationships).[74]

Similarly, while John Locke justified property rights as moral entitlements to the fruits of one's labour provided the valuable goods (or the added value of the good) was produced without violating the rights and basic needs of others,[75] Hegel's *Philosophy of Right* perceives the appropriation, ownership, and alienation of property as expressions of the will, personality, independence, and self-development of the owner in relation to objects and to others who must respect property rights and

---

[72] French Conseil d'Etat (27 Oct 1995), req. nos. 136–727 (Commune de Morsang-sur-Orge), 143–578 (Ville d'Aix-en-Provence).

[73] On defining economic development not only in terms of Pareto efficient satisfaction of utilitarian consumer preferences, but also in terms of individual decisional autonomy, individual 'immunity from encroachment', and substantive 'opportunity to achieve', see A Sen, *Rationality and Freedom* (Cambridge, Mass, Harvard University Press, 2002), chap 17 on 'markets and freedoms'. See also FA Hayek, *The Constitution of Liberty* (London, Routledge 1960), 35: '[e]conomic considerations are merely those by which we reconcile and adjust our different purposes, none of which, in the last resort, are economic (except those of the miser or the man for whom making money has become an end in itself)'.

[74] On different liberal and republican concepts of freedom see P Pettit, *Republicanism. A Theory of Freedom and Government* (Oxford, Oxford University Press, 1997). On the evolving perceptions of the individual in economics see JB Davis, *The Theory of the Individual in Economics* (London, Routledge, 2003).

[75] Cf the 'Essay Concerning the True Origin, Extent and End of Civil Government' in J Locke, *Two Treatises of Government* (Cambridge, Cambridge University Press, 1988), paras

may perceive the personality of individuals (eg, of an artist) in the light of their property (eg, artworks).[76] This often close interrelationship between personality and property illustrates that—just as property may be an existential component of autonomy, personality, and social recognition of an individual—so can deprivation of property imply an attack on the personality of the owner. Joseph Raz and other legal philosophers rightly emphasise the role of property rights as constituent elements of an autonomous life with privacy, material security, and 'positive freedom' that should be limited only to protecting the freedom of others, and requires respect for the fruits of lawful individual choices.[77] As property rights include the right to dispose of one's property, these various moral justifications of legal protection of property rights also justify private rights to supply or demand goods in private markets. As 'enabling devices for individual autonomy',[78] markets and their various social functions (eg, as information, co-ordination and sanctioning mechanisms) can be justified not only on grounds of economic efficiency (eg, as incentives for exertion, creativity, decentralised co-ordination of autonomous actions), but also as preconditions for individual autonomy and for a free, informed, and accountable society, subject to democratic legislation limiting and regulating markets in the public interest.

These various approaches to defining personal and economic development not only in macroeconomic terms but more broadly as freedom, and especially Sen's conception of freedom as empowerment and human capacity for personal self-development, appear more consistent with the universal recognition of human rights (including the so-called 'human right to development') than the state-centred, quantitative conceptions of national income and 'efficiency' cherished by many economists and WTO governments.[79] From a human rights perspective, international economic law should be interpreted as an *instrument* for realising human rights and—like the EU and its Charter of Fundamental

---

27–41, where Locke assumes that, in a state of nature without money, people must limit their unilateral appropriations to a proportional share so that 'enough, and as good' for others is left. Locke's labour theory as the moral justification of just property acquisition raises numerous problems. Notably Locke's views on self-ownership and property rights over one's body appear]inconsistent with respect for human dignity: Individuals have personality rights over their bodies and can claim property rights over body parts only to the extent that such body parts could be separated with the consent of the person (eg, hair) and without violating one's human dignity.

[76] Cf D Knowles, *Hegel and the Philosophy of Right* (London, Routledge, 2002), chap 5.

[77] Cf AR Coban, *Protection of Property Rights within the European Convention on Human Rights* (Aldershot, Ashgate, 2004), 65–77; J Raz, *Morality of Freedom* (Oxford, Clarendon Press, 1986), 372–5.

[78] Cf J Gray, *The Moral Foundations of Market Institutions* (London, The Institute of Economic Affairs, 1992), 3.

[79] Cf Petersmann, n 25 above.

Rights—should be founded 'on the indivisible, universal values of human dignity, freedom, equality and solidarity', 'the principles of democracy and the rule of law'.[80] From such a human rights perspective, many legal questions raised by biotechnology and the modern bioeconomy—such as the legal regulation of trade in body-parts and in biomedical services (eg, to clone human cells, to select the genetic specification of children), and the patenting of inventive work on the human genome—may merit diverse answers depending on the constitutional context, the particular value preferences revealed through 'deliberative democracy', the judicial balancing processes, the scientific evidence, risk assessment procedures and social opportunity costs, and the informed consent or other choice of individuals.

5.2. Human Rights Functions of International Trade Rules?

As shown above (section 3), the human rights approach to international trade law aims at promoting synergies between intergovernmental trade rules and human rights. For example, interpreting the WTO objective of promoting 'sustainable development' (WTO Preamble) in conformity with the UN resolutions on the human right to development could justify interpretations of WTO rules that promote transparency (eg, opening WTO dispute settlement proceedings to the public), participation by citizens (eg, through *amicus curiae* submissions to WTO dispute settlement bodies) and citizens' rights to invoke WTO rules in domestic courts.[81] Interpreting WTO rules and policies as instruments for promoting and protecting human rights could also support arguments in favour of democratic rights to restrict trade in biotech products. Of course, the interpretative relevance of such human rights arguments about the 'object and purpose' and relevant 'context' of WTO rules is limited by 'the ordinary meaning to be given to the terms of the treaty'.[82] Yet, just as human rights law is dynamically evolving in response to new existential problems (eg, by recognising new human rights against abuses of medicine and biology), so should the human rights dimensions of international trade law be seen in a dynamic perspective that must respond to new social challenges.

---

[80] Cf Petersmann, n 13 above.

[81] Cf R Howse, *Mainstreaming the Right to Development into International Trade Law and Policy at the WTO*, ECOSOC document E/CN.4/Sub.2/2004/17 (2004).

[82] According to the customary methods of international treaty interpretation which WTO dispute settlement bodies are required to apply (pursuant to Art 3(2) of the WTO Dispute Settlement Understanding), a 'treaty shall be interpreted in good faith in accordance with the ordinary meaning to be given to the terms of the treaty in their context and in the light of its object and purpose' (Art 31(1) VCLT).

The transformation of GATT's customs union rules into individual 'market freedoms' inside the EC, like the transformation of GATT's free trade area rules into individual 'market freedoms' inside the European Economic Area (EEA), remains the most important example of the potential 'constitutional functions' of WTO rules. The judicial protection of the EC Treaty guarantees of free movements of goods, services, persons, capital (including freedom of establishment) and related payments as 'fundamental freedoms' of producers, investors, traders and consumers, and the related judicial protection of social rights (eg, of migrant workers and their family members), illustrate successful adjustments of national and EC rules to new constitutional challenges that were neglected by human rights and constitutional laws.

Outside the EU, domestic legal systems continue to differ considerably regarding their 'constitutional' and legislative co-ordination of human rights, biotechnology regulations and freedom of trade. The less individual trade rights, intellectual property rights, environmental rights and new 'bio-rights' are related to the protection of personal self-development, the more they may be perceived as matters of mere legislative and administrative rights rather than of human rights or other constitutional rights. Yet, the GATT and WTO guarantees of freedom, non-discrimination, rule of law and social safeguards can ultimately realise their welfare-enhancing objectives for the worldwide division of labour only if they become legally effective for the benefit of the private producers, investors, traders and consumers who determine international production, investments, trade and consumption of goods and services. The incorporation into domestic laws of GATT and WTO obligations to protect individual freedom and other individual rights may complement national constitutional and human rights guarantees of individual freedom, non-discrimination, rule of law and fulfilment of basic needs necessary for a life in dignity.[83]

Human rights appear to leave broad democratic discretion as to whether, for example:

— production and patenting of GMOs and GM products should be promoted (eg, so as to give incentives for research and investment) or restricted (eg, so as to protect traditional farming methods, traditional knowledge, generic medicines);

— whether importation of GM products should be liberalised (eg, so as to offer domestic farmers herbicide-resistant crops and other GM plants with increased yields, and supply domestic citizens with better medicines and food like 'vitamin A rice') or restrained (eg, so as to protect existing

---

[83] This was the central thesis of EU Petersmann, *Constitutional Functions and Constitutional Problems of International Economic Law* (Fribourg, Fribourg University Press, 1991).

biodiversity, promote consumer choice by labelling requirements, monitor GM crops by traceability requirements); and

—to what extent the human right to respect for one's free and informed consent, and against unwilled coercion by others, should be limited by democratic legislation (eg, in order not to compromise the human dignity of other citizens, and to protect all human life from the point of conception onwards).

International economic law must respect this democratic discretion and legitimate regulatory diversity among countries, subject to internationally agreed limits (eg, for the prohibition of genetic engineering of weapons of mass destruction). As biotechnologies (eg, therapeutic cloning producing stem cells for medical research), GM products and related services (eg, gene therapies) are tradable, they are likely to affect ever more areas of WTO law as well as of human rights. So far, there appears to be no evidence for conflicts between WTO rules and universal human rights on the level of international principles. Initiatives by UN bodies to promote respect for human rights and labour standards in the domestic implementation of WTO rules and trade policies are likely also to promote the WTO objectives of 'sustainable development' and of 'raising standards of living', especially if 'development' is being defined—as suggested, eg, by Nobel Prize economist A, Sen—not only in utilitarian, economic terms but more broadly as positive freedom and human capacity for personal self-development.

5.3. What is the Impact of the Indivisibility of Human Rights on International Law?

Just like the 1993 Vienna Declaration of the UN World Conference on Human Rights emphasises that 'all human rights are universal, indivisible and interdependent and interrelated', so are the EU and its Charter of Fundamental Rights 'founded on the indivisible, universal values of human dignity, freedom, equality and solidarity'.[84] Compared with European integration law, however, the about 100 UN human rights instruments and diverse UN human rights institutions remain much less co-ordinated with international economic treaties and institutions (such as the WTO).[85]

---

[84] The quotation is from the Preamble to the EU Charter and is discussed, eg, by J. Kenner, 'Economic and Social Rights in the EU Legal Order: The Mirage of Indivisibility' in T Hervey and J Kenner (eds), *Economic and Social Rights under the EU Charter of Fundamental Rights* (Oxford, Hart Publishing, 2003).

[85] For a comparison of UN and EU human rights law see, eg, D McGoldrick, 'The Charter and UN Human Rights Treaties' in S Peers and A Ward (eds), *The EU Charter of Fundamental Rights* (Oxford, Hart Publishing, 2004). See also EU Petersmann, 'On Indivisibility of Human Rights' [2002] 13 *European Journal of International Law* 381.

The 'indivisibility' of human rights requires the taking into account of all relevant human rights and the promotion, and if necessary reconciliation and 'balancing' of their simultaneous realisation. At national levels of policy-making, such balancing outcomes may legitimately differ among poor and industrialised, importing and exporting countries. For instance, developed countries may invoke human rights (such as freedom of sciences, economic liberties, property rights) in favour of protecting GM crops for better weed and insect control, higher productivity and more flexible crop management, and have promoted biotechnology especially for the production of pharmaceuticals and industrial products by companies in industrialised countries. Less developed importing countries, however, may be concerned by foreign property rights over seeds and GM crops that may be too costly for poor farmers and may adversely affect their production and exportation of conventional crops. At the international level of rule-making in specialised *intergovernmental* organisations, however, protection of 'indivisible' human rights poses more difficult, procedural and substantive problems in view of the 'sovereign equality' of states and the diversity of their national human rights traditions.

The impact of procedural human rights to participate in democratic governance,[86] such as 'the right of everyone to be consulted and participate in significant decision-making processes that affect them',[87] on intergovernmental rule-making in worldwide organisations remains controversial. Specific rules and procedures for the balancing of conflicting human rights, trade and environmental rules exist in a few international trade agreements (eg, the SPS, TBT and TRIPS Agreements), environmental agreements (eg, the 1997 Aarhus Convention on Access to Information, Public Participation in Decision-Making, and Access to Justice in Environmental Matters) and human rights instruments (eg, the 1997 Universal UNESCO Declaration on the Human Genome and Human Rights). Even if such environmental agreements and human rights instruments have not been accepted by all WTO Members, they may reflect 'common concerns of mankind' that may be relevant for interpreting WTO rules (eg, on import restrictions protecting 'public

---

[86] On the emerging 'human right to democratic governance' see, eg, TM Franck, 'The Emerging Right to Democratic Governance' (1992) 86 *American Journal of International Law* 46; GH Fox and BR Roth (eds), *Democratic Governance and International Law* (Cambridge, Cambridge University Press, 2000); E Stein, 'International Integration and Democracy: No Love at First Sight' (2001) 95 *American Journal of International Law* 489. On the recognition and promotion of human rights by international organisations see more recently *Interdependence between democracy and human rights, Report of the Office of the High Commissioner for Human Rights*, E/CN.4/2004/54 of 17 Feb 2004.

[87] Cf *Substantive Issues Arising in the Implementation of the International Covenant on Economic, Social and Cultural Rights*, ECOSOC E/C.12/2001/15 of 14 Dec 2001.

morals' and 'public order' inside the importing country). The universal recognition of the 'inalienable' and 'indivisible' character of human rights calls for a comprehensive, transparent balancing among different human rights also in the law of worldwide organisations (eg, rights to the integrity of each person, consent requirements for medical interventions). Certain provisions in the UNESCO Declaration on the Human Genome and Human Rights (eg, Article 2) and in the 1997 European Convention on Human Rights and Biomedicine (eg, Articles 6, 11)—for example, by prohibiting 'discrimination based on genetic characteristics' and practices that are 'contrary to human dignity, such as reproductive cloning of human beings' and other 'germ-line interventions' (Article 24 of the Convention on Human Rights and Biomedicine)—seem to be more concerned with 'human dignity as legal constraint' than with protecting autonomous choice. Yet, they do not explicitly outlaw biotechnologies (such as sex-based embryo selection) that may be regulated differently in different countries depending on their prevailing bioethics and their respective balancing of human values (such as autonomy, dignity, integrity, vulnerability of human beings, protection of the embryo).

5.4. Respect for the Different National and International Human Rights Traditions

The universal UN human rights instruments tend to include minimum standards and recognise that national legislative, administrative and judicial safeguards and balancing of human rights may differ depending on the individual and democratic value preferences and resources of the people and communities concerned. National human rights guarantees (eg, for indigenous people), regional human rights conventions (eg, the 1997 Council of Europe Convention on Human Rights and Biomedicine), constitutional guarantees of fundamental freedoms and other constitutional rights (eg, in EU law) often complement UN human rights law by offering higher and additional, substantive or procedural guarantees. For example, human rights of access to health services may be protected in developed countries more comprehensively (eg, access to gene therapies) than in poor countries. And modern human rights guarantees (such as the prohibition of eugenic practices in Article II–63 of the TCE) may not be recognised in constitutional democracies whose 'Bills of Rights' were drafted centuries ago (like the US Constitution), and whose legal systems may protect human dignity and social rights by means of legislation rather than by means of human rights.[88]

---

[88] Cf EJ Eberle, *Dignity and Liberty. Constitutional Visions in Germany and the United States* (Westport, Conn, Praeger, 2001). On the 'federal rejection' but 'state protection' of economic

International trade law and international environmental law leave broad 'policy space' for national and regional diversity in the regulation of trade (cf Articles XXIV of GATT, V of GATS) and in the protection of human, animal or plant life or health (cf Articles XX of GATT, XIV of GATS). The diversity of national, regional and worldwide human rights guarantees entails that 'human rights impact assessments' of trade rules at national or regional levels may legitimately differ from those at worldwide levels; they should be undertaken by specialised human rights bodies and national governments rather than by WTO bodies without expertise in human rights. Whether countries want to admit or prohibit commerce in human organs, biotech services and service suppliers (eg, for helping couples to destroy or donate an embryo, terminate a pregnancy, select the sex of their child) depends on the particular conceptions of human dignity and human rights in national societies and may give rise to exports from a more pluralistic society, the importation of which may be prohibited in less tolerant societies.

5.5. States as Primary Duty Bearers for the Protection of Vulnerable Individuals and Groups

Millions of people suffer from unnecessary poverty, hunger, malnutrition, lack of access to medical care, unsafe drinking water, inadequate housing and education, and ineffective protection of human rights. The UN High Commissioner for Human Rights rightly emphasises that:

> 'a human rights-based approach focuses particularly on the needs of the most disadvantaged and marginalized individuals and communities. Because a human right is a universal entitlement, its implementation is evaluated particularly by the degree to which it benefits those who hitherto have been the most disadvantaged and marginalized and brings them up to mainstream level. Thus, in adopting intellectual property regimes, States and other actors must give particular attention at the national and international levels to the adequate protection of the human rights of disadvantaged and marginalized individuals and groups, such as indigenous peoples.'[89]

This focus of human rights on vulnerable individuals and groups differs from a 'constitutional approach' that emphasises the legal equality of all citizens (eg, in the exercise of their democratic rights). The UNHCHR reports also note that 'trade liberalization will create losers even in the long

---

and social rights in the US see JM Woods and H Lewis, *Human Rights and the Global Marketplace. Economic, Social and Cultural Dimensions* (New York, Transnational Publishers, 2004), chap 10.C.

[89] N 87 above.

run and that trade reforms could exacerbate poverty temporarily. Human rights law concerns itself in particular with the situation of the individuals and groups who might suffer during the reform process'.[90] UN human rights law recognises rights of '[e]veryone ... to a social and international order in which the right and freedoms set forth in this Declaration can be fully realized' (Article 28 of the UDHR), as well as corresponding obligations of governments and intergovernmental organisations to respect and protect the human rights not only of domestic citizens but also of foreigners suffering from unnecessary poverty and disregard of their human rights, for example rights 'to a standard of living adequate for the health and well-being of himself and of his family, including food, clothing, housing and medical care' (Article 25 of the UDHR).

Yet, as the persistence of severe poverty is often due to local causes (eg, poor resources, bad governance, capital flight, civil wars), the 'cosmopolitan responsibilities' of governments for the protection of human rights *abroad* remain controversial.[91] The numerous WTO provisions for special and differential treatment of developing countries have been equated by the UN High Commissioner for Human Rights with 'human rights notions of affirmative action as well as international cooperation under the ICESCR, the Convention on the Rights of the Child and the Declaration on the Right to Development'.[92] The controversies over food aid consisting of GMOs and GM food illustrate the complex balancing processes required by human rights, for example between human rights to health services and to adequate food (which may argue in favour of liberalising trade in GM products if they do not pose risks to human health) and democratic rights to restrict imports of GM products that may adversely affect domestic natural products inside the importing country and its export opportunities. Protection of farmers' rights, of traditional knowledge of indigenous people, or of freedom of choice of consumers may likewise justify restrictions of imports of GM products. The UN High Commissioner for Human Rights has criticised WTO rules for recognising the protection of public health and of other human rights objectives only as exceptions to liberal trade rules: '[a] human rights approach ... would explicitly place the promotion and protection of human rights, in particular those in the ICESCR, at the heart of the objectives of intellectual property protection' and of other trade rules.[93] Yet, this criticism ignores the fact that human rights and human health should be protected

---

[90] E/CN.4/2002/54, n 19 above, para 34.
[91] Cf TW Pogge, *World Poverty and Human Rights: Cosmopolitan Responsibilities and Reforms* (Cambridge, Polity Press, 2002).
[92] E/CN.4/Sub.2/2001/13, n 18 above, at para 20.
[93] Ibid, at para 22.

primarily by *non-discriminatory* rules that may be freely applied by WTO members without any recourse to 'WTO exceptions'.

5.6. Interpretation of WTO Rules in Conformity with the Obligations of WTO Members under Human Rights Law, Environmental and Biotechnology Law

All international agreements remain incomplete and include rules the legal meaning of which may need to be clarified by taking into account the 'context, object and purpose' of the treaty terms as well as 'any relevant rules of international law applicable in the relations between the parties'.[94] In the Preamble to the WTO Agreement, all WTO Members commit themselves to 'the basic principles and ... objectives underlying this multilateral trading system.' The WTO case law on the general international *legal principles* underlying the WTO legal system has so far not referred to human rights.[95] The Appellate Body's judicial approach towards balancing the rights and obligations of WTO Members on the basis of general principles of international law (such as good faith, prohibition of abuse of rights[96]) could change if, in addition to general principles for the balancing of reciprocal rights and obligations of states, WTO dispute settlement bodies were to use also the universal human rights obligations of WTO Members as balancing principles that apply *erga omnes* and protect *individual* rights. Whereas trade diplomats tend to view WTO law as a limited trade regime and the WTO institutions as a 'Member-driven framework' for intergovernmental trade bargains, WTO judges—notwithstanding their commitment to 'the principles for the management of disputes heretofore applied under Articles XXIII and XXIIII of GATT 1947' (Article 3(1) of the DSU)—cannot interpret WTO rights and obligations 'in clinical isolation from international law'.[97]

The WTO Appellate Body has repeatedly emphasised the need to take into account, in the interpretation of WTO rules, the 'contemporary concerns of the community of nations about the protection and conservation of the environment'.[98] In the abovementioned WTO dispute over the delays in the EC approval procedures for GM products, the EC

[94] Art 31 VCLT which—according to the WTO Appellate Body—reflects the customary methods of treaty interpretation recognised in customary international law.
[95] See, eg, Petersmann, n 15 above.
[96] WT/DS58/AB/R, *US-Shrimps* (Oct 1998), para 158: '[o]ur task here is to interpret the language of the chapeau, seeking additional interpretative guidance, as appropriate, from the general principles of international law'.
[97] WT/DS2/AB/R, *US—Gasoline* (adopted in May 1996), section III.B.
[98] *US—Shrimps*, n 96 above, para 129.

claimed that GM products (eg, genetically modified tomatoes, ketchup made from such tomatoes) were not 'like' conventional products; the international community had, through the Biosafety Protocol, recognised that GM products require their own, distinct authorisation procedure and are objectively different from (ie, not 'like') non-GM products. The EC further claimed that the precautionary principle had become a fully fledged, general principle of international law that should be taken into account in the interpretation of the autonomous right of WTO members (under Article 5(7) of the SPS Agreement) provisionally to restrict or prohibit GMO imports in the face of scientific uncertainty.

The UNHCHR draws attention to the different functions of non-discrimination requirements under human rights law (which may call for 'positive discrimination') compared with those of WTO law, notwithstanding various common elements (eg, no need to demonstrate discriminatory intention, applicability to both *de jure* and *de facto* discrimination, justifiability of differential treatment based on objective criteria). WTO rules permit subsidies, tax benefits and other preferential treatment in favour of poor people, vulnerable minorities and suppliers of 'essential services'. The 'flexibility' of WTO rules appears to enable countries to avoid legal conflicts between WTO rules and human rights.[99] The customary rules of international treaty interpretation, whose use is explicitly prescribed in the WTO Dispute Settlement Understanding (Article 3), require WTO rules to be interpreted with due regard to the universal human rights obligations of WTO Members and other 'relevant rules of international law' (cf Article 31 of the VCLT). Even though the narrow trade policy perspective of most WTO bodies, and the limitation of the jurisdiction of WTO dispute settlement bodies to the 'covered WTO Agreements' (cf Articles 3, 7, 19 of the DSU), have so far prompted WTO bodies to refrain from discussing the human rights dimensions of WTO rules, it seems to be only a matter of time until human rights arguments are raised more openly in WTO dispute settlement proceedings.

5.7. Human Rights as 'Relevant Context' in WTO Dispute Settlement Proceedings?

In the abovementioned WTO dispute settlement proceeding against the delays and lack of transparency in the EC's approval procedures and import bans on GMOs and GM products, none of the parties appears to have invoked human rights arguments. As the WTO Dispute Settlement Understanding limits the jurisdiction of panels to 'the covered agreements'

---

[99] Cf Petersmann, n 25 above.

(cf Articles 3, 7 of the DSU), WTO complainants may invoke only legal claims based on the WTO Agreements. Yet, human rights may be a relevant context for interpreting WTO rules, for example if human rights have been invoked by defendants as justification for departures from WTO obligations.[100] Some of the human rights arguments presented in trade disputes before the ECJ could similarly be presented in WTO dispute settlement proceedings, notwithstanding the obvious fact that the different legal contexts of EC law and WTO law may require different legal interpretations.

For instance, in a recent judgment on an application by the Netherlands for annulment of EC Directive 98/44 on the legal protection of biotechnological inventions, the ECJ had to decide, inter alia, on the plea that the patentability of isolated parts of the human body provided for in Article 5(2) of the Directive reduced living human matter to a means to an end, undermining human dignity. It was also claimed that the absence of a provision requiring verification of the consent of the donor or recipient of products obtained by biotechnological means undermined the right to self-determination. The ECJ affirmed, without further explanation, that 'it is for the Court of Justice, in its review of the compatibility of acts of the institutions with the general principles of Community law, to ensure that the fundamental right to human dignity and integrity is observed'.[101] This judicial finding was far from obvious in view of the fact that human dignity is mentioned neither in the EC Treaty nor in the ECHR. Even though human dignity had been recognised in Article 1 of the EU Charter of Fundamental Rights, the ECJ did not mention this EU Charter. Nor did the Court mention that only a few national constitutions of EC Member States recognise human dignity as a constitutional principle (eg, in Belgium, Finland, Germany, Greece, Ireland, Italy, Portugal, Spain, Sweden), and that apparently only the guarantee of human dignity in Article 1 of the German Basic Law has been construed by the courts as protecting a justiciable human right to respect for human dignity. Yet, Article 1 of the 1997 Council of Europe Convention on Human Rights and Biomedicine committed all contracting

---

[100] Pauwelyn, n 4 above, chap 8, argues that human rights and environmental agreements may be directly 'applicable law' in WTO dispute settlement proceedings. Most WTO lawyers (like G Marceau, 'Conflicts of Norms and Conflicts of Jurisdiction: The Relationship between the WTO Agreement, MEAs and other Treaties' (2001) 35 *Journal of World Trade* 1081, at 1089–1107, interpret the limited jurisdiction of WTO dispute settlement bodies more narrowly: If WTO dispute settlement bodies conclude that WTO obligations are superseded by non-WTO provisions (eg, a human rights obligation, a regional integration agreement), they may have to decline jurisdiction on the ground that WTO dispute settlement bodies 'cannot add to or diminish the rights and obligations provided in the covered agreements' (Arts 3,19 DSU) if non-WTO rules supersede WTO provisions.

[101] See n 58 above. This judicial finding might have been facilitated by the case law of the ECtHR that 'the very essence' of the ECHR is 'respect for human dignity and human freedom' (*SW v UK and CR v UK*, (1995) 21 EHHR 363, paras 42, 44).

parties to the protection of human dignity in biology and medicine. The ECJ had also long since construed the 'common constitutional traditions of EC member states' in a way enabling a high standard of constitutional protection in EU law (rather than the 'lowest common denominator' of common national constitutional traditions).[102] Like the ECtHR, the ECJ interprets and protects human rights and European integration law as an objective 'constitutional order'[103] that requires recognition of citizens as legal subjects (rather than mere objects) of international law and of fundamental rights.

As every WTO Member has human rights obligations under the UN Charter and UN human rights conventions, legal pleas to interpret WTO obligations with due respect for human dignity and human rights may be raised also in WTO dispute settlement proceedings (eg, in a WTO complaint challenging the WTO consistency of the abovementioned EC Directive, or of import prohibitions of genetically engineered stem cells). The mandate of WTO dispute settlement bodies is, however, much more constrained than that of the ECJ. Neither WTO law nor UN human rights law appears to justify a WTO dispute settlement finding that the references to human dignity in UN human rights conventions entail a general obligation of WTO Members to respect human dignity as a human right (rather than only as a legal principle) in the interpretation and application of WTO rules. The recognition of human dignity as a human right in EU law[104] (but not in UN human rights law which recognises only objective government obligations to respect and protect human dignity), and the much broader constitutional protection of individual freedoms (eg, in the economic field) in EU law than in UN law, illustrate that the judicial interpretation of European economic rules in the light of European fundamental rights may require approaches different from those called for if worldwide WTO rules have to be construed with due regard to the UN human rights obligations of the countries concerned.

5.8. Time for a WTO Commitment to Respect the Universal Human Rights Obligations of WTO Members?

WTO law nowhere refers to the traditional sources of democratic input legitimacy (such as respect for human rights, democratic procedures) and output-legitimacy (eg, promotion of general consumer welfare). The abovementioned UN reports on the human rights dimensions of WTO rules do not suggest incorporating human rights into WTO law,

---

[102] See. eg, Case 44/79, *Hauer* [1979] ECR 3727; Case 155/79, *AM&S* [1982] ECR 1575.

[103] Cf ECtHR judgment on *Loizidou v Turkey* (preliminary objections) of 23 Mar 1995, Series A no 310 para 75, referring to the status of human rights in Europe.

[104] See n 57 above.

presumably for fear that the interpretation of human rights by WTO bodies could deviate from the human rights interpretations by UN human rights bodies. Yet, just as the 1994 Ministerial Decision on 'Trade and Environment'[105] has promoted the mutual coherence of international trade and environmental rules and policies, so could a WTO Declaration on 'Trade and Human Rights' promote synergies between trade and human rights and enhance the democratic legitimacy of WTO rules. By limiting itself to the recognition of existing universal human rights obligations without defining their contested scope, such a WTO Declaration would neither introduce new obligations nor enlarge WTO competencies or interfere with the task of specialised human rights bodies to clarify the relevance of human rights for trade laws and policies. The narrow focus on trade liberalisation and trade regulation has, so far, prevented the political WTO bodies from responding positively to the proposals by the UN High Commissioner for Human Rights to adopt a human rights approach to international trade rules and policies. The (quasi-)judicial WTO dispute settlement bodies, by contrast, might be legally required by the customary methods of treaty interpretation to respond to human rights arguments in WTO dispute settlement proceedings.

At the 2004 Berlin Conference of the International Law Association (ILA), the International Trade Law Committee (ITLC) resolved to elaborate—in co-operation with the ILA's Committee on Human Rights—an ILA Declaration on Human Rights and International Trade Law.[106] Such an ILA Declaration could, inter alia:

(1) recall the universal human rights obligations of all WTO Members, including their obligation to ensure the consistency of WTO rules (eg, on the protection of 'public morals' and *ordre public*) and trade policies with universal human rights, as reflected in the increasing number of 'human rights clauses' in regional trade agreements pursuant to GATT Article XXIV and GATS Article V as well as in the generalised systems of tariff preferences by developed WTO Members for LDCs;

(2) support the legal clarification of the human rights dimensions of international economic law through the competent UN human rights bodies and other institutions;

(3) acknowledge the need for harnessing human rights and WTO rules for welfare-increasing co-operation among free citizens in international trade, in conformity with the worldwide recognition—in paragraph 8 of the 1993 Vienna Declaration of the UN World Conference on Human Rights—that

---

[105] Cf *The Results of the Uruguay Round of Multilateral Trade Negotiations. The Legal Texts* (*Geneva*, GATT, 1994), 469–71.

[106] Report of the 71st Conference of the International Law Association in 2004 at Berlin (London, ILA, 2004), 565.

'democracy, development and respect for human rights and fundamental freedoms are interdependent and mutually reinforcing' (para.8);

(4) recognise that the customary rules of international treaty interpretation may require WTO bodies to take into account human rights obligations of WTO Members in the interpretation of WTO rules; and

(5) urge WTO dispute settlement bodies, if they are requested by WTO Members to take into account human rights in the interpretation of WTO rules, fully to respect the 'margin of appreciation' which every WTO Member may legitimately claim with regard to the domestic implementation of human rights obligations for the benefit of its own citizens.

The 2003 WTO Declaration on access to medicines,[107] the 2005 WTO Decision on amending the TRIPS Agreement,[108] the WTO waiver for the 'Kimberley agreement' on the control of conflict diamonds,[109] and the WTO dispute settlement rulings on the right to make trade preferences for less-developed countries conditional on 'objective standards',[110] bear witness to the flexibility and consistency of WTO law and human rights on the level of principles. It is noteworthy that neither past GATT and WTO dispute settlement reports nor the UN experts' reports on the human rights dimensions of WTO rules identified concrete conflicts between WTO rules and human rights. Yet, this apparent consistency of WTO rules with human rights in no way ensures the absence of conflicts in the judicial interpretation of WTO rules and in the domestic implementation of trade laws and trade policies. Almost all the complaints to the ECJ and in the ECtHR about alleged conflicts between trade practices and human rights could likewise arise in WTO dispute settlement proceedings. For example, like the ECJ and the ECtHR, WTO dispute settlement bodies might be requested to examine:

— whether freedom of transit trade (Article V of GATT) might be restricted in favour of environmental demonstrators invoking their freedom of assembly and freedom of opinion as a justification of blocking transit routes;[111]
— whether the importation of services (laser games simulating killing of human beings) can be prohibited on grounds of respect for human dignity;[112]

---

[107] WT/L/540, Implementation of Pararaph 6 of the Doha Declaration on the TRIPS Agreement and Public Health, WTO General Council Decision of 30 Aug 2003.

[108] WT/L/641, Amendment of the TRIPS Agreement, WTO General Council Decision of 6 Dec 2005.

[109] WT/L/518, Waiver Concerning Kimberley Process Certification Scheme for Rough Diamonds, WTO General Council Decision of 15 May 2003.

[110] WT/DS246/AB/R, *EC—Tariff Preferences*, adopted 20 Apr 2004.

[111] In Case C–112/00, *Schmidberger* [2003] ECR I–5659, the ECJ found that a public demonstration on a major motorway temporarily restricted freedom of trade in goods, but was justified by Arts 10 and 11 ECHR.

[112] See Case C–36/2002, *Omega* [2004] ECR I–9609, para 31.

—whether the legal protection of biotechnological inventions can be rejected on grounds of respect for human dignity;[113] or

—how unfair competition rules (eg, in Articles 2 of the TRIPS Agreement and 10*bis* of the Paris Convention) have to be balanced with human rights to commercial freedom of speech.[114]

In view of their limited mandate and expertise, WTO diplomats prefer avoiding WTO discussions on specific human rights—even if such rights are recognised in universal human rights instruments (eg, human rights to education, food, health care). There is no consensus among WTO negotiators on the justifiability of trade sanctions in response to human rights violations *abroad,* nor on which trade measures may be optimal for protecting human rights *at home.* WTO Members are, therefore, unlikely to agree on extending the mandate of WTO bodies (such as the WTO's Trade Policy Review Mechanism) to discuss or clarify the human rights dimensions of WTO rules and policies. Given the past reluctance of political WTO bodies to address the human rights dimensions of WTO rules, and their reservations *vis-à-vis* the application of general international law rules by WTO dispute settlement bodies, it may be easier for UN human rights bodies and non-governmental organisations (like the ILA) to clarify the legal relevance of human rights for the interpretation of WTO rules and the appropriate methodology for interpreting human rights obligations of governments in the trade policy context. Such UN or ILA Declarations on Human Rights and International Trade could assist WTO dispute settlement bodies in responding to such human rights arguments in WTO dispute settlement proceedings, and foster additional public discussion and better understanding of the human rights dimensions of international economic law.

6. CONCLUSION

Biotechnologies and economic welfare are created by individuals rather than by governments, just as goods and services are traded in and consumed mainly by individuals. Respect for human dignity and human rights requires individuals to be recognised as legal subjects also in international economic law. Human rights offer a more coherent and more legitimate framework for enhancing the coherence and legitimacy of international economic law than the state practice of fragmented, intergovernmental negotiations on separate trade, intellectual property, biotechnology

---

[113] See the EC case referred to in n 48 above.
[114] See the Judgment of the ECtHR of 25 Aug 1998 in *Hertel v Switzerland* (59/1997/843/1049), published in the Court Reports 1998-VI.

and environmental regimes, which tend to view citizens as mere objectives of intergovernmental regulation. The recognition of new 'bio-rights' in human rights law—and their focus on the 'primacy of the human being', ie, that 'the interests and welfare of the human being shall prevail over the sole interests of society and science'[115]—are a reminder for WTO bodies to ensure that 'member-driven' WTO negotiations (eg, on trade in GM products, related biotechnological services, intellectual property rights) should be less one-sidedly driven by producer interests and must take into account the need for democratic rule-making for the benefit of all citizens. WTO rules need to be designed and interpreted with due regard to the human rights obligations of all WTO Members and with due respect for the legitimate diversity of national human rights traditions. Due to the limited trade policy mandate of WTO bodies and the longstanding neglect by UN human rights bodies of welfare-creation through trade, academics and non-governmental organisations should contribute to promoting synergies between human rights and world trade rules.

The state-centred focus of WTO law on rights and obligations of governments and 'member-driven' negotiations has all too often favoured abuses of power (eg, discrimination against cotton, textiles, agricultural and other exports from poor and vulnerable countries). The today universal recognition by all governments of inalienable human rights entails that *intergovernmental* organisations and decisions are also limited by human rights and accountable to civil society. The UN proposals for a human rights approach to international trade should prompt WTO bodies to recognise the universal human rights obligations of all WTO Members as a relevant context for the interpretation of WTO rules and for democratic reforms of international trade governance so as to reduce the existing incoherencies between state-centred international law rules and citizen-centred human rights. As long as non-democratic WTO governments prevent the consensus-based WTO decision-making processes from acknowledging the human rights obligations of all WTO Members, civil society and non-governmental organisations should insist on stronger protection of human rights in the trade policy area.

Justice-oriented changes in international economic law[116] could enhance the democratic legitimacy and economic efficiency of the world trading

---

[115] Art 2, European Convention on Human Rights and Biomedicine.

[116] On the incoherencies between Rawls' *Theory of Justice* (Cambridge, Mass, Harvard University Press, 1971) for democracies and Rawls' theory of justice for the relations among self-governing peoples (cf J Rawls, *The Law of Peoples* (Cambrige, Mass, Harvard University Press, 1999)) see, eg, T Pogge, 'The Incoherence between Rawls' Theories of Justice' (2004) 72 *Fordham Law Review* 1739. On the modern, legal definition of justice by means of human rights see EU Petersmann, 'Theories of Justice, Human Rights and the Constitution of International Markets' (2003) 37 *Loyola Law Review* 407.

system, as well as its contribution to reducing the poverty problems in many developing countries. The constitutional and judicial protection of 'fundamental economic freedoms' and of other newly codified 'fundamental rights' in EU law confirms the 'Kantian moral imperative' that constitutional protection of equal freedoms and human rights in the economy may be no less important for personal self-development in dignity (eg, of the more than one billion poor people living below the poverty line of one dollar per day) than in the polity, at home as well as in international relations across frontiers. Human rights law as well as trade law evolved in response to past abuses of power and must respond to the new social challenges of biotechnologies and 'globalisation' in more coherent ways. Just as economic laws and policies must be guided by human rights, so must human rights help citizens more effectively in benefiting from modern technologies and from the worldwide division of labour through international trade based on WTO law.

11

Genetic Engineering, Trade and Human Rights

THOMAS COTTIER[*]

1. THE FOCUS ON TRADE REGULATION

REGULATION OF GENETIC engineering or biotechnology[1] mainly pertains to domestic law. It is here that the processes of democracy and judicial assessment of fundamental rights and principles produce regulatory frameworks, commensurate with basic attitudes in society. Inherently, regulations vary from country to country. To what extent is there a need to involve international law and treaty-making? To what extent is there a need to achieve common and shared perceptions, and to regulate the interfacing of different regulations? These questions address the proper role of international law, and answers are far from clear, as the different chapters in this volume and the growing literature on the subject indicate.[2]

[*] Professor of European and International Economic Law and managing director of the World Trade Institute at the university of Berne, and of the National Research Programme on trade regulation of the Swiss National Science Foundation. Email: thomas.cottier@iew.unibe. ch. I am most grateful to Michelangelo Temmerman, Research Fellow WTI, for assistance in preparing the chapter.
[1] While the term 'biotechnology' is broader than genetic engineering, also encompassing traditional uses of bacteria, see: http://en.wikipedia.org/wiki/Biotechnology, we use the two terms interchangeably for the purpose of this chap.
[2] See also G Annas, 'Protecting the Endangered Human: Toward an International Treaty Prohibiting Cloning and Inheritable Alterations' (2002) 28 *American Journal of Law & Medicine* 151; A Ayer, 'Stem Cell Research: The Laws of Nations and a Proposal for International Guidelines' (2002) 17 *Connecticut Journal of International Law* 393; S Fiandaca, 'In Vitro Fertilization and Embryos: The Need for International Guidelines' (1998) 8 *Albany Law Journal of Science and Technology* 337; S Greenlee, 'Dolly's Legacy to Human Cloning: International Legal Responses and Potential Human Rights Violations' (2000) 18 *Wisconsin International Science Journal* 537; S Marks, 'Tying Prometheus Down: The International Law of Human Genetic Manipulation' (2002) 3 *Chicago Journal of International Law* 115; S Pridan-Frank, 'Human-Genomics: A Challenge to the Rules of the Game of International Law' (2002) 40 *Columbia Journal Transnational Law* 619 and P Riordan, 'Cloning Consensus: Creating a Convention to Ban Human Reproductive Cloning' (2003) 26 *Suffolk Transnational Law Review* 411.

In international trade regulation, all these questions translate into demands for market access and fair conditions of competition for competing biotechnological or genetically engineered products, on the one hand, and to demands of trade restrictions on the other hand. It is not a coincidence that trade law is the prime area where such divergences are truly felt in international law, and serious conflicts are emerging. It is submitted that a need to co-ordinate and integrate widely diverging attitudes and regulations primarily shows in this field of international law. Again, it is here that a necessary process of co-ordination and eventually of integration of different regulatory traditions and attitudes to biotechnology emerges.

Trade regulation essentially deals with the interfacing of different national legal orders and the different and diverging ways products and processes are dealt with.[3] Commonly accepted products are traded widely, and mutually used. The principle of comparative advantage relies upon the assumption of such acceptance of mutual benefit. Products ideally are traded on the basis of efficient allocation of resources and benefits. International trade, to a large extent, responds to these assumptions. Most products are widely shared and accepted, and thus accessible to open trade. New products, however, may challenge existing structures and patterns of consumption. They may be met with suspicion and diverging assessment by, and within, different jurisdictions. This is particularly true for products and processes based upon genetic engineering and DNA recombination.[4] They are met with diverging attitudes, ranging from acceptance to outright rejection. Within societies and states, this leads to difficult political debates. In international relations, the problem risks triggering trade wars. These tensions need to be addressed. How should international law deal with them? What are appropriate rules? To what extent are shared perceptions and harmonisation indispensable? And to what extent should they be left to domestic jurisdictions?

We note at the outset that few uniform answers can be found to the questions regarding technology since it is employed in very different contexts. Biotechnology and genetic engineering encompass a wide field of human activities and applications. Currently, the main areas of genetic engineering (DNA recombination, properly speaking) are: first, genetic engineering as applied to humans in medical research and treatment of illnesses; secondly,

---

[3] See generally T Cottier and M Oesch, *International Trade Regulation: Law and Policy in the WTO, The European Union and Switzerland* (Berne/London, Staempfli Publishers Ltd and Cameron May Ltd, 2005).

[4] See also EU Petersmann in this volume; G Isaac and W Kerr, 'Genetically Modified Organisms at the World Trade Organization: A Harvest of Trouble' (2003) 37 *Journal of World Trade* 1083. For a historical account see MF Cantley, 'The Regulation of Modern Biotechnology: A Historical and European Perspective' in D Brauer (ed), *Biotechnology vol. 12, Legal, Economic and Ethical Dimensions* (Weinheim, Wiley-VCH, 1995), 505–795.

genetic engineering as applied to crops and animals, both essential to support human life; thirdly, additional applications emerge in the field of energy production; fourthly, genetic engineering has been employed in the field of biological warfare and offers a dangerous and uncontrollable potential for mass destruction. While the technology shares common scientific foundations in all these areas, assessments within societies of these main areas as to their effects vary widely and cannot be dealt with uniformly in law. Genetic engineering for improving health is largely accepted. Genetic engineering for mass destruction is widely rejected. Genetic engineering for food and agriculture is perhaps the most controversial field. This is certainly true in respect of international trade regulation.[5]

The host of areas of international trade regulation affected, and of largely unresolved problems, is impressive when we look at those issues falling under the scope of WTO law relating to the importation and exportation of genetically modified products.[6]

—Importation and exportation of stem cells, either obtained from human tissue of living persons or from human stem cells obtained from non-sustainable human embryos produced by way of artificial insemination: moral exceptions (Article XX (a) GATT)[7];
—Importation and exportation of products obtained on the basis of genetic engineering (food, seeds and plants, animals, and medicines) (Article XX (a), (b), (g) GATT);
—Importation and exportation of biological substances and services relating to warfare and mass destruction: national security (Article XI GATT, Article XIV *bis* GATS)[8];
—The identification of genetically obtained products: labelling (Article III, XX GATT);
—The assessment of the safety of genetically obtained products (GATT, TBT[9] and SPS[10] measures relating to risk assessment and risk management, and corresponding trade restrictions);

[5] Cf the pending WTO dispute resolution on *EC—Measures Affecting the Approval and Marketing of Biotech Products*, WT/DS 291-293. See Isaac and Kerr, n 4 above.

[6] See also R Howse and P Mavroidis, 'Europe's Evolving Strategy for GMOs—The Issue of Consistency with WTO Law: of Kine and Brine' (2000) 24 *Fordham International Law Journal* 348; EU Petersmann in this volume.

[7] General Agreement on Trade and Tariffs (GATT), Annex IA to the Agreement Establishing the World Trade Organization, Marrakesh, 15 Apr 1994, available at www.wto.org/english/docs_e/legal_e/06-gatt.pdf.

[8] General Agreement on Trade in Services (GATS), Annex IB to the Agreement Establishing the World Trade Organization, Marrakesh, 15 Apr 1994, available at www.wto.org/english/docs_e/legal_e/26-gats.pdf.

[9] Agreement on Technical Barriers to Trade (TBT Agreement), Annex IA to the Agreement Establishing the World Trade Organization, Marrakesh, 15 Apr 1994, available at www.wto.org/english/docs_e/legal_e/17-tbt.pdf.

[10] Agreement on the Application of Sanitary and Phytosanitary Measures, Annex IA to the Agreement Establishing the World Trade Organization, Marrakesh, 15 Apr 1994, available at www.wto.org/english/docs_e/legal_e/15-sps.pdf.

—Marketing approval of GMOs: recognition of test results: mutual recognition agreements (MRAs);
—The impact of genetically obtained products on the environment: setting free GMOs and the risk of involuntary cross-breeding and implications on natural habitats: the Cartagena Protocol[11] and the precautionary principle[12];
—Liability for potential damages incurred and the problem of extensive insurance costs: a potential non-violation issue under Article XXIII(1)(b) GATT;
—Patenting life forms: to what extent should DNA recombination be subject to appropriation and exclusive rights under the TRIPs Agreement?[13]
—The implications on social and economic development in developing countries: the problem of Special and Differential Treatment (S&D), and the protection of traditional knowledge in relation to genetic engineering under the TRIPs Agreement;
—The implications for conventional production and processes. To what extent do they need protection by means of tariffs and domestic support in the light of competition by more efficient genetically engineered crops and industrial production? How could an appropriate balance be achieved between conventional crops, including organic farming, and the use of genetically modified crops, all with a view to supporting sustainable development, biological diversity and food security?
—The regulation of therapeutic services employing genetic engineering: GATS Agreement.

2. UNDERLYING ISSUES

Trade disputes arising in these areas must be assessed on the basis of existing international law. They are unlikely to produce widely acceptable results since trade rules, essentially based upon the principle of

---

[11] Cartagena Protocol on Biosafety to the Convention on Biodiversity of 5 June 1992, Protocol available at www.un.org/millennium/law/cartagena.htm; see also The Universal Declaration on Cultural Diversity, United Nations Educational, Scientific and Cultural Organization, Paris, 2 Nov 2001, available at http://unesdoc.unesco.org/images/0012/001271/127160m.pdf. In doctrine, see: R Pavoni, 'Assessing and Managing Biotechnology Risk under the Cartagena Protocol on Biosafety' (2000) 10 *The Italian Yearbook of International Law* 113; R Pavoni, 'Biosafety and Intellectual Property Rights: Balancing Trade and Environmental Security—The Jurisprudence of the European Patent Office as a Paradigm of an International Public Policy Issue' in F Francioni (ed), *Environment, Human Rights and International Trade*, (Oxford, Hart Publishing, 2001); R Pavoni, *Biodiversità e Biotechnologie nel Diritto Internazionale e Communitario* (Milan, Giuffré 2004), 239–484.

[12] Principle 15 of the Rio Declaration on Environment and Development, Report of the United Nations Conference on Environment and Development, June 1992, A/CONF.151/26, available at www.un.org/documents/ga/conf151/aconf15126-1annex1.htm.

[13] Agreement on Trade Related Aspects of Intellectual Property Protection (TRIPS Agreement), Annex IC to the Agreement Establishing the World Trade Organization, Marrakesh, 15 Apr 1994 (1994) 33 ILM 1197, available at www.wto.org/english/docs_e/legal_e/27-trips.pdf. In the literature, see R Pavoni, n 11 above at 67–116.

non-discrimination, do not address the underlying valuational issues or operate on assumptions which were created for commonly accepted, conventional products. Trade disputes, in other words, are only the tip of the iceberg of fundamental underlying problems which are beyond the scope of current international trade rules. These underlying problems include:

— The ethics of genetic engineering: is the technology good or evil? In pluralist societies, and more so in the pluralist international society, it is not possible to seek uniform answers to these basic questions. They are bound to vary in light of largely diverging rational attitudes and interests. Moreover, they vary in different fields of application.

— The utility of genetic engineering: is it useful to the large majority of the population and to society at large, even if it may impair individuals? Again, general answers are unlikely as they do depend on different factual circumstances in different fields of application of the technology.

— How far can science, research and industrial applications be trusted, in the light of the economic constraints and personal agendas of researchers and science based institutions, competing with each other?

— The safety of genetic engineering: what are the long-term implications of the technology? Answers again depend on diverging facts in different settings and fields.[14]

— The economics of genetic engineering: to what extent does genetic engineering lead to de facto monopolies and the exclusion of competition, in particular in the field of seeds?[15]

— The global politics of genetic engineering: what are the implications of the advanced and highly sophisticated technology in terms of relations for competing knowledge based economies? What are the implications on power relations?[16]

— Finally, the trade angle of genetic engineering: to what extent are objections motivated by rent-seeking protectionism, in order to protect conventional

---

[14] Cf Parliamentary Assembly of the Council of Europe, Genetically Modified Organisms, Resolution 1419, 26 Jan 2005, available at http://assembly.coe.int/Documents/AdoptedText/ta05/ERES1419.htm, paras 6 and 7: '6. The Assembly believes that although green biotechnology offers a broad spectrum of potential benefits, many risks—for example horizontal gene transfer—have not been sufficiently evaluated and should continue to be studied. While the health risks associated with current GMOs can be regarded as slight, provided that safety controls prove effective, future developments with modified output characteristics will entail new and different risks that will have to be assessed on an individual basis. 7. Long-term effects on biodiversity are difficult to estimate, particularly as there is no generally recognized definition of "ecological damage". The Assembly emphasizes that there are currently no uniform standards for the assessment of mandatory monitoring of crops in cultivation. Long-term monitoring is obligatory to allow the ecological effects of GMOs to be assessed.'

[15] See J Barton and P Berger, 'Patenting Agriculture' (2001), available at www.issues.org/17.4/p_barton.htm.

[16] See T Bernauer, *Genes, Trade and Regulation: The Seeds of Conflict in Food Biotechnology*, (Princeton, NJ, Princeton University Press, 2003).

products and methods of production? And to what extent are they motivated by truly ethical concerns? Are there unholy alliances?
—In a pluralist society, how can we find the right answer to all these questions? Does majority ruling provide the appropriate answers? How are these
processes informed and led?[17] What is the role of normative principles?

These and other questions entail fundamental problems for different
legal orders and nations. The complexity of these questions is multiplied
once it is transferred to the global level and thus to the realm of international
law. We need to examine carefully to what extent international rules are
necessary, both in terms of harmonisation and in terms of interfacing
different national rules. In areas which are mainly left to domestic
applications short of international exchange of goods and services, best
answers may leave matters to domestic or regional law and regulatory
competition. This would seem particularly appropriate in shaping
framework conditions for basic research prior to the generation of
tradable products. However, as soon as tradable products are placed on
the market, international regulation becomes indispensable if distortions
and misallocation of resources to the detriment of mankind are to be
avoided.

In the process of globalisation, it is evident that harmonisation of
principles and rules offers the best answer, from the point of view of
international trade and legal security.[18] Once tradable goods and services
emerge, once common standards, reflecting shared perceptions, are
achieved, trade problems can be solved. The problem is, of course, that
regulatory harmonisation on substance cannot be readily expected. The
contemporary existence of largely diverging views, essentially based
upon beliefs rather than scientific evidence, cannot be easily overcome.[19]
We are faced with the question, what can we reasonably expect from
international trade regulation in the field of genetic engineering? Second-
best solutions would allow for rational interfacing of diverging attitudes
on the basis of existing techniques addressing the interface. Where does
the law stand today? Can it assist in bringing about common and shared
perceptions in the long run?

At this point, it may be useful to ask whether international law
offers the basis of common perceptions by referring to widely accepted
standards in human rights. Human rights reflect the basic perceptions of

---

[17] See: DJ Galligan in this volume; UN Economic and Social Council, Commission on
Human Rights, Promotion and Protection of Human Rights; 'Human Rights and Bioethics'
Report of the Secretary-General Resolution 2001/71, 10 Feb 2003, E/CN.4/2003/98, 8, available at www.unhchr.ch/Huridocda/Huridoca.nsf/0/a9c6597e2b0571f0c1256cf7005b7ea6?
Opendocument.
[18] EU Petersmann in this volume.
[19] But see S Millns in this volume.

justice, fairness and equity in contemporary society. Many of them, but by no means all, substitute for formerly held religious beliefs. They provide a valuational and ethical system upon the basis of which people may approach and assess underlying problems relating to genetic engineering. This, at least, is true for the Western world, but it is increasingly of global reach in the light of extensive commitments made for the protection of a wide range of human rights.

This chapter thus seeks to explore the relationship of trade regulation and human rights in the field of genetic engineering. It addresses the potential impact of trade regulation in genetic engineering on human rights (in accordance with the title of this volume), and the potential normative impact of human rights on regulating trade in this field. It is submitted that human rights values are an important ingredient to be taken into account in international trade regulation in order to achieve viable and acceptable interfaces between diverging perceptions.[20] They can offer a path towards shared understanding and common rules. We explore to what extent existing trade rules offer portals for human rights considerations. At the same time, we observe that the impact on, and of, human rights cannot be defined in the abstract. It strongly depends on the context of a particular problem and application in the various fields open to genetic engineering. Except for specific areas, there are no uniform answers. Trade rules thus need to be read and shaped in a manner that takes into account the context of a particular constellation. To this effect they should first provide for appropriate procedures.

3. FOUNDATIONS

We first seek briefly to address the potentially relevant substantive human rights in biotechnology and to provide a short survey of the relevant trade regulations in WTO law. The lack of explicit links between the two fields is subsequently addressed.

3.1. Core Human Rights Relating to Genetic Engineering

The following core human rights, drawn from the 1948 Declaration of Human Rights,[21] are of particular importance in the present context and may briefly be touched upon: the protection of human dignity (Article 1)

---

[20] See generally T Cottier, J Pauwelyn and E Buergi (eds), *Human Rights and International Trade* (Oxford, Oxford University Press, 2005).
[21] Universal Declaration of Human Rights, UN GA, Resolution 217 A (III), 10 Dec 1948, available at www.un.org/Overview/rights.html.

provides the overall foundation and purpose of post-World War II international human rights protection. It establishes a core value to be respected in all fields of life. It is of particular relevance in genetic engineering, given the potential of manipulation of genetic information, but likewise of beneficial applications. It is supported by the right to life and the prohibition of slavery (Article 3). The former entails both encouragement and restrictions of the technology, while the latter clearly acts as a barrier to manipulations and oppression. The right to freedom of thought, conscience and religion (Article 18) protects the right to advocate, but also to oppose, genetic engineering and thus to accept and protect diverging views on the subject. It provides the foundation of pluralism. The right to adequate standards of living, including the right to food and the right to medical care (Article 25) offers a potential foundation in support of the technology, while the right to property (Article 17, not reflected in global human rights treaties) and the right to protection for moral and material interests resulting from scientific production (Article 27(2)) raise the issue of proprietary rights in the field. The Declaration sets forth an ambitious right to a social and international order in which the rights of the Declaration can be fully realised (Article 28). The provision encourages the respect for human rights in shaping regulations, in particular international trade regulations, in line with the core standards alluded to by the Declaration. Finally, the generic right to development, often, but disputably, qualified as a human right,[22] must today be understood in the wake of the 1992 Rio Declaration and subsequent instruments as a principle of sustainable development. Legitimacy of genetic engineering therefore also depends upon its long-term impact on social and economic development and ecology, in particular biodiversity.

A detailed account and analysis of all relevant human rights instruments cannot be offered here. Other and further studies need to examine a comprehensive list of rights affected, including procedural rights.[23] It will be argued in this chapter that they are of particular relevance. In addition, it will be important to take into account constitutional rights enshrined in

---

[22] UN GA Declaration on the Right to Development, Resolution 41/128, 4 Dec 1986, A/RES/41/128, CAOR 34th Sess. Res. 66, available at www.un.org/documents/ga/res/41/a41r128.htm.

[23] In particular the International Covenant on Civil and Political Rights (ICCPR), adopted 16 Dec 1966, entered into force 23 Mar 1976, GA Res. 2200, 21 UN GAOR, Supp (No 16) 52, UN Doc A/6316 (1966); International Covenant on Economic, Social and Cultural Rights (ICESCR), adopted 19 Dec 1966, entered into force 3 Jan 1976, GA Res 2200, 21 UN GAOR, Supp (No. 16) 49, UN Doc A/6316 (1966); for a useful comprehensive analysis of these and related instruments see, eg, T Meron (ed), *Human Rights in International Law: Legal and Policy Issues* (Oxford, Clarendon Press, 1984).

main legal systems, such as the freedom to conduct scientific research.[24] At this stage, it is sufficient to note that a number of core individual rights are likely to influence the legal environment of genetic engineering. Human rights essentially require the adoption of a homocentric view while taking into account the needs of environmental protection and the preservation of nature.

3.2. Current WTO Rules Addressing Genetic Engineering

No agreement specifically addressing biotechnology exists in WTO law. Domestic regulations of genetic engineering therefore are subject to the rules of the GATT 1994, the SPS Agreement and/or the TBT Agreement. Other agreements on goods may also be relevant (agriculture, licensing, etc).

Trade regulation in the WTO essentially favours free exchange of goods, subject to tariffs. It assesses whether restrictions imposed on the basis of SPS or TBT measures can be justified. Under the SPS Agreement, domestic standards are thus subject to obligations of risk assessment.[25] Under the TBT Agreement, measures need to respond to the requirements of necessity and proportionality.[26] None of these restrictions is explicitly based upon human rights considerations. To the extent that a measure falls within the SPS Agreement, it will be examined under this agreement, prior to the TBT Agreement and the GATT 1994. The regulatory triangle of these agreements is currently under review by the Panel in *European Communities—Measures affecting the Approval and Marketing of Biotech Products*, and clarification as to their application is expected to emerge in the case law.[27]

Biotechnology is further addressed by the TRIPs Agreement.[28] Standards on trademarks, the protection of undisclosed information in Article 39 and patenting life forms in Article 27 are of prominent importance to the new technology. Finally, general provisions of the GATS apply to services using biotechnology, in particular medical services for therapeutics and diagnostics. They are subject to rules of MFN and transparency, and to national treatment to the extent that Members have made specific commitments, also to national treatment, subject to conditions set forth for different modes of delivery.

---

[24] See M Eibert, 'Human Cloning: Myths, Medical Benefits and Constitutional Rights' (2002) 53 *Hastings Law Journal* 1097.

[25] See eg, Cottier and Oesch, n 3 above, at 778–814.

[26] See ibid, at 750–77.

[27] *EC—Measures Affecting the Approval and Marketing of Biotech Products*, WT/DS 291–293 (report of the panel pending).

[28] Pavoni, n 11 above, at 67–116.

3.3. The Absence of Explicit Reference to Human Rights in Trade Regulation

The present law of the WTO does not contain an explicit reference to human rights.[29] The basic concepts of non-discrimination (MFN and National Treatment) are emanations of the fundamental principle of equality.[30] They primarily relate to the treatment of products (goods and services). To the extent that they also protect people as service providers under GATS or holders of intellectual property rights, they may be conceived in terms of human rights of non-discrimination and equality as applied to a specific context.[31] Structurally, they amount to constitutional principles of the WTO the operation of which, subject to well defined exceptions, could be vaguely comparable to the structure of basic human rights norms enshrining principles of equal treatment. However, the principles of non-discrimination have not hitherto been conceived and understood in terms of human rights in the case law of panels and the Appellate Body. WTO law does not enshrine or recognise a human rights principle of economic liberty, unlike a number of domestic constitutions or in a manner comparable to EC law with its four freedoms.

The same is true for intellectual property rights (IPR), which are of paramount importance in the legal framework for genetic engineering. Despite reference to IPR protection in human rights instruments, it is not appropriate generally to assess intellectual property rights, in particular industrial property rights (patents, trademarks, geographical indications, industrial designs, integrated circuits), in terms of human rights. IPR protection emerged on a functional basis, expanding in accordance with the economic and technological needs of different industries and countries. Intellectual property rights essentially are a utilitarian concept.[32] They are designed by legislators to serve society at large, rather than protecting individuals per se. Consider also the limited duration of most IPRs: how can a limited duration of rights possibly be reconciled with the concept of inalienable fundamental rights? The point conclusively proves the utilitarian nature of the concept.

---

[29] Consequently, there is a call in doctrine, including the ILA Committee on international trade, for an explicit WTO declaration, committing all WTO members and bodies to respect universal human rights obligations. See EU Petersmann in this volume. See also UN Commission on Human Rights, Economic, Social and Cultural Rights 'Liberalisation of Trade in Services and Human Rights', Report of the High Commissioner, 25 June 2002, E/CN.4/Sub2/2002/9, available at www.hri.ca/fortherecord2002/documentation/commission/e-cn4-sub2-2002-9.htm.

[30] Cottier and Oesch, n 3 above, at 347 and 382.

[31] Cf UN Commission on Human Rights, Economic, Social and Cultural Rights, n 29 above.

[32] See F-M Abbott, T Cottier and F Gurry, *The International Intellectual Property System*, (The Hague/London/Boston, Kluwer Law International, 1999) 501–4.

A utilitarian approach can also be observed today in the continuing development of the system, as new technologies and markets arise. The fundamental debates on private versus public ownership of knowledge are informed by perceived economic advantage and disadvantage and the need to balance different interests.[33] These debates are likely to persist, and little ground can be gained for the purpose of foundations by recourse to human rights standards for most of the problems involved. Moreover, the international IPRs system may conversely be argued to amount to a system of limitations on fundamental rights. It is a part of economic law and market regulation which in effect limits economic liberties and trade to the extent that these are recognised in constitutional law. From this perspective, the granting of IPRs amounts to a monopoly, for which a sound legal basis is required in statutory law.

Human rights have been seen as answers to specific and fundamental threats to individuals in human history. They establish specific zones of protection from authoritarian and majoritarian rule and cannot entail the design of an overall economic concept beyond principles. They provide guidance, but cannot regulate complex areas on their own. They do not, and must not, cover the entire range of human activities. There is no need to anchor the entire field of intellectual property protection in terms of human rights. The only area where foundations of IPRs may be contemplated are moral rights in the field of copyright. By protecting the integrity of a work, moral rights indirectly also protect the integrity of the author, producer, performer or actor.[34] In the field of IPRs, this is the most personalised and human rights related dimension. Yet, distortions are equally addressed by human rights guarantees relating to human dignity and personal integrity. Protection does not depend upon proprietary concepts but is based upon the need to protect the integrity of individuals.

Since the international IPRs system exists at a fairly high level of refinement, debates concerning its foundation may seem rather esoteric. However, issues of premise or foundation may appear more relevant as we consider the justification for entirely new types of rights, in particular the emerging right of protection of traditional knowledge (so-called Traditional Intellectual Property Rights (TIP-Rights).[35] These are claims

---

[33] See J-H Reichman and K-E Maskus, 'The Globalization of Private Knowledge Goods and the Privatization of Public Goods, Symposium' (2004) 7 *Journal of International Economic Law* 279; K-E Maskus and J-H Reichman (eds), *International Public Goods and Transfer of Technology Under a Globalized Intellectual Property Regime* (Cambridge, Cambridge University Press, 2005).

[34] See, eg, L Bently and B Sherman, *Intellectual Property Law* (Oxford, Oxford University Press, 2001), 233–52.

[35] T Cottier and M Panizzon, 'Legal Perspectives on Traditional Knowledge: The Case for Intellectual Property Protection' (2004) 7 *Journal of International Economic Law* 371; G Van Overwalle, 'Protecting and Sharing Biodiversity and Traditional Knowledge: Holder and User Tools' (2005) 53 *Ecological Economics* 585.

based on broad concepts of equity and justice. They are sometimes shaped in terms of human rights claims for political reasons, such as rights of indigenous peoples, referred to above.[36] They often address communities, more than individuals, and human rights pertaining to individuals may again not provide an adequate philosophical foundation for these aspirational norms. Rather, the concepts currently emerging emanate from the protection from unfair competition which is part of WTO law by virtue of the incorporation of Article 12 *bis* of the Paris Convention[37] into the TRIPS Agreement. Finally, notions of development and sustainability are addressed in terms of environmental protection. Trade regulation in goods and services relating to biotechnology is dealt with under ordinary WTO rules. They are not inherently based upon specific precepts of human rights protection, unless one adopts very broad and general concepts, such as the right to development, or third generational rights to the environment.

In conclusion, trade regulation and human rights emanate from different legal traditions, and what was generally found is also confirmed when one looks at rules relating to genetic engineering. The two areas emerged in splendid isolation.[38] They need to be brought together. But before addressing implicit linkages, we turn to the problem of the normative ambivalence of human rights in the present field.

4. THE AMBIGUITY OF HUMAN RIGHTS IN GENETIC ENGINEERING

Genetic engineering (DNA recombination) per se, as a technology and independently of its various applications, primarily raises philosophical, ethical, moral or religious issues relating to faith.[39] Main objections relate

---

[36] On the issues of Traditional Knowledge and Indigenous Peoples, see also C Bellmann *et al* (eds), *Trading in Knowledge: Development Perspectives on TRIPS, Trade and Sustainability* (London, Earthscan, 2003); DJ Faye, 'Bioprospecting, Genetic Patenting and Indigenous Populations—Challenges Under a Restructured Information Commons' (2004) 7 *The Journal of World Intellectual Property* 401.

[37] Convention for the Protection of Industrial Property, Paris, 20 Mar 1883, as amended on 28 Sep 1979, and last revised at Stockholm on 14 July 1967, administered by WIPO, available at www.wipo.int/treaties/en/ip/paris/trtdocs_wo020.html.

[38] T Cottier, 'Trade and Human Rights: A Relationship to Discover' (2002) 5 *Journal of International Economic Law* 112.

[39] For an overview of ethical issues see A Burgess and J Walsh, 'Is Genetic Engineering Wrong, "Per Se"?' (1998) 32 *The Journal of Value Inquiry* 393; R Brownsword, 'Bioethics Today, Bioethics Tomorrow: Stem Cell Research and the Dignitarian Alliance' (2003) 17 *Notre Dame Journal of Law Ethics & Public Policy* 15; M Frankel and R Chapman, *Human Inheritable Genetic Modifications: Assessing Scientific, Ethical, Religious and Policy Issues* (Washington, DC, American Association for the Advancement of Science, 2000); E Lucassen, 'The Ethics of Genetic Engineering' (1996) 13 *Journal of Applied Philosophy* 51; UN Educational, Scientific and Cultural Organization, International Bioethics Committee, Elaboration of

to the interference with, and into, naturally given genetic structures. It is on these—non-legal—levels that the technology as such is contested and often refuted as such. Ethical, moral and religious motives strongly influence attitudes, the political process and legislation. However, legally protected human rights standards need to be distinguished from the level of ethical, moral and religious discourse and assessed in their own right and in their own terms. It is submitted that they are essentially inconclusive in relation to the technology as such. Whether or not human rights are affected depends upon the particular application and use of the technology. Clear-cut and absolute statements can, in my view, be made only in relation to biotechnological weapons.

Human rights univocally support a ban on the use of biotechnology relating to mass destruction.[40] There are simply no arguments that legitimately speak in favour of such weapons. More than any other weapon, their impact cannot be controlled. They risk proliferating on their own, once applied. They risk falling into the hands of terrorists. An outright ban should be achievable under international humanitarian law.[41] The

the Declaration on Universal Norms on Bioethics: Third Outline of a Text, Paris, 27 Aug 2004, SHS/EST/04/CIB-Gred-2/4 Rev. 2, available at http://portal.unesco.org/shs/fr/file_download.php/49171e727c7935ea103ffb385a3ae6c3PublicOutline3_en.pdf and UN High Commissioner for Human Rights, Human Rights and Bioethics, Commission on Human Rights, Resolution 1999/63, available at www.unhchr.ch/Huridocda/Huridoca.nsf/(Symbol)/E.CN.4.RES.1999.63.En?Opendocument.

[40] A Red Cross statement aptly summarises the international consensus: '[t]estimony from governments, UN agencies, scientific circles, medical associations and industry provides a long list of existing and emerging capacities for misuse. These include: 1) Deliberate spread of existing diseases such as typhoid, anthrax and smallpox to cause death, disease and fear in a population; 2) Alteration of existing disease agents rendering them more virulent, as already occurred unintentionally in research on the 'mousepox virus'; 3) Creation of viruses from synthetic materials, as occurred this year using a recipe from the Internet and gene sequences from a mail order supplier; 4) Possible future development of ethnically or racially specific biological agents; 5) Creation of novel biological warfare agents for use in conjunction with corresponding vaccines for one's own troops or population. This could increase the attractiveness of biological weapons; 6) New methods to covertly spread naturally occurring biological agents to alter physiological or psychological processes of target populations such as consciousness, behavior and fertility, in some cases over a period of years; 7) Production of biological agents that could attack agricultural or industrial infrastructure. Even unintended release of such agents could have uncontrollable and unknown effects on the natural environment; 8) Creation of biological agents that could affect the makeup of human genes, pursuing people through generations and adversely affecting human evolution itself': International Committee of the Red Cross, Appeal on Biotechnology, Weapons and Humanity, 25 Sept 2002, 1–2, available at www.icrc.org/Web/Eng/siteeng0.nsf/iwpList515/274D020806432963C1256C3E005C4338.

[41] The use of biological weapons is essentially banned under international humanitarian law. It is useful to recall the main sources: In the 1907 Hague Convention IV, Art 23 (a) reads: '[i]n addition to the prohibitions provided by special Conventions, it is especially forbidden (a) To employ poison or poisoned weapons' (Convention (IV) respecting the Laws and Customs of War on Land and its annex: Regulations concerning the Laws and Customs of War on Land. The Hague, 18 Oct 1907, available at www.icrc.

problem, however, does not end here. Arms produced by biotechnology essentially tend to be dual-use products. The product may also be used for beneficial medical purposes in another context, and a clear line is difficult to draw in research.

As for all the other applications of biotechnology, the problem already alluded to above in describing the Universal Declaration, is the ambiguity as well as the vagueness or elusiveness of human rights. The relevant human rights standards are broadly termed precepts of justice and fairness. As applied to new technologies and emerging problems, people read different perceptions into these rights. History, context, wording and even purpose allow for widely diverging interpretations as applied to genetic engineering. Moreover, human rights are not absolute. They are all subject to restrictions in the pursuit of other legitimate policy goals. It is therefore almost impossible to draw any abstract conclusions from human rights protection as to the legitimacy and admissibility of genetic engineering.

org/ihl.nsf/0/1d1726425f6955aec125641e0038bfd6?OpenDocument). More specifically, the 1925 Geneva Protocol reads: 'the use in war of asphyxiating, poisonous or other gases, and of all analogous liquids, materials or devices, has been justly condemned by the general opinion of the civilised world' (Protocol for the Prohibition of the Use in War of Asphyxiating, Poisonous or Other Gases, and of Bacteriological Methods of Warfare, Geneva, 17 June 1925, available at http://fas-www.harvard.edu/~hsp/1925. html). Furthermore, the 1972 Biological Weapons Convention explicitly claims in its Art 1: '[e]ach State Party to this Convention undertakes never in any circumstances to develop, produce, stockpile or otherwise acquire or retain: (1) Microbial or other biological agents, or toxins whatever their origin or method of production, of types and in quantities that have no justification for prophylactic, protective or other peaceful purposes; (2) Weapons, equipment or means of delivery designed to use such agents or toxins for hostile purposes or in armed conflict': The Convention on the Prohibition of the Development, Production and Stockpiling of Bacteriological and Toxin Weapons and on their Destruction, signed at Washington, London and Moscow, 10 Apr 1972, available at www.state.gov/t/ac/trt/4718.htm#treaty. This is restated in Art II § 2 of the 1993 Chemical Weapons Convention, outlawing '[a]ny chemical which through its chemical action on life processes can cause death, temporary incapacitation or permanent harm to humans or animals': Convention on the Prohibition of Development, Production, Stockpiling and Use of Chemical Weapons and on their Destruction, Paris, 13 Jan 1993, available at www.opcw.org/docs/cwc_eng.pdf). Moreover, in 2004, the United Nations Security Council announced that 'proliferation of nuclear, chemical and biological weapons, as well as their means of delivery, constitutes a threat to international peace and security': (UN Security Council, Resolution 1540, 28 Apr 2004, S/RES/1540, available at www.reachingcriticalwill.org/political/SC/1540.pdf). Finally, the use of biotechnology as a method of warfare is not consistent with most 'cardinal principles' of International Humanitarian Law (the prohibition to cause superfluous injury or unnecessary suffering, the prohibition to use indiscriminate weapons and the protection of the environment). We recall here that new weapons are considered lawful only to the extent that they are consonant with the principles underlining the rules governing armed conflicts (Shimoda, Tokyo District Court, 1963 [1964] Japanese Annual of International Law, 212). See L Vierucci in this volume.

Human dignity is at the heart of the debate.[42] It is invoked by all sides alike, and thus to some degree its normative functions are neutralised.[43] Freedom of conscience and religion per se can be invoked both in support and rejection of genetic engineering. The right to health is a strong supporter of genetic engineering in medicine, but may be offset by considerations of human dignity and conscience.[44] A substantial amount of effort has been dedicated to this complex relationship on the international level with a view to achieving common standards and perceptions.[45] The right to food is equally ambiguous. It is supportive of genetic engineering to the extent that it offers options to combat famine and plant disease. It is detrimental to the extent that the technology destroys the conventional food base, biodiversity and food security.[46]

[42] See H Boussard and D-J Galligan both in this volume. See also G Wright, 'Second Thoughts: How Human Cloning can Promote Human Dignity' (2000) 35 *Valparaiso University Law Review* 1; EU Petersmann in this volume; R Brownsword in this volume. See also The Convention for the Protection of Human Rights and Dignity of the Human Being with regard to the Application of Biology and Medicine, Council of Europe, Oviedo, 4 Apr 1997, ETS No. 164 (1997) 36 ILM 821. In Case C–377/98 *Netherlands v European Parliament and Council* [2001] ECR I–7079, 'human dignity' was one argument invoked by the Netherlands before the European Court of Justice in order to annul the application of EU Directive 98/44 on the legal protection of biotechnological inventions [1998] OJ L 213/13.

[43] Cf in particular R Brownsword, in this volume, distinguishing the different traits invoking human dignity.

[44] Cf P Singer and A Daar, 'Harnessing Genomics and Biotechnology to Improve Global Health Equity' (2001) 294 *Science* 87; S Benatar, 'Human Rights in the Biotechnology Era, International Health and Human Rights', (2002), available at www.biomedcentral.com/1472-698x/2/3.

[45] See the UN Declaration on Human Cloning, UN GA, 23 Mar 2005, A/RES/59/280, available at www.pre.ethics.gc.ca/english/pdf/UN%20Nations%20Resolution%202005.pdf; The Human Genome and Human Rights, UN GA, 10 Mar 1999, Resolution 53/152, A/RES/53/152, available at www.unhchr.ch/Huridocda/Huridoca.nsf/0/627818011f6703648025674400567c03?Opendocument. The Universal Declaration on the Human Genome and Human Rights, United Nations Educational, Scientific and Cultural Organization, 11 Nov 1997, available at http://portal.unesco.org/en/ev.php-URL_ID=13177&URL_DO=DO_TOPIC&URL_SECTION=201.html: Implementation of the Universal Declaration on the Human Genome and Human Rights, United Nations Educational, Scientific and Cultural Organization, Resolution 17-29C, Records of the General Conference, Paris, 21 Oct–12 Nov 1997, available at http://unesdoc.unesco.org/images/0011/001102/110220E.pdf; the Convention for the Protection of Human Rights and Dignity of the Human Being with regard to the Application of Biology and Medicine, Council of Europe, Oviedo, 4 Apr 1997, ETS No 164, available at http://conventions.coe.int/treaty/en/treaties/html/164.htm; the Additional Protocol to the Convention for the Protection of Human Rights and Dignity of the Human Being with regard to the Application of Biology and Medicine, on the Prohibition of Cloning Human Beings, Council of Europe, Paris, 12 Jan 1998, ETS No 168, available at http://conventions.coe.int/treaty/en/treaties/html/168.htm; the Additional Protocol to the Convention for the Protection of Human Rights and Dignity of the Human Being concerning Transplantation of Organs and Tissues of Human Origin, Council of Europe, Strasbourg, 24 Jan 2002, ETS No 186, available at www.who.int/ethics/en/ETH_EC_Protocole_transplantation.pdf.

[46] See generally CH Breining-Kaufmann, *Hunger als Rechtsproblem—völkerrechtliche Aspekte eines Rechts auf Nahrung* (Zurich, Schulthess, 1991); JP Mishra, 'Intellectual Property Rights and Food Security—The Efficacy of International Law' (2001) 4 *The Journal of World Intellectual Property* 5; K Mechlem and T Raney in this volume; R Mackenzie and P Newell,

The impact of human rights therefore closely depends upon the particular context. Human rights standards, as well as sustainable development, can be claimed to provide neither unqualified support for genetic engineering nor its unqualified and wholesale refutation. Human rights standards—except for a ban on weapons—cannot serve as a general and conclusive foundation for the regulation of international trade in this field. Relevant human rights standards, however, offer important, legally relevant elements to be considered under the facts of a particular constellation. They need a particular setting, and thus a case-by-case approach. They can be taken into account in interpreting and shaping operational provisions, in particular with WTO law, in making appropriate determinations on import or export restrictions of genetically modified products.

Ultimately, the balance of interest inherent to human rights' assessment will primarily depend upon the utility of the technology in a particular context and on the risks it involves in terms of furthering human rights values and those relating to sustainable development. Currently, judgements about utility and long-term impact of genetic engineering vary. While utility is increasingly demonstrated in the field of medical research, it has not been fully established in the field of plant genetic resources. The long-term impact of genetically modified crops on food security, on social and economic development, as well as in terms of consumer benefits cannot be adequately assessed at this stage. The only area over which we have a consensus concerns the use of biotechnology relating to mass destruction.

It follows that both national and international law needs to be able to cope with major uncertainties, except for the use of weapons of mass destruction. Given the regulatory ambivalence and vagueness of normative guidance under human rights standards, it would seem important to stress procedural requirements in regulating genetic engineering. Perhaps more than in substantive rules, it is here that emphasis should be placed in the context of pluralist societies. Procedural due process, entailing the right to be informed and heard before a determination is made, an obligation to argue and justify decisions in a rational and non-arbitrary manner, a right to appeal and judicial review provide essential elements based upon which procedures involving genetic engineering should be

'Globalisation and the International Governance of Modern Biotechnology: Promoting Food Security?' available at www.gapresearch.org/governance/FINALSYNTHESISPAPER.pdf; P Cullet, 'Intellectual Property Rights and Food Security in the South' (2004) 7 *The Journal of World Intellectual Property* 135 and UN GA, Macroeconomic Policy Questions, 'Science and Technology for Development—Impact of New Biotechnologies, with Particular Attention to Sustainable Development, including Food Security, Health and Economic Productivity' Report of the Secretary-General, 9 May 2003, 17, available at http://daccessdds.un.org/doc/UNDOC/GEN/N03/356/09/PDF/N0335609.pdf?OpenElement.

shaped. In addition, freedom of speech, freedom of the press, freedom of association and the right to obtain information from government (sunshine acts) is of paramount importance with a view to conducting an informed political debate on the subject. It will be necessary to examine to what extent all these guarantees, and possibly additional ones, generally emanating from constitutional and administrative law, also belong to the realm of international human rights protection.

5. THE IMPLICIT RELATIONSHIP OF TRADE RULES AND HUMAN RIGHTS

Since human rights cannot serve as a foundation of regulation, international trade in genetic engineering primarily needs to rely upon trade rules, and thus the principles and norms of the WTO. At the same time, the regime should be able to take into account human rights on a case-by-case basis.

Given the lack of explicit reference to human rights in WTO law, the application and interpretation of trade rules bear the risk that specific human rights concerns pursued by domestic law in regulating biotechnology may be trumped and ignored, unless they can be sufficiently taken into account in the process of applying and interpreting WTO law.[47] To what extent can they inform the application of trade rules? The scope and limits of taking into account human rights concerns of Members of the WTO are ultimately defined by the relevant rules of interpretation of the agreements. The extent to which panels and the Appellate Body do and may take into account human rights thus also defines the portals of human rights concerns for Members in the pursuit of policies relating to biotechnology under WTO law.

5.1. Principles of Interpretation

In applying and interpreting WTO rules, panels and the Appellate Body are barred from adding to or diminishing the rights and obligations under WTO law.[48] Conceptually, interpretation is defined in terms of clarification. Panels and the Appellate Body therefore are prevented from assessing

---

[47] E-U Petersmann in this volume.

[48] Cf Art 3(2) Understanding on Rules and Procedures Governing the Settlement of Disputes (Dispute Settlement Understanding, DSU), Annex II to the Agreement Establishing the World Trade Organization, Marrakesh, 15 Apr 1994 (1994) 33 ILM 1226, available at www.wto.org/english/tratop_e/dispu_e/dsu_e.htm: '[r]ecommendations and rulings of the DSB cannot add to or diminish the rights and obligations provided in the covered agreements'. Therefore G Marceau, 'WTO Dispute Settlement and Human Rights' (2002) 13 *European Journal International Law* 753, para 6 said: 'in all cases, WTO adjudicating bodies cannot "enforce" the human rights if in doing so they add to or diminish provisions of the covered agreements'.

trade disputes on the basis of rules other than those contained in the relevant agreements.[49] They do not have per se jurisdiction genuinely to apply other sources of international law, including human rights.[50] The prevailing view has been challenged, and it has been argued that all pertinent rules of international law are genuinely applicable in a dispute and therefore should also be taken into account by panels and the Appellate Body.[51] This view may be further fostered by findings that WTO law cannot exist in splendid isolation and needs to take into account other areas of international law.[52] Based upon the relevant treaty language and the current functions of dispute settlement, it is, however, difficult to accept this proposition, except for the limited concept of *jus cogens* or peremptory norms of international law.[53] To the extent that norms—such as the prohibition of slavery as a core human right or the ban on apartheid—are recognised as mandatory rules they inherently and genuinely need to

---

[49] Cf Art 3(7) DSU.

[50] Hitherto no explicit human rights arguments have been raised before the Dispute Settlement Body. Even the pending WTO dispute resolution on *EC—Measures Affecting the Approval and Marketing of Biotech Products* (WT/DS 291-293), is based, as far as we can see, on SPS and 'like products' arguments, and not directly on human rights arguments. The ECJ on the other hand, and in line with jurisprudence on human rights developed in case law, explicitly stated in a case concerning Directive 98/44 on the legal Protection of Biotechnological Inventions ([1998] OJ L 213/13) that: 'it is for the Court of Justice, in its review of the compatibility of acts of the institutions with the general principles of Community Law, to ensure that the fundamental right to human dignity is observed': Case C–377/98 *Netherlands v European Parliament and Council* [2001] ECR I–7079, para 70. See also E Righini in this volume.

[51] According to Pauwelyn, general international law and norms binding on all WTO members or reflecting their 'common intentions' form part of the law to be applied in the examination of WTO claims; as reference material for the interpretation of WTO covered agreements and as a valid defence against a WTO claim: J Pauwelyn, *Conflicts of Norms in Public International Law: How WTO Law Relates to Other Rules of International Law* (Cambridge, Cambridge University Press, 2003), 477. Based thereon, Petersmann, in this volume, argues that human rights may genuinely be taken into account in interpreting WTO law and might be invoked by defendants as justification for departure from WTO obligations. Contra: Marceau, n 48 above, at 753.

[52] *US—Standards for Reformulated and Conventional Gasoline*, Report of the Appellate Body, 29 Apr 1996, WT/DS2/AB/R, Section III–B: '[t]hat direction reflects a measure of recognition that the General Agreement is not to be read in clinical isolation from public international law' (available at www.law.georgetown.edu/iiel/cases/US-Gasoline(abr)(ab).doc). We recall that general principles of international law (such as good faith and the prohibition of abuse of rights) were eventually accepted by the Appellate Body in *US—Import Prohibition on Certain Shrimp and Shrimp Products*, Report of the Appellate Body, 12 Oct 1998, WT/DS58/AB/R, Section VI–B, para 158: '[h]aving said this, our task here is to interpret the language of the chapeau, seeking additional interpretative guidance, as appropriate, from the general principles of international law'. Para 129 moreover introduces an interpretation in accordance with contemporary concepts: '[t]he words of Article XX(g), "exhaustible natural resources", were actually crafted more than 50 years ago. They must be read by a treaty interpreter in the light of contemporary concerns of the community of nations about the protection and conservation of the environment.'(available at www.wto.org/English/tratop_e/dispu_e/cases_e/ds61_e.htm).

[53] See Cottier, n 38 above, at 111–15.

be considered, even if contravening rights and obligations under WTO agreements.[54] Other than that, it is submitted that the scope for considering other norms of international law is essentially defined by what we may call portals in the agreements themselves. This is equally true for human rights concerns.

Applying and interpreting WTO norms takes place in accordance with the customary rules of treaty interpretation. Panels and the Appellate Body have relied heavily upon the provisions of Article 31 of the Vienna Convention on the Law of Treaties.[55] They have stressed textual and contextual interpretation. In this process, other sources binding upon the parties should be taken into account under Article 31(3)(c), albeit that it remains controversial to what extent this obligation forms part of the customary international law of treaty interpretation.[56] In the present context, this provision is of paramount importance. To the extent that human rights standards can be established to be accepted universally or among the parties concerned, they need to be taken into account in the process of interpretation. To this effect, a principle to respect human rights in WTO law, as proposed by EU Petersmann, should therefore be recognised as much as general principles of law are recognised and apply to the operation of the WTO agreements.[57] To the extent that human rights standards are not in force for one or both of the parties to the dispute, they cannot be considered as a matter of law, but influence the process as a matter of factual information.[58]

Based upon such premises, WTO law can be found to be open to human rights considerations. The extent to which this is possible depends on specific applicable treaty norms. There is no uniform portal for human rights concerns. Different portals of different size exist, depending on the

---

[54] See also W Weiss and C Hermann, *Welthandelsrecht* (Munich, Verlag CH Beck, 2003), 147–8 and 472.

[55] Vienna Convention on the Law of Treaties (VCLT), Vienna, 22 May 1968, entered into force 27 Jan 1980, 1115 UNTS 331 available at http://untreaty.un.org/ile/texts/instruments/english/conventions/1_1_1969.pdf.

[56] *US—Import Prohibition on Certain Shrimp and Shrimp Products*, n 52 above, Section VI-B, para 158: '[h]aving said this, our task here is to interpret the language of the chapeau, seeking additional interpretative guidance, as appropriate, from the general principles of international law' (thereby explicitly referring in a footnote to Art 31(3)(c) VCLT). Art 31(3)(c) reads: '[t]here shall be taken into account, together with the context: . . . (c) Any relevant rules of international law applicable in the relations between parties'.

[57] EU Petersmann in this volume and the following note.

[58] Others do not seem to make this distinction, see EU Petersmann, 'Time for a United Nations "Global Compact" for Integrating Human Rights into the Law of Worldwide Organizations: Lessons from European Integration' (2002) 13 *European Journal of International Law* 621, at 645: '[t]he universal recognition of human rights requires us to construe the numerous public interest clauses in WTO law in conformity with the human rights requirements that individual freedom and non-discrimination may be restricted only to the extent necessary for protecting other human rights. The non-discrimination and "necessity" requirements in the "general exceptions" of WTO law (e.g. in Article XX of GATT and Article XIV of GATS) reflect these human rights principles'.

structure of the pertinent norm concerned. Importantly, these portals do not exist in isolation. They need to be read in conjunction as panels and the Appellate Body have sought to apply all the pertinent rules under a doctrine of effective interpretation, giving equal effect to all treaty provisions as far as possible under the different agreements.

5.2. Portals of Human Rights Concerns on Biotechnology in WTO Law

A survey of pertinent portals for potential human rights concerns relating to biotechnology shows that they vary greatly. The more specific an agreement, the less the scope for taking them into account. The more general a provision, the greater the possibility, and the need, to insert extraneous elements in the process of interpretation. We start with the more specific provisions and eventually turn to the more general ones under GATT 1994. We then briefly address services (GATS Agreement), and then turn to intellectual property (TRIPS Agreement).

5.2.1. GATT Agreements

To the extent that the SPS Agreement applies to products of biotechnology, it offers the most specific rules on assessing trade restrictions imposed. The protection of food safety inherently relates to the right to food and serves the protection of the right to health and life. The application of food standards stricter than internationally accepted ones is subject to mandatory procedures of risk assessment, scientific evidence of risk, and coherence in risk management.[59] The wording of the Agreement does not lend itself to consider specific human rights concerns other than food, health and life.[60] However, since the level of risk is autonomously defined by Members, human rights as well as other concerns (such as consumer perceptions) can be taken into account at this stage. Policies of zero, low or high risk may be adopted accordingly. The same is true for the application of precaution under Article 5(7) of the Agreement.[61] Lacking scientific

---

[59] See, eg, Howse and Mavroidis, n 6 above, at 323–7.

[60] Art 2(1) of the SPS Agreement reads: 'Members have the right to take sanitary and phytosanitary measures necessary for the protection of human, animal or plant life or health, provided that such measures are not inconsistent with the provisions of this Agreement'.

[61] We recall that in the pending WTO resolution on *EC—Measures Affecting the Approval and Marketing of Biotech Products* (WT/DS 291-293), the EC is claiming that in interpreting Art 5(7) of the SPS Agreement, the precautionary principle should fully be taken into account, as a general principle of international law. In the 'EC—Hormones' case, the Appellate Body however did not consider whether the precautionary principle is to be regarded as international customary law, and restricted its decision to the SPS arguments (*EC—Measures Concerning Meat and Meat Products*, Report of the Appellate Body, 13 Feb 1998, WT/DS26/AB/R, WT/DS48/AB/R, para 123), available at www.wto.org/English/tratop_e/dispu_e/cases_e/ds26_e.htm and ds48_e.htm.

evidence, measures may be supported in view of potential adverse implications on human rights conditions.

To the extent that the SPS Agreement does not apply, restrictions imposed on biotechnology on the basis of scientific and technical standards will be evaluated in accordance with the TBT Agreement.[62] The agreement is based upon the precepts of necessity and proportionality. Restrictions should not be more excessive than is necessary to achieve the regulatory goals. Regulatory goals primarily aim at the protection of human, plant and animal health and the environment.[63] They are thus closely related to the right to life and health. Importantly, the list is not exclusive and Members may pursue further regulatory goals.[64] A Member may thus define goals in terms of human rights policies. Under the definitions of the Agreement, such measures also entail methods of production, at least to the extent that they have a bearing on the quality of the product. Arguably, the protection of pertinent human rights in the field of biotechnology therefore permits the adoption of measures that are necessary and appropriate to protect pertinent human rights. Nothing would thus bar a Member from adopting measures to protect human dignity, for example in the field of medical research, or to foster the right to food. It may also include, under this agreement (and unlike the SPS Agreement), measures that are necessary to protect and honour prevailing consumer attitudes towards particular products and technologies. (The problem here is not so much the definition of such a goal as the ways and means of solidly assessing prevailing views.)

Broad portals to considering human rights exist under the GATT.[65] The protection of public morals under the provisions of Article XX(a) GATT amounts to the most important entrance to considering human rights values in assessing trade restrictions.[66] Human rights values, however, may also be considered under Article XX(d) to the extent that the measure is necessary to support corresponding domestic policies, and environmental rights can be pursued under the precepts of Article XX(g). Human rights concerns, it is submitted, may also lend themselves to playing a role in assessing the competitive relationship of products and in defining whether or not they are like under different provisions

---

[62] Art 1(5) TBT Agreement: '[t]he provisions of this Agreement do not apply to sanitary and phytosanitary measures as defined in Annex A of the Agreement on the Application of Sanitary and Phytosanitary Measures'.

[63] Art 2(2) of the TBT Agreement reads: '[s]uch legitimate objectives are, inter alia: national security requirements; the prevention of deceptive practices; protection of human health or safety, animal or plant life or health, or the environment'.

[64] See for a further discussion Cottier and Oesch, n 3 above, at 750–77.

[65] See Marceau, n 48 above, at 804–7.

[66] Howse and Mavroidis, n 6 above; L Hoe, 'Trade and Human Rights: What's at Issue?' (2001) 35 *Journal of World Trade* 275.

of the Agreement, in particular Article III(4) of GATT 1994.[67] Product differentiation based upon human rights concerns may build upon distinctions made in the *EC—Asbestos* case, taking into account the health risks related to a particular product.[68] Such risks could also be expressed in terms of human rights concerns as they affect the right to health.

The crucial issue here is whether such concerns also permit distinctions on the basis of methods of production (PPMs).[69] Article XX(e) permits the taking of measures against prison labour. Article XX(g) was construed to imply production methods affecting non-renewable resources. Apart from these exceptions, it has remained controversial to what extent PPMs per se can serve as the basis for product differentiation, in particular under national treatment obligations of Article III GATT. As product differentiation solely based upon PPMs tends to render market access more difficult and contains an element of extraterritorial application of the rules of importing countries, most developing countries largely refuse to accept the concept of inherent PPM based restrictions. From a human rights concerns perspective, acceptance of differentiation based upon PPMs amounts to a crucial portal which—except for GSP schemes—has not been opened and remains to be further explored. In relation to biotechnology, it would seem that the PPM issue, however, is of importance in relation to production methods and in particular the problem of release of crops and products in nature.[70] This particular problem, however, was addressed by the Cartagena Protocol and human rights values would need to be considered in applying and interpreting this particular instrument in the first place.

5.2.2. The GATS Agreement

The GATS Agreement applies to services using biotechnology and thus may be important in particular in the field of medical treatment and

---

[67] Depending on the non-discrimination provision at issue and the regulatory context, a wide range of factors (so-called border tax adjustment criteria) are relevant to determining likeness or unlikeness of products. See *Japan—Taxes on Alcoholic Beverages*, Report of the Appellate Body, 4 Oct 1996, WT/DS8/AB/R; *Chile—Taxes on Alcoholic Beverages*, Report of the Appellate Body, 13 Dec 1999, WT/DS110/AB/R; *EC—Measures affecting Asbestos and Asbestos-containing Products*, Report of the Appellate Body, 12 Mar 2001, WT/DS135/AB/R). For a general discussion of likeness see Cottier and Oesch, n 3 above, at 389–411.

[68] Ibid.

[69] See Isaac and Kerr, n 4 above; R Howse and D Regan, 'The Process/Product Distinction—An illusory Basis for Disciplining "Unilateralism" in Trade Policy' (2000) 11 *European Journal of International Law* 249; for a discussion of the problem see also Cottier and Oesch, n 3 above, at 412–8. the case law has dealt with PPM-based environmental objectives on the basis of Art XX GATT; see *US—Import Prohibition on Certain Shrimp and Shrimp Products*, n 52 above; *US—Restrictions on Imports of Tuna*, not adopted, 1994, DS29/R (1994) 33 ILM 842; *US—Restrictions on Imports of Tuna*, not adopted, 3 Sep 1991, DS21/R and *US—Prohibition of Imports of Tuna and Tuna Products from Canada*, Report of the Panel, 22 Feb 1982, L/5198–29S/91.

[70] Howse and Mavroidis, n 6 above, at 333–40.

services. This aspect of the problem has not been extensively researched. Since the GATS is an instrument of gradual and progressive liberalisation, culminating in full national treatment, Members retain the authority to subject all services and service providers to domestically defined standards. Moreover, the agreement allows for extensive conditions and qualifications of national treatment and thus market access. Finally, exemptions in Article XIV, comparable and even more extensive than those recognised in Article XX GATT, allow countries to operate restrictions including the protection of public morals and of human rights.

5.2.3. The TRIPS Agreement

Finally, we turn to intellectual property protection and the TRIPS Agreement. Unlike GATT and GATS, it sets forth minimal normative standards, and the relationship of these norms to human rights is thus structurally different.[71] Portals for taking into account human rights concerns exist in treaty interpretation. In addition, due to the normative standard setting role of the TRIPS Agreement, human rights claims are increasingly influencing the shaping of intellectual property norms per se.[72]

(a) Portals for Human Rights-based IPR Interpretation

Unlike under GATT and GATS, the portals for human rights in intellectual property protection are not to be found mainly in provisions addressing exceptions. They can primarily be found in interpreting and applying intellectual property standards themselves. While IPRs are not founded on human rights, it is submitted that existing IPRs should be construed so as to be consistent with these rights. To the extent that treaty or statutory language allows, the scope and contents of rights should be construed in light of related human rights values and provisions. In specific contexts—in particular when addressing the status of individuals and their personalities—the invocation of human rights may well make a difference. For example, trademark protection may also serve the protection

---

[71] See Report by the UN High Commissioner for Human Rights, 'The Impact of the Agreement on Trade-Related Aspects of Intellectual Property Rights on Human Rights', 27 June 2001, E/CN.4/Sub.2/2001/13, available at www.unhchr.ch/Huridocda/Huridoca. nsf/(Symbol)/E.CN.4.Sub.2.2001.13.En?Opendocument.

[72] Cf, eg, Report of the Expert Consultation on Human Rights and Biotechnology, Geneva, 2002, Annex to Human Rights and Bioethics, Report of the Secretary General, Doc. E/ CN.4/2003/98, 10 Feb 2003, para 17: '[t]he linked issues of the ability to patent genetic material and the sharing of benefits deriving from commercial exploitation of that material to be the most important issues in the area of human rights and biotechnology at this time', available at www.unhchr.ch/huridocda/huridoca.nsf/0/a9c6597e2b0571f0c1256cf7005b7ea6/ $FILE/G0310885.doc.

of a person's reputation, and not just the avoidance of consumer deception and confusion.[73] An inventor may retain his or her rights to some degree, even if the economic potential of the invention in question has been transferred to an employer. Intellectual property laws often allow for the importation of such considerations, for example in assessing fair use exemptions under Articles 13 and 30 of the TRIPS Agreement, as well as in the operation and motivation of compulsory licensing in Article 31 of the TRIPS Agreement. Such effects may flow not just from human rights standards which explicitly refer to IPRs, but from a host of rights, including freedom of the press or freedom of expression. In particular cases, it may be necessary to explore how such rights influence the international IPRs system with a view to preserving basic human rights within a utilitarian concept of IPRs protection.

Articles 7 and 8 of the TRIPS Agreement provide the most important portals to this effect.[74] In shaping and applying intellectual property standards, Members are entitled and obliged to shape the intellectual property system 'in a manner conducive to social and economic welfare, and to a balance of rights and obligations' (Article 7). They may adopt measures 'necessary to protect public health and nutrition, and to promote the public interest in sectors of vital importance to their socio-economic and technological development', albeit with the limitation 'that such measures are consistent with the provision of this Agreement' (Article 8).[75] This qualification implies that human rights concerns compatible with the goals set forth in these provisions need to remain within the bounds and scope of rights and obligations in the TRIPS Agreement. For example, the interpretation of Article 27(3)(b) relating to patenting life forms and thus biotechnological inventions cannot depart from obligations to grant patents on a non-discriminatory basis to genetically modified micro-organisms and non-biological and microbiological processes. Also, they cannot refrain from protecting plant varieties either through patents or a *sui generis* system of protection. In designing such a system, members

---

[73] Cf Bently and Sherman, n 34 above, at 661–5.

[74] T Cottier, 'The Agreement on Trade-Related Aspects of Intellectual Property Rights' in P-F-J Macrory, A Appleton and G Plummer (eds), *The World Trade Organization: Legal Economic and Political Analysis*, (Berlin, Springer Verlag, 2005) 1077–9. See also EU Petersmann, 'Constitutionalism and WTO Law' in D Kennedy and J Southwick (eds), *The Political Economy of International Trade Law: Essays in Honour of Robert E Hudec* (Cambridge, Cambridge University Press, 2002).

[75] We recall here the Doha WTO Ministerial, Declaration on the TRIPS Agreement and Public Health, 14 Nov 2001, WT/MIN(01)/DEC/2, available at www.wto.org/english/thewto_e/minist_e/min01_e/mindecl_trips_e.pdf, stating in its para 4: '[t]he TRIPS Agreement does not and should not prevent Members from taking measures to protect public health. Accordingly, while reiterating our commitment to the TRIPS Agreement, we affirm that the Agreement can and should be interpreted and implemented in a manner supportive of WTO Members' right to protect public health and, in particular, to promote access to medicines for all.'

are allowed to take into account human rights concerns in line with the precepts of Articles 7 and 8 of the Agreement. Importantly, all obligations to grant patent protection are subject to exceptions for inventions, 'the prevention within their territory of the commercial exploitation of which is necessary to protect ordre public or morality, including to protect human, animal and plant life or health or to avoid serious prejudice to the environment', subject to the qualification that such exceptions are not made merely 'because the exploitation is prohibited by their law'[76]. To the extent that biotechnological inventions, such as cloning of embryos, are found to be immoral or putting at risk health and life, Members are entitled to exclude them from patentability and thus from protecting and encouraging investment in the field.

Vice versa, intellectual property may not remain without an impact on construing and applying human rights standards. The relationship is a new field which still requires in-depth analysis. It forms part of the broader agenda of defining the relationship between trade and human rights. Even in domestic law, which is subject to constitutional law and fundamental rights, the matter has hardly come up. Utilitarian commercial law, on the one hand, and fundamental rights based on idealism, on the other hand, have coexisted without major interaction due to a lack of communication among different academic disciplines in law.

(b) The Impact of Human Rights on Rule-making

The impact of human rights, however, does not stop with interpretation. Human rights increasingly influence the shaping and making of rights. This second dimension is complex and controversial. Should the shaping of IPRs be subject to human rights?

On a domestic level, many countries are witnessing a controversial debate on patenting life forms.[77] The debate on patenting human genes is, as much as the debate on cloning, a debate on human dignity.[78] Patent rules are merely used instrumentally to encourage or discourage investment and thus activities in the field. Human rights issues, in particular human dignity and the right to health, therefore, are controversially discussed, focusing on patent law which in wider public opinion is often and unrightfully blamed as such and has suffered in credibility, legitimacy, political acceptance and support.

The second and international problem which brought about discussions on the relationship of human rights and IPRs is access to essential drugs.

---

[76] Art 27(2) TRIPS.

[77] See FM Abbott in this volume; M Herdegen, 'Patents on Parts of the Human Body. Salient Issues under EC and WTO Law' (2002) 5 *The Journal of World Intellectual Property* 145; C Campiglio, 'Human Genetics, Reproductive Technology and Fundamental Rights' (2005) 14 *Italian Yearbook of International Law* 83.

[78] Cf n 42 above.

Importantly, unlike the field of patenting life forms, it directly relates to the expansion of the relatively high intellectual property standards to developing countries by means of the TRIPS Agreement, being a mandatory commitment of all the currently 149 Member States of the WTO. The advent of patent protection for pharmaceuticals gave rise to concerns that this would undermine access to essential drugs at low costs, and thus the right to health of a great number of people around the world. The scourge of AIDS in particular fuelled debate.[79] Tensions arose when industrialised country governments sought the restriction of parallel imports or imports of generic medicaments, invoking the provisions of the TRIPS Agreement.[80] Albeit that the TRIPS Agreement does not legally prevent developing countries from taking such measures in most, if not in all, cases upon an appropriate interpretation of the Agreement,[81] the problem stirred an important political debate in the context of the WTO Ministerial Conference of Doha in November 2001.[82] It led to the adoption of a waiver relating to the scope of Article 31(f) of the TRIPS Agreement, thus allowing Members to grant compulsory licensing on essential drugs also for the main purpose of serving export markets.[83]

At this stage, we merely wish to note that the discussion of the relationship of intellectual property and human rights has been triggered by this very political debate. It has not yet reached deeper levels of the legal problem. To what extent do intellectual property rights, being an emanation of the fundamental concept of private property and taking part in the human rights tradition in relation to the protection of moral rights in copyright,[84] foster or impede the cause of human rights? Are there any general answers to this, or does it always depend on the particular context? What for example, is the economic role of copyright protection for the enhancement of freedom of expression, free speech,

[79] See M Dias-Varella, 'The WTO, Intellectual Property and AIDS: Case Studies from Brazil and South Africa' (2004) 7 *Journal of World Intellectual Property* 523.

[80] See, for instance, T Brennan, 'The United States and Brazil Agree to Disagree over Brazil's Patent Law' (2001) 13(a) *Intellectual Property and Technology Journal* 1–6.

[81] See T Cottier, 'TRIPs, the Doha Declaration and Public Health' (2003) 6 *The Journal of World Intellectual Property* 385–8.

[82] For a comprehensive analysis see FM Abbott, 'The WTO Medicines Decision: World Pharmaceutical Trade and the Protection of Public Health' (2005) 99 *American Journal of International Law* 317.

[83] Doha WTO Ministerial, Declaration on the TRIPs Agreement and Public Health, 14 Nov 2001, WT/MIN(01)/DEC/2, available at www.wto.org/English/thewto_e/minist_e/min01_e/mindecl_trips_e.pdf; Decision of the General Council, Implementation of para 6 of the Doha Declaration on the TRIPS Agreement and public health, 30 Aug 2003, WT/L/540, available at www.wto.org/english/tratop_e/trips_e/implem_para6_e.htm and WTO General Council, Amendment of the TRIPS Agreement, 8 Dec 2005, WT/L/641, available at www.wto.org/English/news_e/pres05_e/pr426_e.htm.

[84] See text accompanying n 34 above.

freedom of the press, freedom of information, freedom of art, and for the right to education? What is the impact of patenting drugs on the right to health in general? What is the impact of patenting life forms for human dignity? It would seem that we can hardly find general and easy answers. Effects may be beneficial in one constellation and detrimental in another. Human rights and patent law coexist with their own legitimacy. It will be important to explore to what extent human rights standards can and should influence the shaping and design of intellectual property norms, in particular the scope, the rights attached and limitations of the intellectual property rights imposed.

From the point of view of the human rights community, the relationship is often seen in terms of a one-way street. The Committee on Economic, Social and Cultural Rights in its *Official Statement on Substantive Issues Arising in the Implementation of the International Covenant on Economic, Social and Cultural Rights* evokes the predominance of human rights standards and calls for the use and application of intellectual property in support of these rights. The basic philosophy is reflected in paragraphs 6–9 of the following statement.[85]

> 6.   The fact that the human person is the central subject and primary ben-
> eficiary of human rights distinguishes human rights, including the right of
> authors to the moral and material interests in their works, from legal rights
> recognized in intellectual property systems. Human rights are fundamental,
> inalienable and universal entitlements belonging to individuals, and in some
> situations groups of individuals and communities. Human rights are funda-
> mental as they derive from the human person as such, whereas intellectual
> property rights derived from intellectual property systems are instrumental,
> in that they are a means by which States seek to provide incentives for inven-
> tiveness and creativity from which society benefits. In contrast with human
> rights, intellectual property rights are generally of a temporary nature, and
> can be revoked, licensed or assigned to someone else. While intellectual prop-
> erty rights may be allocated, limited in time and scope, traded, amended and
> even forfeited, human rights are timeless expressions of fundamental entitle-
> ments of the human person. Whereas human rights are dedicated to assuring
> satisfactory standards of human welfare and well-being, intellectual property
> regimes, although they traditionally provide protection to individual authors
> and creators, are increasingly focused on protecting business and corporate
> interests and investments. Moreover, the scope of protection of the moral and
> material interests of the author provided for under article 15 of the Covenant
> does not necessarily coincide with what is termed intellectual property rights
> under national legislation or international agreements.
>
> 7.   Human rights are based on the equality of all persons and their equal
> standing before the law. For that reason, human rights instruments place

---

[85] UN Economic and Social Council, Official Statement on Substantive Issues Arising in the ICESCR, E/C.12/2001/15,14 Dec 2001, available at www.unhchr.ch/tbs/doc.nsf/0/ 1e1f4514f8512432c1256ba6003b2cc6?Opendocument.

great emphasis on protection against discrimination. Articles 2.2 and 3 of the Covenant stipulate that States parties undertake to guarantee that the rights enunciated in the Covenant must be exercised without discrimination of any kind as to race, colour, sex, language, religion, political or other opinion, national or social origin, property, birth or other status, and to ensure the equal rights of men and women to the enjoyment of all the rights set forth in the Covenant.

8.   A human rights-based approach focuses particularly on the needs of the most disadvantaged and marginalized individuals and communities. Because a human right is a universal entitlement, its implementation is evaluated particularly by the degree to which it benefits those who hitherto have been the most disadvantaged and marginalized and brings them up to the mainstream level of protection. Thus, in adopting intellectual property regimes, States and other actors must give particular attention at the national and international levels to the adequate protection of the human rights of disadvantaged and marginalized individuals and groups, such as indigenous peoples.

9.   International human rights law includes the right of everyone to be consulted and participate in significant decision-making processes that affect them. The right to participate is reflected in numerous international instruments, including the Covenant and the International Covenant on Civil and Political Rights, as well as the Declaration on the Right to Development. Accordingly, the Committee supports the active and informed participation of all those affected by intellectual property regimes.

As standards of justice, equity and fairness, human rights do indeed provide guidance and set goals which law ultimately should achieve and realise. However, it does not necessarily imply a hierarchical perception. It would seem important to consider the relationship of IPRs and human rights, not so much as a relationship of subordination, but as one of co-ordination. Functions and legitimate goals of intellectual property protection equally need to be taken into account. This is true both in applying the law and in negotiating and designing new rules relating to genetic engineering.

It was noted above that the application of existing rules takes place within a particular factual context. Human rights values can best be assessed in that way, and more specific guidance and meaning can be found, ie, as to whether a human right acts in support of, or against, genetic engineering.[86] Such guidance and meaning inform the interpretation of existing trade rules, but cannot overrule them under existing principles of WTO law.

More flexibility to import human rights values exists in the process of negotiations and rule-making. It is here that a human rights-based approach can be further developed. Nothing precludes the establishment of explicit references. Negotiators, however, will face the problem of the

---

[86] See text following n 46 above.

ambiguity of human rights as they deal with general rules applicable to a wide range of different and future factual settings. Moreover, human rights, once identified, need to be translated into economic regulations. While human rights standards, in particular in relation to broad policy goals such as the right to food and the right to health, define essential goals, these goals essentially need to be reached by means of international economic law, in particular international trade regulation. Broadly termed human rights per se offer little guidance in defining complex regulatory issues as to the specific scope and duration of economic rights. These elements depend rather on an assessment of pros and cons in terms of the utilitarian foundation of intellectual property rights, discussed above.[87] In particular, it cannot be argued that human rights persistently argue in favour of reduced protection and a preference for public goods. Reduced protection does not necessarily entail better quality and promotion of health and of human rights in general. The quality of research and pharmaceutical products remains a major concern, quite apart from combating outright piracy. While patent protection and exclusive rights may harm short-term policies, one should also bear in mind that in their absence, innovative drugs of a future generation will not be produced short of extensive subsidisation by the public sector or philanthropical institutions and foundations.[88] The exemption of countries exporting generic drugs from Article 31(f) of the TRIPS Agreement[89] certainly is of short-term and medium-term benefit in bringing down the cost of essential drugs. Yet it will be important to observe the long-term effect of the general waiver on the research based industry and the development of a new generation of drugs. Finally, fundamental principles of WTO law, in particular MFN and national treatment, remain of paramount importance and need to be respected on an equal footing as essential pillars of the multilateral system.

(c) Excursus: the Impact of the Principles of Sustainable Development

Technological advances throughout the last century were facilitated by the emerging international IPRs system. Yet the system also contributed to the depletion of natural resources and to endangering the global environment

---

[87] Ibid.

[88] H Somsen, 'Public Participation in Biotechnology Governance', Paper presented at the Conference on the Impact of Biotechnologies on Human Rights, Florence, 2005 (on file with the author).

[89] Doha WTO Ministerial, Declaration on the TRIPS Agreement and Public Health, 14 Nov 2001, WT/MIN(01)/DEC/2, available at www.wto.org/english/thewto_e/minist_e/min01_e/mindecl_trips_e.pdf; Decision of the General Council, Implementation of paragraph 6 of the Doha Declaration on the TRIPS Agreement and public health, 30 Aug 2003, WT/L/540, available at www.wto.org/english/tratop_e/trips_e/implem_para6_e.htm and WTO General Council, Amendment of the TRIPS Agreement, 8 Dec 2005, WT/L/641, available at www.wto.org/English/news_e/pres05_e/pr426_e.htm.

and common resources.[90] Since adoption of the Stockholm principles in 1972,[91] and the principles of the Rio Declaration in 1992, attention has been directed to the goals of sustainable development,[92] and what the idea of intergenerational equity means to the international IPRs system. How should these concepts and principles affect the evolution of the IPRs system? How would this affect the field of biotechnology? It is important to observe in connection with sources of international law that the international IPRs system cannot be considered in isolation. It has to be responsive to prevailing concepts of justice. It has to adapt to the emerging aspirational norms of international law.

In a manner comparable to the relationship between intellectual property and human rights, the relationship between intellectual property and the principles of sustainable development in environmental law and policies of social and economic development amounts to a new issue which was also brought about by the process of globalisation of law-making in the two fields.[93] This is equally true of genetic engineering. Matters of intellectual property, sustainable development and human rights need to be considered jointly.[94] Again, in domestic law, this has rarely happened before. Intellectual property rights have coexisted with the emerging field

---

[90] Cf T Cottier, 'The Protection of Genetic Resources and Traditional Knowledge: Towards more Specific Rights and Obligations in World Trade Law' (1998) 1 *Journal of International Economic Law* 555.

[91] Declaration of the UN Conference on the Human Environment, Stockholm, 1972, available at www.unep.org/Documents.Multilingual/Default.Print.asp?DocumentID=97&Article ID=1503.

[92] The concept of 'Sustainable Development' was introduced by the so-called 1987 'Brundtland Report', defining the term as '[a] development that meets the needs of the present without comprising the ability of the future generations' (Report of the World Commission on Environment and Development, Note by the Secretary-General, 4 Aug 1987, A/42/427, available at www.un-documents.net/a42-427.htm. Since then, it has been referred to in many international conventions and in the preamble to the Agreement Establishing the World Trade Organization: '[r]ecognizing that their relations in the field of trade and economic endeavour should be conducted with a view to raising standards of living, ensuring full employment and a large and steadily growing volume of real income and effective demand, and expanding the production of and trade in goods and services, *while allowing for the optimal use of the world's resources in accordance with the objective of sustainable development, seeking both to protect and preserve the environment and to enhance the means for doing so in a manner consistent with their respective needs and concerns at different levels of economic development*': Agreement Establishing the World Trade Organization, Marrakesh, 15 Apr 1994 (1994) 33 ILM 15 available at www.wto.org/english/docs_e/legal_e/04-wto.pdf (emphasis added).

[93] See generally T Cottier and P Mavroidis (eds), *Intellectual Property: Trade, Competition, and Sustainable Development* (Ann Arbor, Mich, Michigan University Press, 2003).

[94] Cf Y-G Shim, 'Intellectual Property Protection of Biotechnology and Sustainable Development in International Law' (2003) 29 *North Carolina Journal of International Law and Commercial Regulation* 157; UNGA, 'Macroeconomic policy questions: Science and technology for development—Impact of new biotechnologies, with particular attention to sustainable development, including food security, health and economic productivity', Report of the Secretary-General, 9 May 2003, Doc. A/58/76, 17 available at http://daccessdds.un.org/doc/UNDOC/GEN/N03/356/09/PDF/N0335609.pdf?OpenElement.

of environmental law and no explicit overlaps or tensions can be reported. This began to change in international law with the advent of the 1992 Rio Summit on Sustainable Development, Agenda 21 and, in particular, the Convention on Biological Diversity (CBD) which was adopted during this period and entered into force on 29 December 1993.[95] Negotiations leading to the Convention took place during the closing days of the negotiations of the TRIPS Agreement. Protection required for plant varieties in the TRIPS Agreement was considered to be detrimental for the advancement of important goals of the CBD, in particular benefit sharing and access to technology. Intellectual property was also considered to be detrimental to achieving the protection of traditional knowledge of indigenous peoples and rural populations at large.[96] Indeed, patenting GMOs which are based upon non-protectable traditional varieties or landraces risks creating a significant imbalance. Moreover, it is felt that IPRs would seriously hamper the free flow of gene materials stored in gene banks, and this concern largely stimulated efforts to bring about the new International Treaty on Plant Genetic Resources for Food and Agriculture adopted by the FAO Conference in 2001.[97] Finally, environmental concerns were increasingly discussed in relation to cloning.[98]

Beyond these areas, the debate on IPRs and sustainable development has not really developed. Again, we should ask in a broader context: what is the relevance of patents or trademarks or geographical indication laws for the creation and promotion of environmentally friendly products? Does sustainable development call for a 'greening' of intellectual property? What, beyond the areas of traditional knowledge and genetic engineering, could be pertinent issues relating to other forms of IPR? What could and should be the impact of environmental law on the interpretation of intellectual property? Again, it is difficult to discuss these matters at a high level of abstraction. It takes a particular context to advance insights and arguments. But what can be said is that the globalisation of law-making and the expansion of IPR in developing countries and relations

---

[95] Convention on Biological Diversity, Rio de Janeiro, 5 June 1992 (1992) 31 ILM 818 available at www.biodiv.org/convention/articles.asp.

[96] For an overview of arguments. see: C Correa, 'Traditional Knowledge and Intellectual Property—Issues and Options Surrounding the Protection of Traditional Knowledge—A Discussion Paper', Quaker United Nations Office, Geneva, 2001, availabe at www.geneva.quno.info.

[97] International Treaty on Plant Genetic Resources for Food and Agriculture, Food and Agriculture Organization of the UN, 3 Nov 2001, available at www.fao.org/ag/cgrfa/itpgr.htm.

[98] E Seng, 'Human Cloning: Reflections on the Application of Principles of International Environmental and Health Law and Their Implications for the Development of an International Convention on Human Cloning' (2003) 5 *Oregon Review of International Law* 114; A Wang, 'Regulating Human Cloning within an Environmental and Human Rights Framework' (2001) 12 *Colorado Journal of International Environmental Law and Policy* 165.

of North and South has brought about new issues which were not previously present in the highly specialised world of national intellectual property law. The international system has added new and important dimensions which, more than before, require an integrated approach in addressing these relationships. Efforts to develop new forms of IPRs protecting traditional knowledge and niche products of developing countries amount to essential elements in rebalancing the international intellectual property system.

6. SKETCHING ELEMENTS OF FUTURE TRADE REGULATIONS FOR BIOTECHNOLOGY

Based upon our findings in the fields both of human rights and trade regulation, we may finally seek to identify some essential elements for future regulations for genetic engineering in international trade law. This branch of the law continues to provide the basis and the general framework of regulation, while allowing human rights concerns and sustainable development to be taken into account in a relationship of mutual co-ordination.

International trade regulations dealing with biotechnology in a global context should therefore build upon established principles relating to non-tariff barriers in technology assessment, taking into account human rights values. Relevant rules will be inserted into different existing agreements, such as the TRIPS, SPS and TBT Agreements. Alternatively, the technology could also be addressed in a specific agreement on genetic engineering. Such an agreement could be mandatory. It could also be designed as a plurilateral agreement. From the point of view of trade regulation, market access, conditions of competition and coherence with other international instruments need to be addressed. A number of broad ideas may be offered, sketching core elements of future regulations.

6.1. Market Access Regulation

The WTO provides a solid structural basis for assessing restrictions on imports, exports and competition on the market relating to genetically modified products. Existing instruments can be refined. In the triangle of GATT, SPS and TBT, it is conceivable to define genetic engineering more precisely, and to develop the law partly based upon existing disciplines. The approach essentially builds upon a fundamental distinction between risk assessment and risk management. While the former focuses on scientific assessment, the latter will entail a portal for human rights concerns and those relating to sustainable development. Open questions remain, however, in relation to restrictions based upon public morals.

6.1.1. Scientific Risk Assessment

The main emphasis should be laid on procedural approaches: in order to deal with uncertainties, governments should be authorised and obliged to undertake testing and monitoring activities prior to making determinations on substance. Risk assessment procedures need to be shaped in a manner that secures scientific assessment and excludes protectionist motivations and influence. In order to avoid such effects, risk assessment should be undertaken, or at least monitored, by the international community, and not exclusively within a domestic context. Science is international, and procedures should be structured accordingly. It is therefore submitted that risk assessment should be entrusted to appropriate international organisations which would co-ordinate research activities of domestic research institutions around the world. Specialised laboratories in different areas need not be replicated at high costs in each country, but could be put to work within a global network. Specialisation of this kind helps to bring about high quality research and reliable results.

Risk assessments undertaken by, or under supervision of, the international community should be mutually recognised. There is no reason why risk assessments should be duplicated once undertaken, unless it can be shown that the factual setting is substantially different in a different community. Risk assessments may also be valid for a certain time only, and should be repeated after a number of years, commensurate with progress made in scientific research.

6.1.2. Burden of Proof

A major issue in risk assessment is the allocation of burden of proof. Such burden can be placed upon either the exporter (TBT: eg, medicines) or the importer (SPS: foodstuffs).[99] In other words, the burden can be to demonstrate either that the product is safe, or that the product entails a risk. It would seem that the burden of proof in genetic engineering relating to product safety should essentially be borne domestically by the applicant seeking marketing approval and thus, in international dispute litigation, the exporting country. The costs of testing and risk assessment therefore should be borne essentially by exporting countries. It is important to address and should clarify the allocation of the burden of proof and of costs (including the funding of appropriate international organisations) in future agreements.

---

[99] Cf T Cottier, 'Risk Management Experience in WTO Dispute Settlement' in D Robertson and A Kellow (eds) *Globalisation and the Environment: Risk Assessment and the WTO* (Cheltenham, Northampton, 2001), 41–62.

6.1.3. Risk Management

Due to diverging perceptions, governments need appropriate leeway to undertake the management of scientifically identified risks. Unlike risk assessment, risk management is inherently a matter for decision by political authorities. It remains a domestic responsibility and cannot be conclusively delegated to international bodies. The latter may offer recommendations, but they cannot replace domestic risk management. It is on this level (and not in risk assessment) that human rights values should be mandatorily taken into account in the process on a case-by-case basis, again to the exclusion of rent-seeking protectionism. The law should set out pertinent factors to be taken into account. These factors include the protection of human, animal and plant health and the environment. They should extend to related human rights standards, such as the rights to life, health, food and third generational rights relating to the environment. Risk management offers an appropriate portal for human rights concerns.

6.1.4. The Precautionary Principle

The precautionary principle allows measures to be taken pending scientific risk assessment. It pertains to risk management and remains a responsibility of domestic authorities. The principle should be shaped in such a manner as to avoid protectionist motives. An agreement could build upon the Cartagena Protocol, if incorporation by reference cannot be achieved. The disciplines set forth in the Protocol offer a useful structure upon which WTO rules could be built, improving current conceptual shortcomings of the SPS Agreement.[100]

6.1.5. Principle of Consistency

International assessment and judicial review of domestic risk management should be based upon compliance with procedures set out and consistent with the principle of consistency in the SPS Agreement in Article 5(5) of that Agreement. Risk management must not be selective, but coherent and consistent in responding to all like and comparable risks. National treatment in risk assessment and risk management offers an additional safeguard against protectionist recourse to risk management, as it essentially excludes differential treatment of foreign and domestic products.

---

[100] See T Cottier, 'Implications for Trade Law and Policy: Towards Convergence and Integration' in CH Bail, R Falkner and H Marquard (eds), *The Cartagena Protocol on Biosafety. Reconciling Trade in Biotechnology with Environment and Development?* (London, Earthscan Publications Ltd, 2002), 467.

6.1.6. Labelling Requirements

Labelling of GMOs allows consumers to make informed decisions based upon their own value judgements (moral, ethical, religious). Labelling applies where no risk is being identified or where the risk management opts for consumer choices in terms of running a risk. It is evident that labelling can take place only in respect of products which carry little or no risk.

6.1.7. Reference to Morality?

To what extent should an agreement allow for restrictions based upon prevailing values and preferences, short of those relating to life and health of humans, plants, animals and the environment? Should an agreement include the authority to ban market access independently of risk assessment and risk management procedures on grounds of public morality? Should WTO Members be entitled to ban the technology per se on the basis of prevailing ethical and moral grounds? How does the ambiguity of human rights as applied to genetic engineering influence the legitimacy and scope of such an exception? Is it possible to design a human rights based approach to moral exceptions which avoids outright exclusion of the technology and thus also of its potential benefit to mankind and welfare? It is submitted that efforts to this effect should be made in negotiations. Restrictions which are not based upon risk assessment and risk management should be reviewed by recourse to a set of human rights standards, in particular the concept of dignity (both human and animal), the right to health and the right to life. We do not think that religion per se should determine to acceptable standards, as freedom of religion excludes trade policies based upon a particular belief. Human rights standards offer a rational approach to the problem of ethical restrictions. They specify the concept of morality and seek to inject some objectivity which is important for the acceptance of trade restrictions that are imposed. They assist in avoiding disguised economic protectionism put forth in the name of morality or religion.

6.2. Regulation of Competition

Genetic engineering should be accompanied by principles of competition policy, both relating to IPRs and anti-trust, as well as defined rules on liability.

6.2.1. Intellectual Property

Patenting of GMOs is an essential requirement for market based research and commercialisation of genetically engineered products. The TRIPS

Agreement should be revised to this effect with a view to sufficiently covering the industrial application of DNA recombination. At the same time, it is important to introduce enhanced protection for traditional knowledge, as such knowledge offers an important base for genetic engineering in food and medicines. Rebalancing intellectual property is essential in order to achieve an overall balanced system. It is in this vein that proposals to introduce prior informed consent and the indication of source should be examined as a prerequisite for obtaining patent protection in genetic engineering.[101] In addition, flanking policies in support of conventional products should be enhanced, in particular by effectively protecting geographical indications. Both trademarks and geographical indications are important in supporting biodiversity and food safety.

6.2.2. Anti-trust Rules

Anti-trust rules are essential in the field of genetic engineering due to high levels of concentration in the industry and the risks of expansive intellectual property protection (patents). Bans on hard core cartels, including export cartels, and the abuse of dominant positions should be developed in international fora, in particular the WTO, as such disciplines would also bring about a balance with intellectual property protection, in particular to the benefit of developing countries.[102] Current attitudes, rejecting the elaboration of multilateral disciplines do not withstand a close assessment of interests, and should therefore be reviewed.

6.2.3. Liability Rules

Liability rules form an important part of conditions of competition. They essentially determine costs of production and international trade (insurance). International law should envisage defining both minimum and maximum standards, while leaving Members to develop their own solutions within this range, commensurate with their own traditions of tort and contractual liability rules.[103] Such rules may be developed outside (UNEP) or within a WTO agreement on genetic engineering.

---

[101] See M Girsberger, 'Transparency Measures Under Patent Law Regarding Genetic Resources and Traditional Knowledge: Disclosure of Source and Evidence of Prior Informed Consent and Benefit Sharing' (2004) 7 *Journal of World Intellectual Property* 451.

[102] See T Cottier and I Meitinger, 'The TRIPs Agreement without a Competition Agreement?' in F-M Abbott, T Cottier and F Gurry, *The International Intellectual Property System: Commentary and Materials* (The Hague/London/Boston/New York, Kluwer International, 1999), ii,1750–61.

[103] Cf P Cullet, 'Liability and Redress in the Field of Biotechnology: Towards the Development of Rules at the National and International Levels' (2004), available at www.ielrc/content/w0401.pdf.

They should be sought in order to overcome the risk of challenges under non-violation complaints: as long as common rules are lacking, very strict liability rules may be considered to amount to nullification and impairment of benefits as they may lead to import restrictions or render products excessively costly and thus no longer competitive on export markets.

6.3. Procedural Guarantees

The ambiguity of substantive human rights standards and heavy dependence on factual settings in regulating both market access and competition inherently leads to a focus on procedures. Procedural rights and obligations are crucial to well-informed decisions. International disciplines need to entail the procedural standards which governments and international organisations are obliged to observe. Some elements have already been discussed in the context of risk assessment and risk management, as well as in the context of patenting (prior informed consent). Building upon existing general procedural obligations in WTO law, in particular transparency and an obligation to provide judicial review,[104] further research should identify procedural human rights to be incorporated as minimal standards in international agreements regulating genetic engineering. It is submitted that they may will amount to the most important contribution of human rights to the field. Due process of law, entailing the right to be heard, to be informed, the corresponding obligation to give reasons for and justify decisions need to be further explored. Finally, genetic engineering calls for particular attention to be given to an open political debate. Freedom of speech, freedom of the press and the right to obtain information from government are necessary ingredients for dealing with the controversial topic of genetic engineering in an open society. Equally, further research is required in order to assess to what extent these guarantees should and can be made part of international trade regulation of genetic engineering. We see at this point that genetic engineering is part of a broader debate on trade and human rights.

6.4. Coherence

Finally, we need to address the interaction of different fora. International regulation of biotechnology pertains to the domain of different organisations

---

[104] Art X(1), (3) GATT; Art III GATS Agreement, Art 63 TRIPS Agreement, and corresponding provisions in specialised agreements of the WTO.

and treaties. Treaties should ensure mutual consistency by means of mutual referencing, incorporation, and procedural participation in making determinations. Appropriate procedures for cooperation of different international organisations need to be developed, as much as cooperation between different domestic agencies with responsibilities in the field. The latter is key to resolving coherence problems on the international level. We need to examine to what extent obligations to co-ordinate among different governmental agencies should be introduced as a matter of international law and no longer be left entirely to domestic constitutional or administrative law. This is yet another field of research to be explored.

Enhanced co-ordination is also necessary in international dispute settlement. WTO dispute settlement is bound to remain the centre of trade-related disputes. Instead of seeking alternative dispute resolution mechanisms, efforts should be made to ensure that policy interests of other international organisations can properly be heard and taken into account in the process of interpreting and applying trade rules. Procedural rules therefore should be developed accordingly. The current opportunities to hear such organisations and to obtain factual information from them should be further developed into genuine participatory rights. Organisations should be fully integrated in the process of interpreting and applying relevant instruments which may emerge in the field.[105]

7. CONCLUSION

Diverging attitudes towards genetic engineering or biotechnology in different societies primarily play out in international trade and thus need to be addressed by trade rules in the first place. It has been argued that human rights are important factors in assessing specific applications of genetic engineering, both in terms of promoting and restricting related products (goods and services). They should influence regulations and decisions on a case-by-case basis. WTO law offers a number of portals that allow human rights to be taken into account, despite the absence of explicit recourse in current WTO law. They are able to offer guidance in assessing the justifications of policies and restrictions imposed or supportive measures applied by governments, having distorting effects on international trade. In particular, they may replace current and overly broad recourse to morality, rendering ethical considerations more rational and thus acceptable. This chapter also argued, however, that human rights—except for the ban on biological warfare and weapons—do not offer a general foundation for the regulation of the technology. Substantive human rights, as applied to genetic engineering, are ambivalent. Their application may

---

[105] See Cottier, Pauwelyn and Buergi, n 20 above, at 205–42.

be both beneficial and detrimental. They need a particular factual setting which also needs to be assessed in light of the goals of sustainable development. Trade regulation for genetic engineering therefore needs to build upon strong procedural disciplines, based upon the fundamental distinction between risk assessment and risk management, human rights, coherence, precautionary rules, labelling, intellectual property protection, and anti-trust law and liability rules, and by achieving greater coherence between different international fora dealing with the multiple and complex aspects of biotechnology or genetic engineering.

12

Patents, Biotechnology and Human Rights: The Preservation of Biodiverse Resources for Future Generations

FREDERICK M ABBOTT*

1. INTRODUCTION

T HE PATENTING OF biotechnological inventions potentially affects human rights in a number of ways. Human rights to identity and the practice of religion may be affected by the availability of patents on genetically modified human beings (or elements of the human body). Patents as mechanisms for market exclusion affect access to new medicines, including those based on biotechnological innovation. Access to medicines and health care are part of the panoply of human rights.

During the past decade, the international community has focused significant attention on the protection of biological diversity and the potential impact of patents and other intellectual property rights on that protection. All of mankind benefits from the preservation of biological diversity. Genetic resource stocks likely will be the source of future agricultural, medicinal and other innovations. The preservation of plant and animal species is important to the functioning and continuing evolution of the Earth's ecosystem, and therefore to the preservation of human life. While the maintenance of biological diversity is not part of the traditional catalogue of protectable human rights, a generalised human interest in the preservation of such diversity might be considered part of the common human interest in the wellbeing of future generations.

* Edward Ball Eminent Scholar Professor of International Law at the Florida State University College of Law, consultant to the UNCTAD/ICTSD projection TRIPS and Development, to the World Bank, the WHO Department of Essential Drugs and Medicines Policy and to the Quaker United Nations Office (Geneva). Email: FAbbott@law.fsu.edu. This chapter includes excerpts from a more extensive study prepared for the Ministry of Foreign Affairs of Norway which has given its permission for their publication here.

Differentiated biological resources are concentrated in a group of megadiverse countries, almost all of which are developing countries. The geographic territories in which such resources are located are often populated by poor indigenous peoples. The exploitation of biological resources from territories inhabited by these individuals has the potential substantially to affect their economic wellbeing. The maintenance of basic human rights, including rights to security, food and shelter, are dependent on a minimum level of economic welfare. The Convention on Biological Diversity (CBD) recognises sovereign rights over biological resources located within national territories, in part with a view towards assuring that individuals benefit financially from biotechnological inventions derived from such resources.

The potential for conflict between the objectives and terms of the CBD and the rules governing the international patent system has been debated since the conclusion of negotiations on the WTO Agreement on Trade-Related Aspects of Intellectual Property Rights (TRIPS Agreement) in late 1993. In 2006 this subject is on the active agenda of the WTO TRIPS Council, and it is being considered at the World Intellectual Property Organisation (WIPO).[1] This chapter analyses the relationship between the CBD and the rules governing the international patent system with a view to making a recommendation regarding whether a multilaterally agreed mandatory requirement for disclosure of the source and origin of genetic resources in patent applications would aid in achieving greater complementarity. The chapter concludes that adoption of such a requirement would be useful.

This chapter does not expressly address information referred to as 'traditional knowledge' except to the extent that such information is relevant to evaluating applications for patents on inventions under the generally applied criteria of patentability. Traditional knowledge may itself be protected as intellectual property distinct from patentable invention.[2]

---

[1] See, eg, Decision VII/19 of the COP of the CBD requesting technical assistance from WIPO on matters relating, inter alia, to the relationship between the CBD and international patent system disclosure which, 'invited WIPO to examine, and where appropriate address, taking into account the need to ensure that this work is supportive of and does not run counter to the objectives of the CBD, issues regarding the interrelation of access to genetic resources and disclosure requirements in intellectual property rights applications'. See WIPO Secretariat, Patent Disclosure Requirements Relating to Genetic Resources and Traditional Knowledge: Update, WIPO/GRTKF/IC/7/10, at para 11. The subject matter also is also discussed in the context of negotiations on a substantive patent law treaty.

[2] See, eg, T Cottier and M Pannizon, 'Legal Perspectives on Traditional Knowledge: the Case for Intellectual Property Protection' in K Mascus and J Reichman (eds), *International Public Goods and Transfer of Technology* (Cambridge, Cambridge University Press, 2005).

2. DEFINING THE INTERESTS AT STAKE

2.1. The Convention on Biological Diversity

The CBD has the primary objective of preserving diversity of genetic resources found in nature, including in animals and plants.[3] There are various reasons for promoting such preservation, including to allow continuity in the natural evolution of species (including adaptation to new environmental conditions), for use in research and development as a source of primary material for direct and recombinant use (taking advantage of natural development and adaptation of biological systems), and maintaining the quality of life from the presence of a diverse biological environment.

The second objective of the CBD is to recognise state ownership and control over genetic resources located within territorial boundaries. This basic objective has at least two grounds: first, to provide an economic incentive to countries for preserving genetic resources by assuring compensation for their use, and secondly, to enhance economic welfare in countries that house existing stocks of genetic resources by assuring compensation for genetic assets.

Determining the 'economic value' of genetic resource stocks is a problematic exercise because it involves anticipating what technological capacities will evolve to exploit such resources, as well as what technological capacities will evolve as alternatives to the exploitation of genetic resources. The world community remains at early stages in assessing the economic value of genetic resource stocks and strong assumptions concerning their future value should be avoided.[4] Anecdotal references to a comparatively small number of 'biopiracy' cases are not a proper framework for evaluating the economic value of genetic resource stocks. The economic value of genetic resources may remain stable, or increase or decrease dramatically in the future. Notwithstanding caveats regarding indeterminacy in valuing genetic resources, there is a reasonable likelihood that such resources are of 'material' value.[5]

---

[3] See, eg, Secretariat of Convention on Biological Diversity, 'Sustaining life on Earth', Apr 2000 ('CBD Secretariat Summary').

[4] See, eg, T Cottier, 'The Protection of Genetic Resources and Traditional Knowledge: Towards More Specific Rights and Obligations in World Trade Law' (1998) 1 *Journal of International Economic Law* 555 and JP Rosenthal, Fogarty International Center, National Institutes of Health, USA', A Benefit-sharing Case Study for the Conference of Parties to Convention on Biological Diversity' The International Cooperative Biodiversity Groups Program (ICBG).

[5] See G Dutfield, 'Legal and Economic Aspects of Traditional Knowledge' in Mascus and Reichman (eds), n 2 above, at 495, 504–5, for references to literature with economic estimates.

Developing countries are the preponderant owners of diverse genetic resources.[6] The special interest of the international community in encouraging development—evidenced, inter alia, in the Preamble to the WTO Agreement and in the United Nations Millennium Development Goals—suggests that a presumption in favour of recognising rights in genetic resources on the part of developing countries is appropriate. In other words, to the extent that developing countries are able effectively to exploit economic interests in genetic resources it is in the interests of the wider international community to support this.

2.2. The International Patent System

The objectives of the international patent system derive from the basic objectives of the patent which are:

1. To encourage innovation;
2. To encourage investment in the commercialisation of innovation, and
3. To promote the dissemination of technical knowledge.[7]

Multilateralisation of the international patent system effectively commenced with conclusion of the Paris Convention for the Protection of Industrial Property in 1883. That Convention adopted the principles of national treatment, right of priority and independence of patents. The disparate interests of developed and developing countries were not given special attention in the Paris Convention.[8] However, the TRIPS Agreement was adopted in 1994. It significantly expanded the scope of multilateral rules applicable to patenting, and expressly acknowledged developmental objectives, for example, in its preamble which '[r]ecogni[ses] the underlying public policy objectives of national systems for the protection of intellectual property, including developmental and technological objectives', and in Article 7, which provides:

Objectives

The protection and enforcement of intellectual property rights should contribute to the promotion of technological innovation and to the transfer and

---

[6] See FIC—Economic Development and Biodiversity, Table—Economic Development and Biodiversity, available at www.fic.nih.gov/programs/countries.html and UN list of Megadiverse countries.

[7] See F Machlup, *An Economic Review of the Patent System*, Subcomm. on Patents, Trademarks and Copyrights, of the Committee on the Judiciary, 85th Congress, 2nd Sess. (excerpts reprinted in F Abbott, T Cottier and F Gurry, *The International Intellectual Property System: Commentary and Materials* (The Hague/London/Boston, Kluwer Law International, 1999), 224–46. Another school of thought considers patents to protect a human right in a person's creative efforts.

[8] See ET Penrose, *The Economics of the International Patent System*, (Baltimore, Md Johns Hopkins University Press, 1951), chap XI.

dissemination of technology, to the mutual advantage of producers and users of technological knowledge and in a manner conducive to social and economic welfare, and to a balance of rights and obligations.

Furthermore, the objectives of the TRIPS Agreement must be understood in the context of the WTO Agreement, the Preamble to which includes the objectives of:

allowing for the optimal use of the world's resources in accordance with the objective of sustainable development, seeking both to protect and preserve the environment and to enhance the means for doing so in a manner consistent with their respective needs and concerns at different levels of economic development . . .'

The objectives of the international patent system which is grounded in the Paris Convention and the TRIPS Agreement should therefore be considered to include not only the three basic objectives of the patent, but also the objectives of promoting economic development, social welfare and environmental sustainability.

2.3. Mechanisms for Achieving Objectives

The CBD accomplishes its objectives by (i) broadly recognising sovereignty over genetic resources,[9] (ii) requiring prior informed consent (PIC) of the host country as a condition of access to genetic resources,[10] and (iii) providing for the equitable sharing of benefits from the exploitation of

---

[9] The CBD provides:

'Article 15. Access to Genetic Resources

1. Recognizing the sovereign rights of States over their natural resources, the authority to determine access to genetic resources rests with the national governments and is subject to national legislation.'

Because some countries are not parties to the CBD, and, particularly from an economic standpoint, the US which has signed but not ratified the agreement, it is important to clarify that the principle of sovereignty of states over resources located within their territory did not arise in the CBD but was only codified in that agreement. The US has in multilateral fora acknowledged its acceptance of this principle. Therefore, to the extent that the rules of the international patent system are reviewed for promoting compliance with the objective of national sovereignty over genetic resource stocks, the review is not directed only to countries that are party to the CBD.

[10] The CBD provides:

'Article 15. Access to Genetic Resources

5. Access to genetic resources shall be subject to prior informed consent of the Contracting Party providing such resources, unless otherwise determined by that Party.'

such resources.[11] Methods for implementation of PIC and equitable benefit sharing are elaborated in the Bonn Guidelines on Access to Genetic Resources and Fair and Equitable Sharing of the Benefits Arising Out of Their Utilisation.

The CBD is implemented in national and regional legislation in various ways. Explicit legislation has been adopted by the Andean Community, Brazil, Costa Rica and India, among other countries.[12] In addition, a number of countries through regulatory guidance offer some form of protection for genetic resources.

The international patent system accomplishes its objectives by prescribing in the TRIPS Agreement a set of rights in favour of patentees to exclude third parties from the market when their claimed inventions have met four criteria: (1) novelty, (2) inventive step, (3) capability of industrial application, and; (4) enabling disclosure. The international patent system is not a unitary mechanism. Patents are granted and enforced by national and regional authorities. The granting of patents in multiple jurisdictions is facilitated by multilateral agreements, including the Paris Convention, Patent Cooperation Treaty and Patent Law Treaty administered by WIPO, and the TRIPS Agreement.

2.4. Complementarity and its Limits

At a fundamental level the objectives of the CBD and international patent system are complementary. The international patent system should facilitate the objectives of the CBD by allowing states legally to protect their recognised interests in genetic resources, including inventions derived from genetic resources, including through commercialisation.

However, to recognise that rule systems are conceptually complementary does not mean that they are properly aligned so as to achieve that complementarity. For example, if the international patent system as currently implemented facilitates circumvention of the CBD by allowing patent applicants to secure patents based on incomplete or misleading information, this may undermine the objectives of the CBD. Similarly, if

---

[11]   The CBD provides:

'Article 15. Access to Genetic Resources

"7. Each Contracting Party shall take legislative, administrative or policy measures, as appropriate, and in accordance with Articles 16 and 19 and, where necessary, through the financial mechanism established by Articles 20 and 21 with the aim of sharing in a fair and equitable way the results of research and development and the benefits arising from the commercial and other utilization of genetic resources with the Contracting Party providing such resources. Such sharing shall be upon mutually agreed terms."'

[12]   See WIPO database of CBD implementing legislation, available at www.wipo.int. See also references in Cottier and Pannizon, n 2 above at 575–76.

the CBD is implemented in a way which adds an unnecessary level of insecurity to patent rights, this may undermine the commercial value of patents and incentives for the development of new products. The CBD and the international patent system have not been subject to 'conscious alignment' and that is the reason for the present international dialogue.

3. TOWARDS ACHIEVING COMPLEMENTARITY

3.1. The Patent Problem

3.1.1. Inventorship

Patents on inventions derived from genetic resources present some unique issues. Under the jurisprudence of the US Court of Appeals for the Federal Circuit (CAFC),[13] and pursuant to the EU Biotechnology Directive,[14] a biotechnological invention may consist of a purified or isolated form of genetic material as found in nature. Traditionally, patent law has distinguished between discoveries of natural phenomena, on one hand, and inventions which solve a technical problem, on the other. Discoveries are not patentable. Inventions are. The decision by the CAFC and EU to allow patenting of purified or isolated genetic material is an industrial policy decision which effectively modifies traditional patent law and policy.

---

[13] See, eg, *Amgen v Chugai Pharmaceutical*, 927 F 2d 1200 (CAFC 1991).

[14] Directive 98/44/EC of the European Parliament and of the Council of 6 July 1998 on the legal protection of biotechnological inventions provides:

'Article 3

1. For the purposes of this Directive, inventions which are new, which involve an inventive step and which are susceptible of industrial application *shall be patentable even if they concern a product consisting of or containing biological material or a process by means of which biological material is produced, processed or used.*
2. *Biological material which is isolated from its natural environment or produced by means of a technical process may be the subject of an invention even if it previously occurred in nature.*'

'Article 5

1. The human body, at the various stages of its formation and development, and the simple discovery of one of its elements, including the sequence or partial sequence of a gene, cannot constitute patentable inventions.

2. *An element isolated from the human body or otherwise produced by means of a technical process, including the sequence or partial sequence of a gene, may constitute a patentable invention, even if the structure of that element is identical to that of a natural element.*

3. The industrial application of a sequence or a partial sequence of a gene must be disclosed in the patent application.' [Emphasis added]

A patent is granted only to the true 'inventor' of a product or process. The various patent treaties, regional and national patent laws, require an oath of ownership of the invention by the person claiming it. Normally, this will be the individual(s) who first conceived of the invention (and reduced it to practice).[15] But, who is the 'inventor' when the claimed invention is a product of nature, or so closely related to a product of nature as to be essentially indistinguishable? The CBD recognises national sovereignty over genetic resources. If a business bioprospects within a national territory and files a patent application claiming as invention an isolated form of a genetic resource obtained from that territory, who is the owner of the invention? Does a CBD state lose its rights in the genetic resource because the bioprospector took it home and isolated it? Such a result might seem to defeat the purpose of recognising sovereign rights in genetic resources. In light of this, information concerning the source and origin of genetic resources may be directly relevant to determining ownership of the invention, or 'inventorship'.

3.1.2. Criteria of Patentability

The TRIPS Agreement provides that patents should be made available for inventions which meet the criteria of novelty, inventive step and capability of industrial application. In determining the novelty of an invention claiming or based on a genetic resource, whether the claimed invention is found in nature or derived from a material found in nature is relevant in establishing the prior art. The Rules of the US PTO frame the inquiry as follows when identical products are sought to be patented:

> Where the claimed and prior art products are identical or substantially identical in structure or composition, or are produced by identical or substantially identical processes, a *prima facie* case of either anticipation or obviousness has been established. *In re Best*, 562 F.2d 1252, 1255, 195 USPQ 430, 433 (CCPA 1977). 'When the PTO shows a sound basis for believing that the products of the applicant and the prior art are the same, the applicant has the burden of showing that they are not.' *In re Spada*, 911 F.2d 705, 709, 15 USPQ2d 1655, 1658 (Fed. Cir. 1990).[16]

The patent applicant must disclose information relevant to the question of novelty. Claims may be made with respect to genetic resources as found in nature or genetic resource-based inventions derived from nature. Determining whether such inventions are anticipated by the prior art requires identifying the inventions in their natural state; that

---

[15] European law allows a corporate entity to claim inventorship. Under US law, only natural persons may be inventors, although natural persons may assign their rights in patents to corporate persons.
[16] MPEP Rule 2112.01.

is, if a material in its natural state constitutes anticipating prior art it must be identified in order to determine whether the claimed invention is different from that prior art. Similarly, a determination as to inventive step is predicated on an appreciation of the distance between the prior art and the claimed invention.[17] Determining whether a newly claimed invention would be obvious to a person skilled in the art necessitates a determination as to what constituted the prior art. For reasons with respect to both determinations of novelty and inventive step information regarding genetic resources in their natural state is important to the patent examiner.

It is possible for a patent applicant to disclose the composition and structure of genetic material claimed in an invention either by description of the composition and structure or by deposit of the genetic material.[18] It is possible for a patent applicant to describe the prior art as a form of genetic material found in nature without specifically disclosing the country of source or origin of that material. It does not follow, however, that a requirement of disclosure of the source and origin is not reasonably related to a determination of novelty and inventiveness even if genetic material might be described in writing or by deposit without such information. The patent applicant may, for example, claim that he or she isolated or purified the genetic material, or identified a use for the material which was previously unknown. Or, the patent applicant may claim an invention derived from a material described as found in nature. In any of these cases, the work of the patent examiner may be facilitated by access to information regarding the source or origin of the materials. If the genetic materials are unique to a particular geographic location, the most likely prior art with respect to uses of or derivatives from that material may be found in sources from that geographic territory. A genetic material which is well-known in one geographic territory may be unknown in another. If a particular country or countries has chosen not to require disclosure of source or origin, this does not imply that other countries may not reasonably choose to do so within the meaning of the concepts of novelty and inventive step.

Article 27(1) of the TRIPS Agreement does not preclude WTO Members from requiring the disclosure of source and origin of genetic resources. Article 27(1) provides that 'patents shall be available for any inventions, whether products or processes, in all fields of technology, provided that they are new, involve an inventive step and are capable of industrial application'. A requirement to disclose the source or origin of genetic resources does not inhibit the grant of patents for biotechnological

---

[17] See F-K Beier, 'The European Patent System' (1981) 14 *Vanderbuilt Journal of Transnational Law* 1 and MPEP Rule 2141.
[18] See also the Communication from the European Communities, IP/C/W/383, 17 Oct 2002, at para 47.

inventions. It is rather a requirement reasonably related to determinations of novelty and inventive step and, as discussed below, inventorship. The same Article requires that patents should be available 'without discrimination as to the place of invention'. A requirement to disclose source and origin of genetic resources does not discriminate as to place of invention because, inter alia, the inventor is not prejudiced in any way by the disclosure. If in fact some WTO Members consider that a national obligation to disclose source and origin in a patent application is inconsistent with Article 27(1) of the TRIPS Agreement, this would be good grounds for seeking an amendment of the agreement.

Article 29(1) of the TRIPS Agreement obligates Members to require sufficient enabling disclosure and Article 29(2) allows members to require information concerning foreign applications and grants. Article 62 of the TRIPS Agreement allows Members to impose reasonable procedures and formalities with respect to the grant and maintenance of intellectual property rights.[19] The provisions of these Articles do not by their terms preclude Members from requiring the disclosure of information relevant to patentability and inventorship.

3.2. Measures to Promote Complementarity between the CBD and the TRIPS Agreement

3.2.1. CBD-Based Enforcement

Parties to the CBD could adopt a supplemental agreement on enforcement obligating each state to take steps to investigate and pursue violations

---

[19] The TRIPS Agreement permits imposition of reasonable procedural requirements to further substantive compliance, and provides that procedures should not unreasonably interfere with grant of patents (Art, 62.1–2). This would not preclude imposition of mandatory disclosure obligation for CBD compliance so long as an exceptionally cumbersome process was not put in place. Issues of the effect of introducing mandatory disclosure requirements on administrative efficiencies should be addressed, but placing affirmative obligation on an applicant is not inherently inefficient. Patent examiners do not typically verify the accuracy of all information provided by patent applicants (eg, patent examiners do not independently test whether the invention is enabled by repeating invention in patent office). This is the role of third party opposition and litigation. The patent office is protected by rules precluding provision of misleading information and applicable penalties.

Article 62 provides:

'1. Members may require, as a condition of the acquisition or maintenance of the intellectual property rights provided for under Sections 2 through 6 of Part II, *compliance with reasonable procedures and formalities*. Such procedures and formalities shall be consistent with the provisions of this Agreement.

2. Where the acquisition of an intellectual property right is subject to the right being granted or registered, Members shall ensure that the procedures for grant or registration, subject to compliance with the substantive conditions for acquisition of the right, permit the granting or registration of the right *within a reasonable period of time so as to avoid unwarranted curtailment of the period of protection.*' [Emphasis]

of CBD-based obligations. Such an agreement could impose an obligation to pursue enforcement measures against persons who obtain genetic
resources within the territory of CBD members in the absence of PIC or an
agreement on equitable benefit sharing. The steps to be taken could include
intervention with respect to patent applicants and patent holders, including potentially de-recognition of patent rights or the grant of compulsory
licences to remedy compliance failures. One particular obstacle to effective
implementation of such a supplemental CBD agreement is that the United
States is not a party to the CBD. Therefore, one of the largest economic markets would be outside the territorial scope of the agreement.

3.2.2. Need for a Multilateral Solution

The present situation under the TRIPS Agreement is that WTO Members
may require the disclosure of the source or origin of genetic resources, but
are not obligated to do so. This situation is unsatisfactory from the standpoint of countries housing substantial genetic resource stocks because
the enforcement of CBD-based rights of these countries depends on third
country co-operation and enforcement. An enterprise that bioprospects in
a country without complying with that country's PIC requirements may
not be able to patent an invention derived from the wrongfully acquired
genetic resources in that country, or in other countries which require adequate disclosure for compliance purposes, but it may well be able to patent the invention in third countries (which may include the major markets
for its product). So, for example, a bioprospecting enterprise which fails
to comply with the Andean rules on PIC may elect not to seek a patent
in the Andean Community, but it may seek patents in the United States,
European Union, Japan and Switzerland without a disclosure that would
aid in identifying its compliance failure.

In the absence of a multilaterally agreed requirement of disclosure or other
means for encouraging compliance with obligations arising from the CBD
and public international law, a country in which a patent application is filed
may have no basis for determining whether the applicant has complied with
obligations, and consequently whether the applicant is the legitimate owner
of the claimed invention and whether the invention is novel or inventive.

Existing international patent rules require a country seeking to enforce
rights acknowledged under the CBD to initiate claims either before the patent
office or the courts in each country where a patent is sought or has been
granted to a third party. The pursuit of such claims in this way is inefficient,
costly and time-consuming. It places an enormous burden on the limited
resources of developing countries. A multilaterally agreed requirement
to disclose source and origin would not remedy this situation on its own.
However, to the extent that it reduced the frequency with which problematic
patents are granted it would reduce the burden on these countries.

3.2.3. Mandatory disclosure of source and origin

Patent-system-based measures could be adopted as a means to promote compliance with CBD-based obligations. A requirement to disclose the source and/or origin of genetic resources in the context of the patent application process would serve several purposes. First, it would aid in providing patent applicants with effective notice of their obligation to comply with national legal requirements with respect to PIC. Patent attorneys and agents preparing patent applications would be aware of the legal basis for the requested information and would communicate this information to their clients. Secondly, patent examiners would be directed to prior art from those countries most likely to have information relevant to the issues of novelty and inventive step. Thirdly, when patent applications are published notice would be provided to countries where genetic resources were obtained of the claims to inventions based on such resources, allowing them to intervene in the application process by providing third-party information. Fourthly, a disclosure requirement would provide an independent basis for action by the patent office or third party with respect to the patent applicant. The nature of such action could vary depending on the nature of the compliance deficiency, for example, it might vary depending on the state of knowledge of the applicant or the economic consequences of the deficiency.

Also, a strong argument in favour of an existing requirement to disclose the source or origin of genetic resources is the necessity to demonstrate ownership of the invention. Starting with the premise that each country owns the genetic resources located within its territory, the development of an invention by a third party based on the use of such resources would appear to presuppose a transfer of ownership or consent to use of such resources. The addition of a clarifying rule to existing patent office rules requiring applicants to demonstrate the means by which they acquired the right to own or use genetic resources, including by disclosure of source and origin of such resources, would appear to be consistent with existing patent office rules and procedures.

A combination of obligations to disclose source 'and' origin of the genetic resources in a patent application would appear preferable to an obligation to disclose either source 'or' origin because this would provide alternative routes for tracing the sovereign owner of the resources.[20] It

---

[20] There are technical issues with respect to identifying the 'source' or 'origin' of genetic resources. A patent applicant may not be able to identify the 'origin' of a genetic resource with assurance from a technical standpoint. A plant genetic resource may reflect hundreds of thousands of years of evolution across a wide geographic expanse. The country from which the genetic resource is obtained may not be the true or only 'country of origin' of the resource from an evolutionary perspective. Cases may arise in which states parties to the CBD dispute the origin of a genetic resource. Therefore, a requirement to disclose the origin of the genetic resource may involve a best effort on the part of the patent applicant.

may be that no single CBD contracting state is able to claim ownership of a genetic resource, but it may be possible to identify a group of countries that share ownership. In such cases, equitable benefit sharing may involve distribution to more than one country.

3.2.4. Disclosure of PIC and Equitable Benefit Sharing

In addition to disclosure of the source and origin of genetic resources, it has been suggested that patent applicants might also be required to disclose compliance with PIC and equitable benefit sharing.[21] The disclosure of compliance with CBD PIC requirements would involve certification of compliance with national laws implementing such requirements.[22] Such

Generally speaking, a patent applicant will be aware of the person from whom a genetic resource was obtained. Therefore, the last person in the 'chain of custody' of the genetic resource should generally be identifiable by the patent applicant without substantial effort. It is, of course, possible that one or more intermediaries may be involved in a chain that moves from securing the genetic resource from a particular geographic location to the patent applicant. A disclosure obligation limited to the person from whom a genetic resource was directly obtained may be too narrow to provide a CBD party with an adequate basis for intervening to protect its rights. It may therefore be necessary to impose on the patent applicant an obligation to identify the originating source of the genetic resource, which in turn would require those supplying such materials to maintain adequate records of the chain of custody. Alternatively, as has been suggested, some form of 'certificate of provenance' could be introduced as a means for identifying the source of genetic resources, without which a patent applicant would not be entitled to pursue its application.

In addition, there is a technical issue regarding the relationship between genetic resources in their natural state and inventions 'derived from' such genetic resources. A patent applicant may have obtained a genetic resource within a national territory and have used that resource as the basis for experimentation which ultimately yielded a product substantially different from the genetic resource. In such case, the CBD would have required that the patent applicant obtain PIC with respect to the basis for the research, and requiring disclosure of the source and origin of the genetic resource upon which experimentation was based would remain useful for promoting compliance with the CBD. There are likely to be cases in which an invention is sufficiently remote from the genetic resource, and the inventor (and patent applicant) is sufficiently remote from the person who obtained the genetic resource, that it may be unreasonable to hold the patent applicant (who acted in good faith without notice) responsible for disclosure as to the source or origin of the resource.

[21] See Submissions from Bolivia, Brazil, Cuba, Ecuador, India, Pakistan, Peru, Thailand and Venezuela, IP/C/W/438, 10 Dec 2004 and IP/C/W/429/Rev.1, 27 Sept 2004.

[22] IP/C/W/438 specifically proposes:

'10. To fulfil the requirement of furnishing evidence of prior informed consent, the applicant will have a positive obligation and would therefore have to discharge a positive burden in this regard. This means that the applicant will have to provide evidence that he or she accessed the genetic resources and/or traditional knowledge used in the invention for which a patent is sought through approval or consent of the national authorities of the country of origin and/or the local or indigenous community, as applicable.

11. It is foreseen that the applicant will be deemed to comply with the requirement of furnishing evidence of prior informed consent if the patent application contains

requirements will vary from country to country.[23] Some countries may have a procedure pursuant to which a certificate of compliance is issued by the regulatory authorities. A certified contract between the patent applicant and the national authorities might be submitted.

The proposal to extend the disclosure requirement to certification of compliance with PIC and equitable benefit sharing appears to be based on recognition that disclosure of source and origin, standing alone, will not prevent the grant of patents when genetic resources have been taken and used without the consent of the host country. Action by the host country to prevent the issue of the patent or to seek its revocation would be required.[24] The additional certification requirements would elevate the burden of proof on the patent applicant and, theoretically, provide an unambiguous signal of compliance with CBD obligations to the patent office. The patent examiner could delay processing the application until the necessary certification and evidence is provided.

A new system requiring certification of compliance with PIC and equitable benefit sharing would be substantially more complex than a system requiring disclosure of source and origin of genetic materials. The adoption of a system of certification of compliance with PIC and equitable benefit sharing would require substantial working through of details. It can be argued that more experience in the implementation of national legislation on PIC and equitable benefit sharing would be useful before attempting to introduce a certification requirement into the international patent system. An internationally adopted system of certification might be a useful adjunct to the CBD. Such a system might reduce potential conflicts involving evaluations of foreign law. On the other hand, it can be argued that a system relying solely on disclosure of source and origin will not be adequate to remedy the problem of misappropriation or misuse of genetic resources. Moreover, the adoption of an international framework

and/or is accompanied by a declaration, in the prescribed form, indicating that prior informed consent was obtained from the relevant national authorities (and local and indigenous communities, where applicable). Further, the declaration would be accompanied, where relevant, by the actual evidence of prior informed consent, for example, in the form of a certificate or duly certified contract between the applicant and the national authorities of the country of origin. In this regard, it should be noted that it may be possible that a single declaration with the necessary evidence could be furnished to cover the requirements on disclosure of source and country of origin, evidence of prior informed consent as well as evidence of equitable benefit-sharing. . . .'

[23] The Bonn Guidelines suggest general principles for implementing PIC and equitable benefit sharing requirements, but these are not binding obligations.

[24] Such as by submitting evidence that materials were acquired without PIC, or by submitting evidence of anticipation and lack of inventive step, during the examination phase. Alternatively, ex post facto proceedings for revocation or invalidation of the patent would be necessary.

for certification is far in the future and not a practical solution for present purposes. Finally, even though legal questions will arise if a certification system is adopted without further international agreement, courts and administrative bodies are capable of working through such questions.

3.2.5. Remedies for Non-compliance

The objective of the disclosure system would be to assure compliance with the basic objectives of the CBD. There are several different approaches that states may take with regard to non-compliance with mandatory requirements to disclose the source and origin of genetic resources in patent applications.

Patent offices typically impose on applicants a duty to deal with the office in good faith. The United States Patent and Trademark Office ('US PTO'), by way of illustration, maintains in its Manual of Patenting Examining Procedure ('MPEP') a fairly extensive set of rules regarding the 'Duty of Disclosure' on the part of patent applicants.[25] Since the source and origin of genetic resources are relevant to determinations of novelty and inventive step, as well as to the question of patentable subject matter (ie, discovery or invention), it would be consistent with existing Rules of the US PTO to require applicants to disclose the source and origin of such resources.

US PTO Rules describe very broadly the information as to which a duty, of disclosure is owed. This includes, for example, information regarding 'prior invention by another, inventorship conflicts, and the like'.[26] The US PTO recommends that patent attorneys ask inventors whom they represent questions about 'the origin of the invention and the point of departure from what was previously known and in the prior art'.[27]

The remedy for fraud, inequitable conduct and/or a violation of the duty of disclosure is to render all of the claims by the inventor unpatentable or invalid.[28] A determination of inequitable conduct involves an evaluation of the 'intent' of the patent applicant. The applicant must have intended to mislead the patent office through its action or omission. As the US PTO Rules note, 'inequitable conduct is not set by statute as a criteria for patentability but rather is a judicial application of the doctrine of unclean

---

[25] Chapter 2000, Duty of Disclosure, Manual of Patenting Examining Procedure, 8th Edition, Aug 2001, Latest Revision, May 2004. The Chapter 2000 Rules of the US PTO are based on several sections of the Patent Act prescribing the duties of the Director of the Patent Office (35 USC s 2, 3, 131, and 132), which are further elaborated in the Code of Federal Regulations, at 37 CFR s 1.56.

[26] MPEP, Rule 2001.04.

[27] Ibid, at Rule 2004.

[28] Ibid, at Rule 2016. See, eg, *Bristol-Myers v Rhone-Poulenc*, 2003 US App LEXIS 7103 (CAFC 2003).

hands. . .'[29] An applicant who by inadvertent error files an incomplete application is generally permitted to correct that application.[30] Similarly, if an error is discovered subsequent to the grant of the patent, the patent holder may request correction and reissuance of the patent.[31] As a general rule, claims with respect to fraud, inequitable conduct and/or a violation of the duty of disclosure must be determined by the federal courts and not by the US PTO.[32] The patent office does not consider itself equipped to evaluate evidence as to intent.

A finding of invalidity and/or revocation of a patent is not the only potential remedy even in cases of intentional misconduct. Conceptually remedies may be fashioned that would provide equitable benefit sharing to countries from which genetic resources were obtained in the form of royalties.[33] Also, the patent holder may be required to license its invention to third parties as is sometimes the remedy in competition cases. Remedies may well be fashioned to address the specific circumstances of cases. The key point, however, is that a system of purely voluntary compliance is unlikely to have any real effect on market participants. There must be some material risk to patent applicants for failure to comply with their obligations.

The international patent system does not recognise the concept of a 'central attack' on a patent. The principle of independence of patents recognised in the Paris Convention establishes that a determination of patent invalidity or the revocation of a patent in one member state does not affect the validity of parallel patents in other member states. Judges in patent cases may be cognisant of findings of foreign judges and take them into account as evidence, but are not bound by such findings. Therefore, a requirement to disclose source and origin of genetic resources combined with a potential remedy of invalidity or revocation will still require action in more than one forum by a country from which genetic resources have been improperly appropriated. Nonetheless, if the patent holder is subject to legal proceedings in the major market countries this is likely to have a significant compliance and deterrence effect.

---

[29] Ibid, at Rule 2010.

[30] Ibid, at Rule 2004, para 11.

[31] Ibid, at Rules 2012 and 2022.05. The use Rules are based on 35 USC §251 which provides:

'Whenever a patent is, through error *without any deceptive intention*, deemed wholly or partly inoperative or invalid, by reason of a deceptive specification or drawing, or by reason of the patentee claiming more or less than he had a right to claim in the patent, the Director shall, on the surrender of such patent and the payment of the fee required by law, reissue the patent for the invention disclosed in the original patent, and in accordance with a new and amended application, for the unexpired part of the term of the original patent…' [emphasis added]

[32] Ibid, at Rules 2010, 2012, 2013, 2014 and 2022.05.

[33] Ibid, eg, Communication from the European Communities, IP/C/W/383, 17 Oct 2002, at para 55.

3.2.6. Locus of Amendment

The international patent system is effectively regulated by two multilateral institutions—the WTO and WIPO—with complementary and overlapping rules and institutional mechanisms.[34] As a consequence of this relatively unique multilateral institutional framework, rules to implement a mandatory multilateral disclosure requirement with respect to source and origin of genetic resources will be needed in each forum. Therefore, an amendment to the TRIPS Agreement, presumably situated at Article 29, will be required. Also, rule changes reflecting the mandatory nature and the disclosure obligation will be required for the Patent Cooperation Treaty and the Patent Law Treaty. Finally, such a requirement should be reflected in the draft text of the Substantive Patent Law Treaty.

4. THE CBD, THE INTERNATIONAL PATENT SYSTEM AND HUMAN RIGHTS

The objective of the CBD is to preserve biodiversity and the global genetic heritage. If preservation of biodiversity is necessary or important to the future wellbeing of the human race, it is reasonable to include its preservation as part of the broad panoply of human rights. The CBD has recognised sovereign rights in genetic resources in part with a view to assuring the equitable distribution of benefits from the exploitation of those resources. If such distribution aids individuals in developing countries to obtain the necessities of food, shelter and health care, it too is important to the promotion of human rights.

The international patent system can and should be consciously aligned with the CBD to promote the preservation of biodiversity. The adoption of a modest change to the international patent system to incorporate a disclosure rule would be a good step in this direction.

---

[34] See FM Abbott, 'Distributed Governance at the WTO–WIPO: An Evolving Model for Open-Architecture Integrated Governance' (2000) 3 *Journal of International Economic Law* 63.

Part V

Participatory Rights and Remedies

13

Citizens' Rights and Participation in the Regulation of Biotechnology

DJ GALLIGAN*

1. INTRODUCTION

BIOTECHNOLOGY REPRESENTS THE rapid advance of science in a range of complex and often controversial activities. It is defined as any technological application that uses biological systems, living organisms or derivates from them in order to make or modify products or processes for specific use.[1] It includes such matters as human reproduction, genetic enhancement, research involving human subjects, organ transplants, stem cell research and genetically modified organisms (GMOs).[2] It is the controversial character of these activities, together with their scientific complexity, that draws them to the attention of legal and moral analysis. They are controversial in opening up questions of a fundamental moral kind about the ethics and limits of genetic engineering and their consequences for individual persons and cultures; they also raise more practical questions, such as the effects of genetic engineering on the environment through the release of organisms. While it is the nature of science that it discovers new horizons and creates fresh possibilities, both of which lead in uncharted directions towards unknown destinations, biotechnology appears to add a new dimension not just of unimaginable possibilities, but also concerns and risks. The task for ethical theorists is to bring these matters within the moral traditions of Europe and elsewhere; it is then for others to devise legal structures that enable practical controversies to be resolved and decisions to be made.

* Professor of Socio-legal Studies at the University of Oxford and Jean Monnet Chair, Università degli Studi di Siena . Email: denis.galligan@csls.ox.ac.uk. With assistance from Ms Kasia Zalanowska, Doctoral Student, University of Oxford.
[1] Convention on Biological Diversity (United Nations, 1992), available at www.biodiv. org/convention/default.shtml.
[2] For a full description of biotechnological processes, see Report of the International Bioethics Committee on the Possibility of Elaborating a Universal; Instrument on Bioethics, SHS/EST/02/CIB-9/S(Rev3) (UNESCO, 2003).

Here we enter the domain of regulation, which means simply the use of law and legal techniques to supervise an area of activity. A regulatory system is generally thought of as having three elements: the formulation of standards that direct and guide the activity, the implementation of the standards in practical situations, and the monitoring of the process to ensure that it works properly.[3] Within the contours of this general approach, the scope for variation is considerable. (i) As to the first element, standards may be set by the legislative authority, the parliament or assembly, or the task may be delegated to the regulatory body itself, although in practice it is usually a combination of the two. The regulatory authority generally has a fair amount of discretion in formulating standards or in rendering general standards into more operational guidelines. How that standard-setting function is performed is one of the important issues for a regulatory regime. (ii) The second element, the implementation of standards, requires the making of decisions concerning practical cases and situations. One part of this process is the application of the standards to specific cases; this requires not only settling the facts and circumstances, but also judgement as to how the standards apply in a given situation. This again involves elements of discretion in giving content and substance to standards that are often deeply open-textured. Application may be a straightforward process of applying standards to a clear set of facts, but it is often more complex, involving a range of groups and interests whose voice should be heard and whose concerns should be taken into account. Another part of implementation concerns the action that may be taken to ensure compliance by the party subject to regulation. This sometimes means enforcement mechanisms, such as prosecution, although it more often refers to persuasion, negotiation, and accommodation, with coercive action retained as a last resort. (iii) The final element, the monitoring of regulation, draws on various processes to ensure that the goals and purposes of the regulatory regime are achieved. This takes such forms as parliamentary scrutiny, complaints mechanisms and judicial review or other court proceedings.

The purpose of this presentation is to consider selected issues that arise in the creation and functioning of regulatory systems for biotechnology. Since a complete account would be a large undertaking well beyond my present scope, my concern here is with a very particular part, namely, the scope for individuals, groups and entities to be involved in each stage of the regulatory process. Here one set of questions involves such matters as: what rights and interests do these parties have with respect to each element of the regulatory process; what is the basis for such rights and interests; and what practical steps are necessary and what difficulties are

---

[3] See generally A Ogus, *Regulation: Legal Form and Economic Theory* (Oxford, Clarendon Press, 1994).

faced in making them effective. A second set of questions concerns the current regulatory environment regarding biotechnology, which leads us to enquire into the international, European and national approaches to it. Here the object is to provide a sketch of the regulatory environment, to asses its progress in dealing with biotechnology and to isolate some of the issues that have not yet been addressed or which leave room for further consideration and development.

2. THE REGULATORY FRAMEWORK OF BIOTECHNOLOGY

2.1 Introduction

On a first approach to understanding the current position on regulation of biotechnology, one is struck by a bewildering array of discussion documents, reports, international covenants, and EU and national laws. Indeed, the various jurisdictional levels are a good point at which to begin. Since biotechnology raises issues that transcend national boundaries in both theoretical and practical ways, it has attracted the attention of international bodies, particularly the United Nations (UN) and UNESCO; it has proved equally enticing to regional bodies which for European purposes are the Council of Europe and the European Union (EU). And of course it has long since found its way into the law and administration of national states. That biotechnology should occupy so many different levels and organisations is wholly understandable, although that does mean that different bodies are approaching the issues in different ways, which adds to the complexity and diffusion. Nevertheless, the motivation common to many participants appears to be the need for universal standards and common approaches to biotechnology, although whether that is wise or feasible is another matter.

2.2 Nation States

If we start with national states and work upwards, the first thing one notices is that European states are at very different stages in addressing biotechnology and in developing regulatory regimes for its different aspects. To say that most have made little progress may be an over-simplification, although it is hard to obtain an accurate overall picture from the evidence available. A survey by the European Commission of one aspect of biotechnology, namely stem cell research, shows that EU Member States take very different positions on the standards to be applied and that many have not yet taken steps towards regulation. Italy, for instance, has a law prohibiting research on embryos unless it is for therapeutic or diagnostic purposes that are beneficial to the embryo, while the *Istituto Superiore di Sanità'* has the task of approving protocols for that purpose.[4] In the UK, research is permitted for a wider range of purposes and a regulatory system has been created with authority invested in the *Human Fertilisation and Embryology Authority*.[5] On the other hand, a random sample shows that Poland, Portugal and Slovakia, for example, have taken only minimal steps towards dealing with the issue. It is likely that a comprehensive survey of European states, even those within the EU, would demonstrate similar patterns in other areas of biotechnology. National states are central to the regulatory process since the implementation of international covenants and EU laws both depend largely on them.[6]

2.3 The European Dimension

2.3.1 European Union

The European Commission has been active in initiating an EU position which has resulted in regulations or directives on matters such as: GMOs, stem cell research and cloning. There are several Directives regulating GMOs (Genetically Modified Organisms), the most important dealing with the use of genetically modified micro-organisms both for research and commercial purposes.[7] Another covers the deliberate release into the environment of GMOs.[8] There is also a Regulation on trans-boundary movements of genetically modified organisms[9] and one concerning the traceability and labelling of GMOs and the traceability of food and feed products produced from GMOs. The most extensive Directive of the European Parliament and the Council of Ministers concerns the legal protection of biotechnological inventions.[10]

A significant role is played by the *European Group on Ethics* in science and new technologies which is an independent, pluralist and multi-disciplinary body which advises the European Commission on ethical aspects of science and new technologies concerning the preparation and implementation of Community policy and law. In 1997, the European Group on Ethics was created to succeed the Group of Advisers on

---

[4] *Norme in materia di procreazione medicalmente assistita* (Ddl Senato 1524, 2001) Art 13.

[5] Established in Aug 1991 following the passing of the Human Fertilisation and Embryology Act 1990 (c 37).

[6] On implementation of international law standards and their inter-relation with national systems, see DJ Galligan and D Sandler, 'Implementing Human Rights' in S Halliday and P Schnidt (eds), *Rights Brought Home* (Oxford, Hart Publishing, 2004).

[7] 90/219/EEC [1990] OJ L 117/1.

[8] 90/220/EEC [1990] OJ L 117/15.

[9] 1946/2003/EC [2003] OJ L 257/10.

[10] 98/44/EC [1998] OJ L 213/13.

the Ethical Implications of Biotechnology.[11] During its first mandate the *European Group on Ethics* provided opinions on subjects as diverse as human tissue-banking, human embryo research, personal health data in the information society, doping in sport and human stem cell research. At the request of the President of the Commission, the Group wrote the Report on the Charter on Fundamental Rights related to technological innovation. The Group on Ethics has issued opinions on topics such as products derived from human blood or human plasma, on ethical questions arising from the Commission's proposal for a Council Directive for legal protection of biotechnological inventions, the ethical implications of gene therapy, ethical aspects of the labelling of the food derived from modern biotechnology, and the ethical aspects of cloning techniques.

2.3.2 *Council of Europe*

The most significant document of the Council is the *Convention for the Protection of Human Rights and Dignity of the Human Being With Regard to the Application of Biology and Medicine (Convention on Human Rights and Biomedicine)* adopted in 1997. The Convention is the first internationally binding legal text designed to protect against the misuse of biological and medical advances, while protocols have been added to clarify, strengthen and supplement the overall Convention on medical research, organ transplants, the protection of the human embryo and genetics.

The Convention sets out to preserve human dignity, rights, and freedoms through a series of principles and prohibitions. In terms of genetics it bans discrimination on the grounds of a person's genetic make-up and allows the carrying-out of predictive genetic tests only for medical purposes. It allows genetic engineering only for preventive, diagnostic or therapeutic reasons and only where it does not aim to change the genetic make-up of a person's descendants. It also bans the use of *in vitro* fertilisation to help choose the sex of a child, except where it would avoid a serious hereditary condition. Medical research is regulated according to detailed and precise conditions, especially for people who cannot give their consent. The Convention prohibits the creation of human embryos for research purposes and requires adequate protection of embryos where countries allow *in vitro* research. There is also a prohibition on financial gain from the use of any part of the human body. Removal of organs and other tissues which cannot be regenerated is banned from people not able to give consent.

---

[11] GAEIB (1991–97).

2.4 The International Sphere

Moving then to the UN itself, the most important documents here are the *Universal Declaration on the Human Genome and Human Rights*[12] *and the International Declaration on the Human Genetic Data* 2003. The former Declaration raises issues related to the human genome and states that 'research, treatment or diagnosis affecting an individual's genome shall be undertaken only after rigorous and prior assessment of the potential risks and benefits pertaining thereto and in accordance with any other requirement of national law'.[13] The prior, free and informed consent of the person concerned has to be obtained, and no one shall be subjected to discrimination based on genetic characteristics. States should take measures to encourage other forms of research, training and information dissemination conducive to raising the awareness of civil society as to the risks created to human dignity by research in biology, genetics and medicine. States should recognise the value of promoting the establishment of independent, multidisciplinary and pluralist ethics committees to assess the ethical, legal and social issues raised by research on the human genome and its applications. The International Bioethics Committee (IBC) of UNESCO is expected to contribute to the dissemination of the principles of the Declaration and to examine further the issues raised by the evolution of the technologies in question. Also relevant is the 2003 International Declaration on Human Genetic Data which, as its title suggests, is concerned with the collection, processing, use and storage of human genetic data. The Declaration aims at establishing principles to guide states in the formulation of their policies and laws.

Of special importance from the point of view of civil society is the *Convention on Access to Information, Public Participation, and Access to Justice with Respect to Genetically Modified Organisms*. Known as the *Aarhus Convention*, it recognises the difficulties that arise from the release of GMOs and the need for public information and public participation concerning them. This has been added to since by way of additional protocols, culminating in 2003 in a set of detailed guidelines formulated by the Economic and Social Council of the UN and adopted by the parties to the Convention. These are considered more fully in the later discussion of civil society participation.

---

[12] The Universal Declaration on the Human Genome and Human Rights was adopted unanimously and by acclamation at the 29th session of UNESCO's General Conference on 11 Nov 1997. The following year, the UNGA endorsed the Declaration: Gen Conf Res 29 C/Res16, reprinted in Records of the General Conference, UNESCO, 29th Sess., 29 C/ Resolution 19, at 41 (1997) (adopted by the UN General Assembly, GA Res 152, UN GAOR, 53rd Sess., UN Doc. A/RES/53/152 (1999)).

[13] Art 5.

2.5 Conclusions

What conclusions may be drawn from this complicated jigsaw of laws, conventions, declarations and recommendations? One conclusion is that at the national level European states vary greatly as to the steps they have taken to regulate biotechnology. Some are reasonably well advanced, others have hardly begun even to consider the ethical issues, let alone devise regulatory regimes; many fall somewhere in between. The EU and the Council of Europe have both been active in recognising the need for the formulation of standards and rendering them into law. The number of EU Directives and Regulations is growing, although in general the EU is limited jurisdictionally as to the matters on which it has authority to legislate. The UNESCO proposal for a universal code has considerable merit, although it must be remembered that, even if the proposal becomes a reality, such a code is inevitably general in its terms, with considerable scope left to nations states as to how it should be implemented. That does not detract from the importance of a code but is a reminder that the practical issues of regulation remain with each nation state.

3. THE INVOLVEMENT OF CITIZENS, GROUPS, AND ORGANISATIONS: CIVIL SOCIETY

3.1 Issues for Civil Society

For ease of reference, citizens, groups and organisations are referred to here as *civil society*.[14] Civil society is sometimes used as a term of art, referring to non-governmental organisations as opposed to those forming part of government; its use here is for convenience and is to be understood in an extended sense to include all non-governmental persons, entities and organisations. The main issue for our present purposes concerns the involvement of civil society in the regulation of biotechnology. This in turn divides into several more particular issues: how is civil society dealt with in current approaches; what is the theoretical and social basis for civil society's involvement; what should be the practical expression of that basis; and, finally, are there obstacles that are likely to hinder its realisation.

3.2 The Theoretical and Social Basis for Civil Society's Involvement

The case for civil society's involvement in the formulation, implementation and supervision of standards concerning biotechnology is based

---

[14] The European Commission offers a definition of civil society in its White Paper on Governance: *European Governance: A White Paper*, COM(2001)428 at 14.

partly on the need to reach good outcomes and partly on the rights of the representatives of civil society to participate in the process for reaching those outcomes. By outcomes is meant having a regulatory system the standards of which are based on sound policies and principles, which is implemented fairly and effectively, and where suitable forms of recourse are available to ensure that it is implemented. The involvement of civil society therefore has dual purposes: to contribute to good outcomes and as an expression of the rights of individuals, groups and entities to participate in the policy process. The first is plainly instrumental to good outcomes; the second an expression of the rights of members of civil society. The case for each part of this approach has to be made more fully.[15]

The ideal for any society is that laws and policies should be for the common good. Since we do not trust claims that there is an objective way of determining the common good on any matter, some mechanism has to be deployed for doing so. In democratic societies, the guiding principle is that the common good is forged from considering and accommodating the different interests and viewpoints within the society. The usual mechanism for making that determination consists in electing representatives and empowering them to decide matters. Legitimacy is gained, not so much by the content of the outcomes, but from the fact that those who decide are elected by the people.

But while this is a fundamental feature of democratic societies, normally it is not enough. In the *first* place, elected representatives are prone to making foolish or unwise decisions; there is also no guarantee that different viewpoints genuinely will be taken into account; and there is a risk that more powerful interests will exert excessive influence. Moreover, decisions are often delegated, as much by necessity as by choice, to officials, agencies and regulatory bodies who are neither elected nor directly accountable to the democratic process. The democratic principle plainly needs to be supplemented by other mechanisms that draw on and reinterpret it in terms of participation in administrative processes. This also has the effect of educating and informing officials with a view to better quality decision-making. Mechanisms can be devised to inform and educate decision-makers, to make them aware of the various aspects of an issue and the different viewpoints with respect to it. The British public inquiry is one mechanism, the American notice-and-comment procedure another; while several Australian states have adopted quite complex and exhaustive approaches. The range of fact-finding and consultative approaches available is extensive.[16] The common factor, despite the differences, is the concern to provide decision-makers with information,

---

[15] For a fuller account of these issues, see DJ Galligan, *Due Process and Fair Procedures* (Oxford, OUP, 1996), Part IV.

[16] For discussion of these approaches, see ibid, chap 16.

analysis and a plurality of views. This can be linked to the common good and to the idea running through the liberal tradition represented by, for instance, John Stuart Mill, namely, that in order to make good decisions one needs to be well-informed, and that listening to the competing views in society is a good way of being informed.

This leads to a *second* consideration: in a democratic society it is important that all elements of civil society have their interests and views taken into account. This means ensuring that decisions of policy, at whatever level they are made, are responsive to the range of ideas and positions held within a society. We saw above how decisions are likely to be improved if individuals, groups and entities are able to insert their views into decision-making processes at various points. However, involvement is valued not just for its instrumental effect on outcomes; it is also important in the relationship between the state and the people. That relationship is mediated by rights of various kinds, where to have a right is to have an interest that has special protection.[17] The rights we are concerned with here protect the interest each person has in being able to participate in the political process, but participation, while a basic element of the relationship, does not fully capture the range of interests at issue. It is not only to participate and to have one's interests represented, but also to have those interests considered and taken into account. This second aspect, which is often harder to achieve, I have described elsewhere as the *right to consideration*.[18]

The pluralist debate, if properly structured, will reveal the different facets of an issue and ensure that the decision-maker is well-informed. There are, however, other kinds of information vital to good policy-making that may not emerge from it. In some fields of endeavour, of which biotechnology is surely a prime example, expert, scientific knowledge is an essential ingredient of policy-making. This is a *third* factor to take into account in devising regulatory regimes for biotechnology: how to ensure that scientific analysis and research are available, not only to the regulatory officials but also to interested parties. The scientific community itself divides in ways with different interests, and it may be that it should be represented in the regulatory process as an interested party. But quite apart from its interest as a group or number of groups, the common good has its own need for ensuring that scientific knowledge is on the policy table.

To sum up, the objective here is to identify a framework within which to understand the place of civil society in the regulatory process. The need for a theoretical foundation is plain in evaluating the current approaches to the issue and making proposals for the future. Although it is feasible

---

[17] This definition is adopted from J Raz, 'On the Nature of Rights' (1984) xciii *Mind* 18.
[18] DJ Galligan, 'Rights, Discretion, and Procedure' in C Sampford and DJ Galligan (eds), *Law, Rights, and the Welfare State* (London, Croom Helm, 1986).

here to sketch only its outline, and while a fuller account is plainly needed, we have enough to guide the discussion. There are basically four propositions: *first,* that the regulation of biotechnology should be guided by a concern for the common good; *secondly,* that in a democratic order, representatives and officials are empowered to determine the common good with respect to policy matters; and *thirdly,* that they should do so by hearing and taking account of the different elements of civil society and their different viewpoints, partly in order to respect the rights of those parties and partly in order to be well-informed; while, *fourthly,* steps should be taken to ensure that scientific knowledge is also available. There is interplay between processes *and* outcomes, between the rights of civil society to participate and to have diverse interests considered *and* the common good in having sound policies.

3.3 The Engagement of Civil Society in Current Approaches

Within the general regulatory framework for biotechnology set out earlier, we now focus on the provisions made for the participation of civil society. Participation is often used as a short-hand expression for several more specific activities which, taken together, express the extent of civil society involvement and the forms it takes. It is clear from the following discussion that the involvement of civil society in regulation and policy-making is widely recognised as a social good that ought to be encouraged. At all three levels of law-making—international, European, and national—the signs of a normative and cultural commitment to such involvement are plainly visible. What is less plain is the extent to which that commitment is translated into rights with clear legal rules and processes to back them, without which it will remain an aspiration rather than a reality.

3.3.1 International Level

The international level, as the first level of regulation, contains several conventions and declarations relevant to the participation of civil society. These tend to be in the nature of general principles rather than detailed proposals, but they are important in shaping the way international organisations and nation states regard the role of civil society. To begin, the *Convention on Biological Diversity*, adopted in 1992, envisages an active role within national states for public participation in the setting of standards, their implementation and their monitoring.[19] It also makes extensive reference to the need for raising public awareness. This concern appears in the later *Convention on Human Rights and Biomedicine* which urges the

---

[19] *Sustaining Life on Earth: How the Convention on Biological Diversity promotes Nature and Human Well-being* (Secretariat, 2004).

parties to it 'to create greater public awareness of the fundamental questions raised by the application of biology and medicine'.[20] To that end public discussion and consultation are recommended.

The *Universal Declaration on the Human Genome and Human Rights*, adopted in 1997, the same year as the preceding Convention, contains similar provisions which are spelt out more fully. First, the need for public awareness again is emphasised; in this case the concern is to protect human dignity with respect to issues raised by research in biology, genetics and medicine, and its applications.[21] Member states are encouraged to facilitate open discussion at an international level, while ensuring the free expression of different socio-cultural, religious and philosophical opinions. A second theme concerns the creation of ethics committees. This idea is repeated in various later legal instruments and is one practical measure that has been widely implemented. Committees are encouraged as vehicles for assessing the ethical, legal and social issues raised by research on the human genome and its applications.[22] They provide opportunities for the involvement of civil society. A third related proposal is for the free exchange of scientific knowledge concerning biotechnology, a matter relevant to our present concerns since responsibility for much of the proposed exchange is likely to rest with civil society.[23] In the implementation of these provisions, member states are asked to take various measures, including assessments of the risks and the benefits relating to genome research, a process which could only be properly conducted with the active engagement of civil society.[24]

The Declaration delegates the task of further developing its ideas to the *International Bioethics Committee of UNESCO*[25], urging it in the process to 'organize appropriate consultations with the parties concerned'. One of the IBC's main responses has been the proposal of a set of universal standards for bioethics.[26] In formulating the declaration, the IBC was expected to work closely with all relevant organisations, including 'the scientific community and representatives of civil society'.[27] In its report on the adoption of a universal declaration on biotechnology, the IBC emphasises the key role of 'specialists, civil society, and the international community'.[28] It refers to the building of 'alternative frameworks for

---

[20] Art 28.
[21] Art 9.
[22] Art 16.
[23] Art 18.
[24] Art 19.
[25] Art 24.
[26] *Report of the IBC on the Possibility of Elaborating a Universal Instrument on Bioethics*, SHS/EST/02/CIB-9/S (Rev 3) (2003).
[27] *Bioethics: International Implications* (Paris, Roundtable of Ministers of Science, 2001) para 7 (viii).
[28] Ibid, para 12.

arriving at a broad consensus regarding important issues', which means taking account of the 'cultural sensitivities of human communities'. The report returns again and again to the idea that biotechnology cannot be understood separately from the cultural and social setting in which it is applied, with the consequence that fundamental ethical and practical questions can be addressed and resolved only partly at the universal level, and must also be considered at the local level within local communities. Since the growth of ethical standards depends on understanding and acceptance within communities with varying traditions, the role of civil society in helping to bridge the two is obvious. The report endorses the idea that 'bioethics must be based on the practice of democracy and the active participation of all citizens'.[29]

Ideas about biotechnology must begin somewhere, and international conventions are a good breeding ground of values, attitudes and aspirations.[30] The conventions mentioned so far may be important in fostering a cultural approach to biotechnology which has civil society at its centre, the hope being that it will gradually seep down to the regional and national levels where cultural aspirations become legal principles. A serious step towards that goal was taken at the international level with the adoption of the Aarhus Convention which deals with access to information, public participation, and access to justice with respect to environmental issues, such as the formulation of plans, programmes and policies, and the setting of legislative and administrative standards.[31] The Aarhus Convention introduces the concept of rights into the discourse; it states in the Preamble that every person has 'a right to live in an environment adequate to their health and well-being'. From that general right, more particular ones follow: the right of access to information, the right to participate in decision-making, and the right of access to justice in relation to a wide range of environmental matters.[32] Extensive provision is made for the practical application of each of these rights,[33] including detailed procedures to be adopted by member states for that purpose. Three types of processes at the national level are identified: decisions on specific activities, decisions concerning plans, programmes and policies, and the preparation of executive regulations and other normative instruments.[34]

---

[29] Ibid, para 50.

[30] Also relevant here is the Cartegna Protocol on Biosafety to the Convention on Biological Diversity: note Art 23 which provides that the parties to the convention shall promote public awareness and participation. In particular, the parties 'shall consult the public in the decision-making process regarding living modified organisms'.

[31] The full title of the convention is: Convention on Access to Information, Public Participation in Decision-Making, and Access to Justice in Environmental Matters.

[32] The range of activities to which the Convention applies is stated in Annex 1.

[33] Arts 2, 3, 6, 7, and 8.

[34] Arts 6, 7, 8.

While national governments are naturally allowed discretion as to how they give effect to the provisions, the Convention could prove to be of major significance in setting a general legal framework within which that discretion must be exercised, and in specifying certain rights in the public that must be respected.

At the same time, many of the rights remain at a high level of generality with generous escape routes for national governments. To take an example, the right of civil society groups to participate is fairly firm with respect to decisions about specific activities threatening the environment, since here there must be the opportunity to submit in writing, or at a public hearing, comments and submissions. That right is weaker with respect to plans, programmes and policies where the duty on national governments is only 'to make appropriate practical and/or other provisions for the public to participate', while its duty in preparing regulations is merely 'to strive to promote effective public participation'. Each right furnishes civil society with some means for influencing and swaying government actions, and is a basis for criticising those that do not measure up, although ultimately their moral and legal force is variable.

At its meeting in Kiev in 2003, the Economic Commission for Europe adopted guidelines pursuant to the Convention with respect to GMOs. The guidelines, however, apply only to certain decisions concerning specific activities, and do not cover the more policy-oriented processes, such as plans, programmes and regulations.[35] It may be that guidelines will appear in the future concerning these other matters, but for the moment national governments have only the general provisions of the Convention to guide them. Bearing in mind that they apply only to decisions about specific activities concerning GMOs, the guidelines are nevertheless of interest and importance. They emphasise public participation in decision-making, which is to be achieved by 'open, transparent, efficient, and accountable decision-making' with particular reference to the release, marketing and use of GMOs.[36] The guidelines offer detailed advice for national governments in providing for public participation, which includes making provision for such matters as: early notice of planned activities; the provision of information as to the decision procedure and as to the circumstances surrounding the GMO; the opportunity to present evidence and argument, in writing or at an oral hearing; and ensuring that representations made by concerned parties are properly taken into account in the decision.[37] The information provided to civil society should include details of intended uses, an environmental impact assessment,

---

[35] The scope is stated in Art 3.
[36] Arts 1 and 3.
[37] Arts 6–21 and Annexes II and III.

and the measures proposed to be taken in order to limit adverse effects. The text of each decision should be published and should include the reasons on which it is based and an explanation of how 'due account has been taken of the outcome of the public participation'.[38] Provision should also be made within a national legislative framework for civil society groups to have access to a review of the decision before a court of law.[39]

These guidelines are a measured attempt at the international level to provide a framework of principles and procedures for the involvement of civil society in one limited area of regulation of biotechnology. Their purpose is to give practical guidance on how rights to information, participation and review should be implemented by national authorities. To offer one or two notes of caution should not undermine their importance. Being guidelines, they are not legally binding but represent good practices which states are encouraged to adopt 'in a flexible manner', a disclaimer that may seem like giving with one hand and taking away with the other. The next step should be to determine whether the guidelines have been adopted by the parties to the Convention or whether they remain idle aspirations. Under the Convention, a Secretariat was created, but it is not clear that it has any role in encouraging, facilitating, or monitoring compliance by member states. Without adequate attention to these matters, there is little reason for confidence in the Convention or the Guidelines having a real impact on state practices. The final point is to repeat that the Convention's coverage of the more policy-oriented decisions and processes concerning biotechnology and the environment is severely and unjustifiably limited, while guidelines on such matters do not yet exist. The approach of international law of regarding the policy process as somehow beyond regulation, as outside the ambit of civil society's interests, is embedded in the legal and political culture of Europe, unlike other places such as Australia and the United States. I return to this point later on to suggest that a paradigm shift is needed on this matter if civil society is to have a full role in the future regulation of biotechnology.

3.3.2 The European Dimension

Moving from international law to the EU, it is noticeable that the *Charter of Fundamental Rights of the European Union*, signed in Nice in 2000, makes no mention of rights to participate in policy matters. The EU does, however, take a close interest in biotechnology, as noted above, and over the last 25

---

[38] Art 19.
[39] Art 32 of the Guidelines which follows the Convention Art 9.

years has adopted directives and regulations governing such matters as the patenting of inventions and the licensing of pharmaceutical products, although the main reference is to GMOs.[40]

Of special importance in the EU context is the Commission's vision of how biotechnology should be approached within Europe.[41] In its *Strategy for Europe*, the European Commission sets out how Europe should 'develop sustainable and responsible policies' to deal with issues of biotechnology, in particular to deal with the revolution that is taking place in the life sciences and biotechnology, with new applications appearing in health care, agriculture and food production, as well as in new scientific discoveries. Policies have to be devised to deal with the many issues that will arise, while amongst the Commission's objectives is winning the confidence and support of the public for its policies and those of Member States. The idea that policies should be 'people-centred' recurs throughout the strategy and requires that there be an 'open, collaborative, and sustained process to develop coherent and credible policies'.

The *Strategy for Europe* makes a range of proposals for the regulatory systems that need to be developed, and specifies the principles that ought to govern them. These include: risk assessment, the precautionary principle, safeguarding the internal market, proportionality and consumer choice, transparency and monitoring. The importance of public confidence in European regulation is emphasised; there is talk of 'a structured dialogue' among all interested parties including civil society, while the Commission is made responsible for encouraging public debate between scientists, industry and civil society.[42] A Stakeholders' Forum is to be established and the need for a 'multilateral consultative forum' is also proposed.[43]

And yet on the precise role and participation of civil society in the regulatory system, the *Strategy for Europe* is strangely silent; the many expressions of concern for civil society involvement are not translated into practical or justiciable principles. That is taken up in a later progress report, where the Commission states that the formulation of sound and coherent policies 'lies with all stakeholders in Europe—public authorities, science, economic operators, and consumers as well as the general public'.[44] The case for setting up a *multilateral consultative forum* to facilitate open and balanced dialogue is considered, the purpose of which would be to foster 'a wide spectrum of interests, including scientists and a

---

[40] Council Directives 90/219/EEC, [1990] OJ L 117/1 and 90/220/EEC [1990] OJ L 117/15.

[41] *Life Sciences: A Strategy for Europe* (European Commission, 23 Jan 2002, COM (2002) 27 final).

[42] Ibid, at 28.

[43] *Progress Report and Future Orientations* (European Commission, 5 Mar 2003, COM(2003) 96 final, 21).

[44] Ibid, at 4.

cross-section of civil society'.[45] In another document published at the same time, the Commission goes a little further in declaring that '[d]emocracy depends on people being able to take part in public debate. To do this, they must have access to reliable information on European issues and be able to scrutinize the policy process in its various stages.'[46]

Here reference is made to an earlier 2001 White Paper of the Commission's entitled *European Governance*,[47] in which attention was drawn to the need to open-up policy-making in general, and to make it more inclusive and accountable. It is noted that there needs to be a 'stronger interaction with civil society' and it is proposed that the involvement of civil society in shaping policy should be increased and a link made with the demands of democracy. This means in practical terms establishing and publishing minimum standards for consultation in relation to EU policy-making.[48] The participation of civil society is put forward as one of the five principles of good governance, with the exhortation that there should be 'wide participation throughout the policy chain—from conception to implementation'.[49] The White Paper then proceeds to specify several practical measures for enhancing public participation. Public involvement also requires access to reliable information and the capacity to scrutinise the policy process at each of its stages. Consultation is regarded as a key concept the function of which is 'to help the Commission and other institutions to arbitrate between competing claims and priorities' in developing policy.

While the White Paper expresses firm support for developing a culture of consultation in all policy matters, and while it recognises that consultation can lead to better policies, it then asserts that it cannot be achieved by legal rules.[50] Legal rules, it is claimed, would create 'excessive rigidity and risk slowing the adoption of particular policies'.[51] The solution instead is a code of conduct that sets minimum standards. Not only does this avoid the rigidity of law, but it will supposedly reduce the risk of some civil society groups having privileged access to the policy process, which is said to be a weakness of the present ad hoc arrangements. Although this may seem to be a sensible first step towards a more effective and fairer system of civil society participation, it is surprising that no mention is made of how similar issues are dealt with in other systems, not least those of Member States. The merits of a more

---

[45] Ibid, at 21.
[46] *Commission Staff Working Document* (Brussels, 5 Mar 03, SEC(2003)248, 4.
[47] COM(2001)428 final of 25 July 2001.
[48] Ibid, at 4.
[49] Ibid, at 10.
[50] Ibid, at 17.
[51] Ibid.

legal approach are not considered, while the claim that such an approach would lead to rigidity is so bluntly asserted, without the slightest concern for any supporting evidence, that we may wonder whether the issues of participation have been seriously considered. One of the main reasons, indeed, for a more legal approach, an approach that guarantees to all groups, no matter how weak or disparate, the right to have their voices heard in policy matters, is equity and fairness. In the absence of firm legal entitlements, larger, more powerful groups are likely to form relationships with policy-makers that continue over time and which leave little room for more occasional and fragmented groups to be involved.[52] The ideas in the White Paper are fluid and will change and evolve in future years, so that any concerns about its proposals should not be overstated, although an opportunity to formulate a rigorous, properly researched approach to governance, and to the participation of civil society in particular, was plainly lost.

3.3.3 National Dimension

At the national level, the practices of Member States of the EU regarding regulation of biotechnology generally, and the involvement of civil society in particular, vary greatly. A common factor across European jurisdictions is the absence of general procedural codes for the making of delegated legislation or deciding issues of general policy. Those countries, such as Germany, Poland and Hungary, which have a tradition of administrative procedure codes do not normally include within them detailed procedures on standard-setting or policy-making by administrative agencies or government departments. There is usually a separate law dealing with general normative acts, but these on the whole are rather formal and furnish little procedural guidance on matters such as the involvement of civil society. Other countries, such as France and the United Kingdom, do not have administrative procedure codes, and, as would then be expected, do not have general laws specifying the procedures for standard-setting or policy-making. This is not to say that countries adopting this approach lack interest in procedures concerning the policy-or rule-making process, but means that each case depends on the particular legislative framework, where variation is extensive, with procedural rules varying from those that are well-developed to those that are effectively non-existent.

It was suggested earlier, using the regulation of stem cell research as an example, that European countries have been slow, in general, to devise regulatory systems for biotechnology. Although examples can be found of

---

[52] Research on these matters shows how policy communities form and how influential they are on policy-outcomes: see J Richardson and G Jordan, *Governing Under Pressure* (Oxford, Martin Robertson, 1979) and GK Wilson, *Interest Groups* (Oxford, Blackwell, 1990).

very different approaches, from almost no legal provision on biotechnology to reasonably mature regulatory approaches, the general conclusion must be that Europe is still at an early stage of regulatory development. This also helps to explain why the legal approach to participation by civil society is both variable and relatively undeveloped.

3.4 Building a Model of Civil Society Involvement

Now that we have a theoretical approach to the involvement of civil society in regulating biotechnology, and a sketch of the current context in which regulation occurs at the three levels of national, European and international, the next task is to consider what model or models should be designed for the future. That in turn depends on the context in which policy issues occur; several reasonably distinct cases can be separated. *One* is an individualised decision such as whether a licence should be granted according to fairly clear criteria to an applicant for the release of a product or organism; here the policy aspects are likely to be fairly minimal. *Another* is where a specific decision has to be made in which the policy implications are extensive; for example, a general decision to license a new product for release where novel considerations arise or where there are few guidelines to follow. Since the policy element here is prominent, various civil society groups are likely to have an interest in participating in the decision process. *A third* situation is where more general policies are being determined to guide future practice, a process which results in new laws, often in the form of secondary legislation, administrative rules, new regulations or general guidelines. *Finally*, the process may be the formulation of new laws, whether by way of statutes in Member States or EU regulations and directives.[53] Just as the policy element varies in its scope and importance in each of these cases, similarly the level of civil society engagement varies, with a plain case for a wide range of groups in the more general policy issues and a more restricted group where policy is more narrowly confined. Elsewhere I have developed the *principle of selective representation* as a means of determining the scope of participation.[54]

In a fuller analysis it would be necessary to explain this principle in detail, but, that not being possible here, it is enough to suggest a general model of civil society involvement in policy decisions concerning biotechnology. While the model needs to be tailored according to the principle of selective representation, its general features are as follows below. It should be noted also that the same goal of civil society

---

[53] For elaboration of these categories, see DJ Galligan and others, *Administrative Justice in the New European Democracies* (Budapest, 1998).

[54] DJ Galligan, *Discretionary Powers* (Oxford, Clarendon, 1986), 345 ff.

participation can be achieved through different means; procedures are primarily means to ends, ensuring that certain desired outcomes are achieved, an undertaking that can usually be achieved in different ways, according to different procedural regimes. This general principle applies to the procedures for policy-making, as can be seen by comparing the way different jurisdictions approach the matter. The US model for federal agencies, which is perhaps the best developed, has two main approaches at its disposal: one makes frequent use of the *notice-and-comment* procedure while the less frequently invoked alternative is to settle policy through a *trial-type* procedure.[55] European countries do not generally have a clear or uniform approach and use different techniques in different contexts. In Britain the public inquiry is often used on matters ranging from local issues to those of national importance, their attraction being that all interested parties have the chance to put their case and to question the evidence and cases of others. They are well regarded as a way of dealing with difficult issues, not least because the chairman of the inquiry is independent of government, although the fact that he recommends a decision rather than takes it himself means that ultimate political control is retained by the government.[56] Other approaches are found in other European countries. Often, however, in the British system consultation with civil society groups is conducted through structured procedures that are sometimes formally provided for by law, at other times a matter of informal arrangement, while several Australian jurisdictions have devised a highly exacting and elaborate regime of consultation, assessment and reason-giving, before new regulations may be adopted.[57]

Similarly with regard to biotechnology, there is no reason to think that there should be just one approach. Nevertheless, it may be useful to identify the elements that each approach should include, even though the way they come together is likely to vary according to local and contextual factors. We begin by adopting the general stance that the object in making decisions concerning biotechnology is to act in the common good and that the common good, while being more than the good of any one group or interest, should take account of the views and concerns of all groups and interests.[58] From that stance, the following elements need to be included in any regulatory regime: (i) notice of the issue is widely displayed; (ii) the issue is fully defined; (iii) relevant scientific knowledge is made available in forms intelligible to the layman; (iv) relevant evidence and facts are

---

[55] For a fuller account of these, see M Asimow, 'Delegated Legislation: United States and United Kingdom' (1983) 3 Oxford Journal of Legal Studies 253.

[56] For this and other approaches in the UK see G Ganz, *Administrative Procedures* (London, Sweet and Maxwell, 1974).

[57] For an account, see Galligan, n 15 above at 501 ff.

[58] A fuller account of senses of the common good is contained in ibid at 465 ff.

made available; (v) an assessment of the consequences of the proposed course of action is carried out and made available; (vi) opportunities are provided for interested parties to present their cases, either orally or in writing; (vii) the reasons for decisions are given; (viii) the decision is based on the common good and is shown to be in the common good; (ix) the process respects relevant rights under international, European and national jurisdictions; (x) the reasons must show that all cases put forward have been considered and taken into account; (xi) the whole process is conducted in conditions of openness and transparency; (xii) subject only to defined situations of confidentiality.

Plainly, adjustments have to be made to the general approach depending on various factors, such as whether it is directed at formulating general standards or at resolving a specific issue, perhaps the grant of a licence to release a genetically engineered organism or to patent a mechanism or product. Similarly the role of civil society should take different forms depending on the nature of the process. The most widespread involvement should occur where general standards are being formulated, while it will be narrower in individualised cases where the standards to be applied are clear. In between the two points, numerous variations are possible. The complexity of the matters at issue will also be a significant variable, so that more complex matters, or those that create high uncertainty as to their consequences, should be subject to more stringent scrutiny from civil society. Similarly, it should not be assumed that all elements or representatives of civil society are entitled to engage in the same way and on the same terms. Some individuals or groups or interests may have more clear and immediate interests than others, and therefore should have rights to participate in different ways and levels of intensity. The conclusion to be drawn from this brief sketch is, first, that the regulatory architecture suitable for biotechnology varies and, secondly, that the exact manner and form of civil society involvement depends on variable factors of the kind noted, and on the different combinations they form in different contexts.

The model outlined here should be flexible enough to accommodate the various situations in which decisions have to be made concerning biotechnology. The key ideas are information, participation and explanation. Once a decision has been made, a different but related set of issues arise as to monitoring its implementation. Monitoring includes such matters as whether the decision is put into practice, whether subsequent action is in accordance with the decision, and the capacity to revise a decision in the light of new scientific knowledge or other evidence. This gives rise to another set of questions concerning the standing of civil society in the monitoring process: it could be directly involved in monitoring implementation; more indirectly, avenues should be available for expressing concerns and complaints, whether through the regulatory process or in the courts; for questioning the implementation process at each of its stages; for having matters re-opened where there is new evidence or knowledge. The issues arising in relation to monitoring are again diverse and need to be resolved according to the contours of different biotechnological contexts. The guiding principle from the perspective of civil society is that monitoring is an essential element in the regulation of biotechnology, particularly given the levels of complexity and uncertainty, and that civil society's involvement is essential both for the common good and as an expression of its rights.

3.5 Obstacles to Involving Civil Society

Now that we have a normative model of civil society involvement in the regulation of biotechnology, we should attend briefly to some of the obstacles to making it work in practice. So far nothing has been said of the social context of biotechnology or its regulation. The first general point to note is the gap between a set of normative principles, no matter how well designed, and their realisation in practice. Put simply, the central issue is how to make the normative standards governing the involvement of civil society effective in practice. This is part of the wider question of how to ensure that a regulatory regime achieves its purposes, which in turn means asking how the different participants, the regulators, those who are regulated and any other parties, comply with its requirements. It is plain from the extensive socio-legal research on the matter that there is no straight path to success; rather, implementation is a complex process that is never perfectly achieved and will often be obstructed by competing factors.[59] Perfection is not the goal, however, and regulation can achieve tolerable levels of effectiveness. The regulation of biotechnology inherits all the traits of any other area of regulation, with extras to add. The same general problem applies to that aspect of regulation with which we are concerned: how to ensure that civil society is able to be involved according to the model proposed.

The analysis begins with a general outline of what is necessary for effective involvement, and then considers some of the obstacles. Civil society involvement depends on several elements. The first is a satisfactory legal regime or what might be referred to as the legal architecture. Architecture suggests design, which is an apt notion since regulatory regimes have to be designed in order to realise their purposes. Poor design of the regulatory system prejudices its effectiveness from the outset, a

---

[59] For discussion of implementation issues see F Haines, *Corporate Regulation* (Oxford, OUP, 1997).

matter that is often overlooked in studies of regulation.[60] Here good design means that the legal framework makes provision for involvement that is coherent and suited to its purposes, where the members of civil society are able to participate and, in case of default, enforce or otherwise secure their right to do so. The point need not be laboured that the better the design, the better the chances of its being effective. The second matter concerns the organisations that are responsible for enabling participation. To a degree this overlaps with design, but it also raises questions as to the kinds of organisations that can make sure participation is possible. This embraces not just decision-making bodies but also those that are capable of monitoring and supervising participation. An inadequate organisational structure is likely to have negative effects on participation. The third, and possibly most difficult, aspect of effective regulation is how to encourage the parties engaged in the regulation of biotechnology to internalise the norms of civil society involvement. It is plain from studies of regulation that enforcement through coercion or the threat of it, although a necessary element, will never be a secure or sufficient basis for effective implementation. Persuasion, negotiation and accommodation are often more effective both at securing compliance and at encouraging the internalisation of standards. The same analysis applies here: the norms of civil society participation need to be seen as right and natural in the resolution of matters of biotechnology.

This leads on to an initial obstacle to engaging civil society in regulatory processes: European administrative law systems are not generally committed to the full inclusion of civil society in the administrative process. That the degree of inclusion is variable and lacks a central commitment reflects a deeper cultural tendency to treat administrative systems, whether departmental or agency-based, as bureaucratic and routine, rather than policy-based and discretionary. Perhaps it is the shade of Max Weber's account of law and administration in advanced systems that has led us to think that the hard questions of policy are settled in the political process and that, once settled, it is then for administrators to implement them. Although this analysis has some plausibility in some areas of administration (the distribution of social welfare, for example, can be fairly routine), it does not present an accurate picture of modern administration. It is now well recognised that elements of policy occur at all levels of government and administration, and that administrative and regulatory agencies have discretion, whether expressly conferred or implicitly assumed, to decide a great range of matters. And it is to be expected that the way they use discretion unavoidably depends on

---

[60] For a fuller account of the notion of legal architecture, see DJ Galligan, *Law in Modern Society* (Oxford, OUP, 2006).

the influences brought to bear on them. If that holds true for the more pedestrian subjects of administration, it must a fortiori be true for complex and difficult matters like biotechnology.

The consequence of this general cultural feature is that the case for the involvement of civil society, not just at the level of high policy-making, but at the various levels of administration and regulation, is not fully realised. Administrative law has reached heights of sophistication in recognising the rights of the individual to be involved in decisions that concern him or her, through doctrines of notice and hearing, the disclosure of information and the giving of reasons. But when it comes to groups, interests and organisations that do not have that same personal right or interest at stake, administrative law doctrines are often inadequate. Similarly, when the issue is not how an individual should be treated, but what policy ought to guide government and administration, either by way of a general rule or the disposition of a complex matter, the tools of administration are very limited. Civil society has its sphere of action and influence at the macro-political level rather than at every level of government and administration. To generalisations of these kinds, there are of course many exceptions; nevertheless the suggestion here is that the European approach to administration and the legal framework within which it operates, and in particular the place of civil society within it, has to transform itself, has to undergo a paradigm shift in order to catch up with and reflect the reality.[61] The Commission's white paper on *European Governance* and its *Strategy for Europe* are steps towards the better involvement of civil society in policy matters, and in recognising that such matters occur in different contexts and at different levels of government and administration, from the macro level of high politics to the micro level of decisions about very particular issues. It also recognises the elements that good *governance* must include in the participation of civil society.

The EU approach fails, however, in taking the next major step towards a legal structure that reflects the role of civil society. It was noted earlier that its proposals stop short of legal implementation and that they do so for reasons that are unconvincing. The concept of rights is largely missing from the discussion, which suggests that the EU authorities still regard good governance as a matter of policy to be achieved through informal negotiations and actions, rather than recognising the need for a legal structure. We have seen how international covenants concerning biotechnology use the concept of rights, but only in an attenuated sense, so that practical measures for the participation of civil society are proposed

---

[61] The approach to these matters varies within Western Europe but deteriorates rapidly within the new democracies of post-communist Europe.

with great caution, leaving the serious working out of the arrangements to national governments. If there is one general theme running through the analysis of both EU and international approaches to the regulation of biotechnology, and to the involvement of civil society, it is that they are still at an early stage of development.

Another obstacle flows from the nature and weight of the groups and interests of civil society. Some are more powerful than others; some establish relationships with the policy-making bodies that give them a special place in consultation and allow them to exert influence informally rather than through formal processes. Research into *policy communities*, that is to say, the network of informal relationships that develop and continue, shows how they are the real centres of policy-making.[62] Those within the community are often engaged in a continuing dialogue with the policy-maker, while those outside make little impact. The divide between insiders and outsiders is prone to undermining efforts to have all views heard and considered. It may not be surprising to learn that the stronger economic interests are adept at becoming insiders, while those groups representing a cause rather than their own financial advantage find it more difficult to gain entry and often are left on the outside. One of the main points of a formal set of procedures for hearing and considering the views of interest groups is to contain the natural imbalances among them.

Here again the gap between the legal design and the social realities is hard to bridge. The spectacle of formal procedures being followed conscientiously, knowing well that real influence has already been exerted and positions already taken elsewhere, is surely familiar. In the context of biotechnology, where the stakes may be high and the interests powerful, the problem is acute, leading to the tendency towards *capture* of the regulatory body.[63] Precisely to what extent there is capture remains unclear. But here, as elsewhere, practical steps can be taken to reduce the levels of informal influence through policy communities and capture, and to raise the efficacy of the legal framework. Following the model put forward, the three elements are: sound legal architecture, suitable organisations, and acceptance or internalisation of the legal rules. By patient attention to each of these elements, it is suggested, the legal rules guiding civil society participation can be made reasonably effective in practice. They will never hold full sway because the natural tendency is towards informality based on common interests and power relations, but law can be made to work reasonably well if the variables affecting compliance are understood and then addressed through practical measures.

[62] For further discussion, see W Grant, *Pressure Groups, Politics, and Democracy in Britain* (London, Philip Allan, 1989), and Galligan, n 15 above chap 15.

[63] For discussion of capture of regulatory bodies by powerful interests see R Baldwin and C McCrudden (eds), *Regulation and Public Law* (London, Weidenfeld and Nicolson, 1987).

4. CONCLUSION

The object of this chapter has been to consider the place of civil society in the regulation of biotechnology. In the background are the special issues biotechnology poses for the setting of standards and their implementation. One is a high level of complexity and uncertainty as to the nature and consequences of bio-products, another the moral and social dilemmas they create. Against this setting, the reasons for civil society involvement in policy-making, implementation and monitoring are considered both as to the theoretical basis and their practical application. The need for regulation has caught the attention of international affairs, the European dimension and nation states, although it is hard not to conclude that the current regulatory environment is patchy and piecemeal, while at the same time overly complicated. This should not be too surprising, considering the nature of the issues and their trans-national effects, but the need for a simplified and more coherent approach is plain. The main point, however, has been to take forward the idea of civil society engagement in regulation by proposing a model approach that is both compelling and practicable. Some support for the model can be seen in the current regulatory environment, particularly within the contours of the Aarhus Convention and the strategies of the European Commission. Both offer positive and imaginative ways forward, although, as the final section of the chapter shows, the practical difficulties of making regulatory norms concerning civil society effective in practice are considerable. At the same time, it is suggested that the effectiveness of legal norms is not an arcane art but a reasonably rational process that now needs to be applied to biotechnology.

Part VI

International Humanitarian Law

14

Offensive Military Applications of Biotechnologies: Loopholes in the Law?

1. INTRODUCTION: RELATIONS BETWEEN BIOTECHNOLOGIES AND WARFARE

THE DEVELOPMENTS IN biotechnology[1] that have taken place in the last few years and that can be reasonably expected in the near future are so momentous as to be often equated to the industrial revolution of the eighteenth century or the communication evolution of the last century. Nobody doubts the immense potential that developments in life sciences, eg, in the area of prevention of genetic diseases or increasing food production, may have for humanity. The industrial application of living organisms to the military field, often in conjunction with nanotechnology,[2] may also offer an array of instruments and devices capable of affording enhanced protection from the adversary. Lighter armour made of bioengineered protein molecules, biological markers to distinguish friendly soldiers, edible vaccines designed to fit a soldier's individual genetic make-up are only a few examples of the advantages that defensive military applications of biotechnologies may bring to modern armed forces.[3]

---

* M Stud (Oxon), PhD (EUI), researcher in international law at the University of Florence. She has worked for the Red Cross in Uganda and Uzbekistan. Email: luisa.vierucci@unifi. it.

[1] The definition of the term `biotechnology´ would deserve discrete treatment, but we refer to the definition set out in Art 2 of the Convention on Biological Diversity of 1992 ('any technological application that uses biological systems, living organisms, or derivatives thereof, to make or modify products or processes for specific use'). Compare this definition with the one offered by the International Committee of the Red Cross (ICRC): 'the industrial application of technologies researched, developed or used in the biological sciences, particularly those associated with genetic engineering': see `Biotechnology, Weapons and Humanity—FAQ and Reference´, available at www.icrc.org/web/eng/siteeng0.nsf/html/ 1F00E7A (last visited 7 Jan 2004).

[2] For a definition of nanotechnology and prospects of its use by the armed forces cf. RD Pinson, 'Is Nanotechnology Prohibited by the Biological and Chemical Weapons Conventions?' (2004) 22 *Berkeley Journal of International Law* 282.

[3] US National Research Council, *Opportunities in Biotechnology for Future Army Applications* (*Washington, DC*, National Academy Press, 2001).

However, history shows that technological advances also lend themselves to possible hostile uses. Biotechnology is no exception. Costing little and being very difficult to detect, military applications of biotechnologies could be turned into offensive instruments that deliberately spread diseases or modify body functions in an irreversible manner.

The high potential of risk that certain applications of biotechnology to warfare may pose to humanity, especially if handled by terrorist groups, is well illustrated by two significant initiatives which have recently been taken at the national and international levels. Following the anthrax attacks that occurred at the end of 2001 on US soil, the US Congress has passed two Acts[4] aimed at facilitating the purchase, stockpile and distribution of vaccines and drugs that may be used in case of attacks using biological, chemical or radiological agents, as well as at expediting research in and development of medicines against bioterrorist attacks.

At the international level, for the first time in 2004 the UN Security Council resolved not only to define the proliferation of weapons of mass destruction (nuclear, chemical and biological) as a threat to international peace and security, but also to urge states under Chapter VII of the Charter to take a series of measures to prevent non-state actors from manufacturing, acquiring, possessing, developing, transporting, transferring or using such weapons or their means of delivery.[5]

This alarm may seem misplaced as the use of biological agents in warfare in modern times has been extremely rare.[6] Nonetheless, the prospects that the biotechnology industry offers to the military is far more appealing than those connected with already known biological agents. Biotechnology allows manipulation of potentially all fundamental human

---

[4] Respectively the Biological, Chemical, and Radiological Weapons Countermeasures Research Act of 2002, s 3148, 107th Cong 2 (2002), and the Project Bioshield Act of 2004, s 15, HR 2122, 108th Cong (2003) PL 108–276. It is also to be noted that the US budget for biodefence has increased by 15 times between 2001 and 2004.

[5] S/RES 1540 (2004). Also the UN High-level Panel on Threats, Challenges and Change has addressed the question of the new threats posed by biological warfare and stated: '[u]nlike anthrax, which can be treated by antibiotics, ricin has no antidote and is lethal to humans in quantities smaller than the size of a pinhead. Use of similar materials to cause deliberate outbreaks of infectious disease could prove equally if not more lethal than a nuclear detonation. Under worst-case assumptions, an attack using only one gram of weaponized smallpox could produce between 100,000 and 1,000,000 fatalities': *A More Secure World:Our Shared Responsibility*, UN doc A/59/565, 2 Dec. 2004, para 115, available at www.un.org/secureworld/.

[6] The rare recourse to biological warfare in history is mainly due to the risks that such weapons pose for the personnel handling them and their limited tactical value because of unpredictability and delayed effects. The 'germ warfare' the infamous Japanese Unit 731 experimented with in Word War II by spreading bubonic plague agents on Chinese population centres via bombing missions is one of the very few examples of recourse to biological agents in warfare: see generally JP Zanders, 'International Norms against Chemical and Biological Warfare: an Ambiguous Legacy' (2003) 8 *Journal of Conflict & Security Law* 391.

features such as cognition, memory, equilibrium, fertility, as well as life processes in animals and plants. The potential for changing the basic features of life as we know it today is not only immense, but is proceeding at a pace that is unexpectedly rapid especially in the field of molecular biology and genetics.

In particular, biotechnology may enhance the interest in offensive military uses of biological agents and toxins—and to a limited extent also chemicals and their precursors—in two principal ways: (1) through the *creation* of new pathogenic agents; and (2) through the *modification* of already known bio-chemical agents.[7]

The *creation* of new pathogens may be either the unintended or the intentional result of research. The likelihood of such occurrence increases in parallel with fear of bioterrorist attacks, which may cause an uncontrolled proliferation of experiments capable of multiplying opportunities for release of pathogenic substances the effects of which are difficult to foresee.[8] For example, scientists have accidentally created a more dangerous version of the 'mouse pox' virus, which is similar to smallpox, and have then warned against the dangers of such research.[9] The greatest risk in this area is the possible creation of a microbial or biological agent resistant to antibiotics, the effects of which might be more devastating than any other known disease.

Equally worrisome is the intentional artificial creation of viruses. It is well known that a polio virus has been synthesised from segments of DNA ordered by mail and on the basis of genetic information available on the Internet.[10] This unprecedented creation of a virus from synthetic materials is most unsettling, given the easy availability of information to researchers who do not have good intentions.[11]

---

[7] Drawing on a variety of sources, mainly scientific studies carried out in the UK, these risks are identified by the ICRC: see www.icrc.org/web/eng/siteeng0.nsf./html/1F00E7AC049A4758C1256C300054AD. See also M Dando, *Biological Warfare in the 21st Century: Biotechnology and the Proliferation of Biological Weapons* (London, Brasseys, 1994). Biotechnology advances may also have the indirect effect of inducing states or other actors to use biological warfare agents where they possess a vaccine against a newly produced biological agent, as the vaccine would protect their own troops or population against the effect of those weapons.

[8] J Knight, 'Biodefence Boost Leaves Experts Worried over Laboratory Safety' (2002) 415 *Nature* 719.

[9] On the prospects of the production of new agents for biological warfare cf E Croddy, *Armi Chimiche e Biologiche* (Turin, Bollati Boringhieri, 2004), esp. at 262–3.

[10] J Cello, A V Paul and E Wimmer, 'Chemical Synthesis of Poliovirus cDNA: Generation of Infectious Virus in the Absence of Natural Template' (2002) 309 *Science* 1016.

[11] In addition, accidental infections of researchers are likely to occur, especially if biodefence laboratories expand without the necessary control. Recently, three researchers were accidentally infected with tularaemia, a deadly bacterium which could be used in bioterrorism, at Boston University: cf R Dalton, 'Infection Scare Inflames Fight against Biodefence Network' (2005) 433 *Nature* 344.

The *modification* of known biological or chemical agents may make those agents more attractive for use as weapons. For example, it seems that the most poisonous toxin, botulin, and the most dangerous bacillus, anthrax, may be modified so that their effects develop faster. Manipulation may also make such agents dangerous or deadlier, thus making them more powerful warfare agents. An example of this hypothesis is the development of hidden biological agents in safe vaccines capable of infecting only a particular ethnic group.[12]

Not only pathogens but also chemical agents may be relevant to our analysis. For instance, a particular kind of virus may be used as a vector for a chemical which attacks only specific receptors of the central nervous system.[13] In addition, the use of manufactured 'bioregulators'[14] for attacks aimed at altering body functions offers one of the most recent possibilities for the use of chemicals modified through genetic engineering as offensive instruments. Because of the totally disruptive effects that weapons capable of altering vital body functions would have and the ease in manipulating them to make them more potent and controlable, the ability to manufacture these chemicals for offensive purposes is possibly the most disquieting prospect among those that have so far been evoked.

Although not necessarily anti-personal, the creation or manipulation of microbial, biological or chemical agents, used in isolation or in conjunction with chemical agents, could adversely affect either or both human life and the environment. All these broadly depicted potential abuses of biotechnologies can lead to the industrial development of weapons (`weaponisation´) to be employed in armed conflict or for hostile acts.[15] The question then arises of the legality of these military (mis)applications of biotechnology under international humanitarian law (IHL). In other words, would one or all of the above examples of biotechnology use

---

[12] This prospect is not so futuristic. Already in the 1980s South African scientists carried out similar research in order to reduce fertility within the target (black) population, although such agents were never produced: see ICRC study, n 7 above. The ability to create this ethnic weapon flows from our knowledge about the particular genetic characteristics of an ethnic group: M Dando, *The New Biological Weapons: Threat, Proliferation, and Control* (London Lynnee Rienner, 2001), 125 ff.

[13] M Dando, 'Incapacitants', *Open Forum on the Chemical Weapons Convention*, 1 May 2003, 39, avilable at www.sussex.ac.uk/Units/spru/hsp/OpenForum CWC.pdf.

[14] Bioregulators are chemicals which are naturally produced by our body in order to regulate functions such as consciousness, behaviour and body temperature.

[15] There is very little detailed information on the capacity to build weapons using such agents, but it seems that the high risks of contamination that exist both for the scientists and manufacturers, in addition to difficulties in production, make the prospects of mass manufacture of such weapons quite remote. For an extremely interesting account of the USSR's ability to produce similar weapons see the famous book by K Alibek and S Handeman, *Biohazard* (New York Random House, 1999).

breach IHL rules or principles? Is IHL well equipped to face the dramatic challenge that misuses of biotechnology may cause to humanity?

While the question has recently come to the attention of renowned international institutions in the field of IHL and disarmament, some of them have hastily concluded that `biotechnologies in warfare are prohibited by IHL´. Although from a logical viewpoint such a conclusion is certainly justified, the lack of an express treaty provision and practice does not allow us simply to discard the possibility that at least *some* offensive military applications of biotechnologies may be consonant with IHL. The question deserves closer scrutiny in light of IHL rules and principles.

2. ARE OFFENSIVE MILITARY APPLICATIONS OF BIOTECHNOLOGIES PROHIBITED BY IHL?

The legality of offensive military applications of biotechnologies needs to be reviewed in light of two distinct sets of IHL rules. On the one hand, the question arises of the legality of the new 'weaponised' agents according to existing arms control treaties; on the other, an assessment relating to the existence of prohibitions or limitations on the modality of their use under general IHL rules or principles is called for.

2.1 Existing Arms Treaties relevant to Offensive Military Applications of Biotechnologies

There is no specific treaty or explicit treaty provision banning any offensive military application of biotechnology *as such*. This is not surprising, considering the newness of those applications and the still scant information on whether some of them have actually achieved production stage.[16] The analysis must therefore turn to the possibility that existing arms control treaties are able to cover such new applications as well. This is not a new type of exercise: the question of the impact of new technological developments on the law has often arisen both in international and domestic law. Relevant to our purposes are in particular those treaties which ban biological and chemical weapons.

---

[16] Although practice shows that some weapons have been the object of a treaty ban subsequent to their manufacture and use, at least in one instance a category weapon has been prohibited before it reached production stage: blinding laser weapons: see Additional Protocol to the Convention on Prohibitions or Restrictions on the Use of Certain Conventional Weapons which may be Deemed to be Excessively Injurious or to have Indiscriminate Effects (Protocol on Blinding Laser Weapons), opened for signature on 13 Oct 1995 and entered into force on 30 July 1998, available at www.icrc.org/ihl.nsf/FULL/570?Open Document.

2.1.1. *Biological weapons*

Although the Regulations annexed to the 1907 Hague Convention IV Respecting the Laws and Customs of war on Land already prohibited the use of poisonous weapons (Article 23(a)), a more specific and extensive prohibition on the use in war of certain means of warfare was laid down in the Geneva Protocol of 1925 on Asphyxiating, Poisonous or other Gases and Bacteriological Methods of Warfare.[17] The Protocol, which was the fall-out of the use of gases and bacteriological warfare during World War I,[18] was revolutionary because it was the first treaty banning the use of an entire category of weapons. Such weapons were deemed to be so abhorrent to human conscience that there was little doubt about their use, at least in war, being prohibited. Although the significance of the Protocol for the times cannot be overstated, the major shortcomings of the Treaty soon became apparent. It does not contain any definition of the prohibited means and methods of warfare, with the consequence that, for example, certain agents were used in the Vietnam war because they were claimed not to fall within the purview of interdiction.[19] In addition, the Protocol is limited to the prohibition of only 'use' of certain means and methods of warfare, thus leaving other activities such as production and transfer legal; and it prohibits such use exclusively 'in war', meaning international armed conflict. Finally, it sets up no verification regime.

The limits of the 1925 Protocol and the increased awareness that biological weapons were comparatively cheaper to manufacture,[20] many times deadlier than other weapons,[21] and almost anonymous because symptoms appear after the incubation period needed by the agent to establish itself in the host and replicate,[22] spurred the adoption of what was at the time considered the most comprehensive ban on weapons ever adopted: the Biological Weapons Convention of 1972 (BWC).[23]

---

[17] The Protocol was opened for signature on 17 June 1925 and entered into force on 8 Feb 1928.

[18] N Ronzitti, 'Le désarmement chimique et le Protocole de Genève de 1925' [1990] *Annuaire Français de Droit International* 149.

[19] RR Baxter and T Buerghenthal, 'Legal Aspects of the Geneva Protocol of 1925' (1970) 64 *American Journal of International Law* 853.

[20] The cost of a large arsenal of biological weapons is reportedly as low as US $10 million.

[21] Biological agents are many times deadlier than chemical agents. For instance, 10 grammes of anthrax spores could kill as many people as a ton of the nerve agent sarin.

[22] The stealthy nature of these weapons, hence the difficulty in identifying the employer, clearly leaves questions of responsibility unresolved and entails violations of one of the basic premises on which IHL is grounded.

[23] The Convention on the Prohibition of the Development, Production and Stockpiling of Bacteriological (Biological) and Toxin Weapons and on their Destruction of 10 Apr 1972 entered into force on 26 Mar 1975, available at www.opbw.org. 151 states are party to it.

The BWC is wide-ranging in that it bans the 'production, stockpile, or otherwise acquisition' of 'microbial or other biological agents or toxins'.[24] Although the BWC provides no definition of the term 'other biological agent', the *travaux préparatoires* indicate that the term was chosen so as to be wide enough to cover not only those agents that had already been discovered but also other agents that could have been identified in the future.[25] Today it is generally agreed that any agent that depends for its effects on multiplication within the target organism is biological.[26] The ban on this broad category of agents has been further extended in the BWC by the prohibition of pathogens 'whatever their origin or method of production'.[27] This means that the agents are proscribed regardless of their natural or artificial synthesis. Furthermore, the BWC prohibits only those agents and toxins 'of types and in quantities that have no justification for prophylactic, protective or other peaceful purposes'.[28] Therefore activities such as research or defence use are allowed under the Convention. Specifically, the Convention bans only 'weapons, equipment or means of delivery designed to use such agents or toxins for hostile purposes or in armed conflict'.[29]

It is remarkable that the ban applies both in times of armed conflict and for 'hostile purposes'. This suggests that it extends also to non-international armed conflict and situations in which the threshold of the armed conflict has not been reached (otherwise there would be no reason for couching the sentence in the alternative—using 'or'). This provision is crucial in the time of the War on Terror, whose legal qualification as an 'armed conflict' is not

---

[24] Art I BWC. Note that toxins (eg botulinum, which is the most toxic substance known to man) are a product of living organisms but behave in a way, and yield effects, similar to chemical agents.

[25] J Goldblat, *Arms Control: The New Guide to Negotiations and Agreements* (2nd edn, London, Sage, 2002) at 138; G Fischer, 'Chronique du désarmement: la Convention sur l'interdiction de la mise au point, de la fabrication et du stockage des armes bactériologiques (biologiques) ou à toxines et sur leur destruction' [1971] *Annuaire Français de Droit International* 102. It is to be noted that while the definition of what substance constitutes 'other biological agents' does not appear to have given rise to major objections, the determination of what constitutes weapons, equipment or means of delivery has instead been the object of debate in the past but seems now settled (see the reservation made by Switzerland to the effect that it would decide unilaterally what constitutes weapons, equipment and means of delivery and the US objection that only those items whose *design* indicates that they could have no other use than that specified in the Convention, or that they are intended to be capable of the use specified, would fall under the treaty ban: see also J Goldblat, 'The Biological Weapons Convention —an Overview' (1997) 318 *International Review of the Red Cross* 254).

[26] Those agents whose classification as biological or chemical is debatable, such as 'bioregulators', fall under the ban: see M Wheelis, 'Biotechnology and Biochemical Weapons'(2002) 9: *The Non Proliferation Review* 52, available at http://cns.miis.edu/pubs/npr/vol09/91/ 91whee.htm.

[27] Art I(1) BWC.

[28] Ibid.

[29] Art I(2) BWC.

clear-cut. The absolute ban relating to those agents for offensive purposes is reinforced by the commitment undertaken in Article I of the BWC 'never in any circumstances' to put in place the prescribed activities.[30] Finally, there seems to be no room left in the Convention for the legality of the use of non-lethal agents. While some states claim that those agents are not expressly prohibited, in our opinion the comprehensiveness of the ban leaves no doubt as to their inclusion when non-lethal agents would be used for other than peaceful purposes.[31]

In brief, the major advances of the BWC lie in the broad spectrum of prohibited agents as well as its application below the armed conflict threshold.

Nevertheless critical shortcomings are left in the Convention.[32] In particular, 'peaceful applications' of biological agents and toxins are expressly authorised under the BWC as well as biological weapons research, provided that they are 'designed' for use only for peaceful purposes (eg training). The functional definition of the prohibited agents is troublesome because pathogenic agents are inherently 'dual-use'. Their nature makes it difficult, if not impossible, to distinguish between the peaceful and hostile military application of the agents as long as they are not actually employed.[33] Similarly, the obligation to destroy stockpiles does not apply to biological agents that are 'diverted to peaceful purposes' (Article II), thus leaving states an important degree of discretion in deciding what activities may be classified as peaceful. For example, a state may lawfully decide not to destroy stockpiles of polio agent because of its prospective protective use as a vaccine by its armed forces in a specific armed conflict scenario. However, the stockpile may easily be used to enhance the effects of the virus and later be employed

---

[30] However these advances should not mask an important reason which led to their outlaw, namely the fact that, because of their unpredictable effects—mainly due to their delayed action linked to the incubation period—they are of limited value on the battlefield. Actually, their threat and use are aimed at terrorising rather than defeating the enemy.

[31] The definition of 'peaceful purpose' may arguably be interpreted as encompassing law enforcement purposes as the latter do not usually require the use of the armed forces. However this gimmick is aimed only at evading the BWC obligations which exclude this interpretation by specifically referring to 'hostile acts' among the circumstances where the use of the agents are prohibited (as we have seen, the preamble supports this conclusion).

[32] The fact that the BWC does not specifically ban the 'use' of the agents does not appear to be a serious loophole because the BWC complements the 1925 Protocol, which explicitly prohibits the 'use' of the agents, although the limited number of ratifications of the 1925 Protocol and the more limited scope of application of the Protocol compared to the BWC may cause a reduced coverage of the ban on the 'use' of bacteriological weapons compared to their production, stockpile or otherwise acquisition. In any case, the ban on production coupled with the obligation to destroy the existing agents and weapons obviously minimises the risks of their 'use' and is to be construed as implying prescription of 'use'. See also the Final Declaration of the Fourth Review Conference (1996), BWC/CONF.IV/9, available at www.opbw.org, 13, where the states parties affirmed that the convention encompassed also prohibition of 'use' of the agents.

[33] The UNSCOM experience in Iraq is quite illustrative of this difficulty.

as a means of delivery in armed conflict without great possibilities of identifying the illegal use of the research facility.

In addition, the precise scope of the ban on pathogens set forth in Article I (1) is uncertain. The provision generally outlaws pathogens 'of types and in quantities that have no justification for prophylactic, protective or other peaceful purposes', but the BWC contains no list of the types or quantities of an agent which may pose concerns relevant to the Convention.

On account of these shortcomings, it is most regrettable that no verification regime is attached to the Convention. Where a state party suspects a breach of the Convention by another state, the only enforcement mechanism available under the Convention is lodging a complaint with the UN Security Council (Article VI(1)). The question is a thorny one because any control mechanism would hardly be effective, given the ease with which a very small amount of a banned agent can be transformed into an arsenal which is simple to conceal. Indeed, the regularly held review conferences of the BWC have so far failed to address this major loophole, despite repeated attempts to introduce a monitoring system mainly on those very grounds.[34]

2.1.2 Chemical weapons

The other treaty relevant to our analysis is the Chemical Weapons Convention of 1993[35] (CWC), a complex and elaborated legal text which complements both the 1925 Protocol and the BWC (Article XIII). The CWC outlaws toxic chemicals and their precursors, namely 'any chemical which through its chemical action on life processes can cause death, temporary incapacitation or permanent harm to humans or animals',[36] regardless of their origin or method of production, unless they are used for purposes that are not prohibited by the treaty.[37] The definition of toxic chemical is to be interpreted in light of the annexes to the Convention, which list

[34] Negotiations have been carried out since the 1990's on a Protocol to the BWC establishing a verification mechanism. Negotiations have now stalled as there is no agreement, not only on the type of monitoring mechanism, but also on the body (the World Health Organisation, an organ purposefully set up under the BWC, etc) that should be in charge of control. The UN High-level Panel on Threats, Challenges and Change has also recommended in its report that states parties 'without delay return to negotiations for a credible verification protocol, inviting the active participation of the biotechnology industry': n 5 above, para 126.

[35] Convention on the Prohibition of Development, Production, Stockpiling and Use of Chemical Weapons and on their Destruction, adopted in Paris on 13 Jan 1993 and entered into force on 29 Apr 1997 available at www.opcw.org. 167 states are parties to it.

[36] Art II, (2), CWC.

[37] According to Art II (1), CWC, '[c]hemical weapons means the following, together or separately: (a) Toxic chemicals and their precursors, except where intended for purposes not prohibited under this Convention, as long as the types and quantities are consistent with such purposes; (b) Munitions and devices, specifically designed to cause death or other harm through the toxic properties of those toxic chemicals specified in subparagraph (a),

chemical products according to their level of toxicity and potential use in combat (so-called 'schedules').[38] As emphasised by the states at the negotiating conference, this definition of chemical weapon is intentionally broad so as to be able to encompass future discoveries due to progress in technology.[39] However, the precise scope of the ban, for instance the question whether it covers weaponisation of 'bioregulators', remains to be addressed in light of the provisions of the Convention dealing with scientific and technological developments (see section 2.2 below).

The scope of the CWC is broad also as far as the range of prohibited activities is concerned. The Convention introduces a ban applicable 'under any circumstances'[40] on the development, production, or other acquisition, retention, stockpiling and use of chemical weapons and chemical weapons production facilities, as well as on the assistance or encouragement to engage in any of the prohibited activity (Article I). It further requires the destruction of all such weapons and facilities which are located either on the territory or 'under jurisdiction and control' of a party not later than 10 years after entry into force (Articles I(2) and IV(6)).

Similarly to the BWC, a functional definition of the prohibited chemicals is given, so that the use of those substances is not prohibited, inter alia, for research and protective purposes, or for 'military purposes not connected with the use of chemical weapons and not dependent on the use of the toxic properties of chemicals as a method of warfare'.[41] As we have seen for the BWC, the general purpose criterion introduces a subjective element in the determination of the scope of the Convention that is regrettable because of

which would be released as a result of the employment of such munitions and devices; (c) Any equipment specifically designed for use directly in connection with the employment of munitions and devices specified in subparagraph (b)'. Paras (2) and (3) of the same Art define respectively the terms 'toxic chemical' and 'precursor'.

[38] M Bothe, 'The Chemical Weapons Convention: a General Overview' in M Bothe, N Ronzitti and A Rosas (eds), *The New Chemical Weapons Convention—Implementation and Prospects* (The Hague, Kluwer Law, 1998), 4.

[39] See Pinson, n 2 above, at 7, and N Ronzitti, 'La Convention sur l'interdiction de la mise au point, de la fabrication, du stockage et de l'emploi des armes chimiques et sur leur destruction' [1995] *Annuaire Français de Droit International* 887.

[40] The ban therefore appears to apply in all types of armed conflicts as well as in time of peace. Such interpretation is warranted also in light of the preamble, which specifies that one of the aims of the Convention is 'to exclude completely the possibility of the use of chemical weapons' and to achieve a 'complete and effective prohibition of the development, production, acquisition, stockpiling, transfer and use of chemical weapons and their destruction'. However, Art II(9)(c) lists '[l]aw enforcement including domestic riot control purposes' among the purposes not prohibited under the Convention. This provision leaves doubt as to the applicability of the CWC for law enforcement activities, especially at the domestic level: cf A Gioia, 'The Chemical Weapons Convention and Its Application in Time of Armed Conflict' in Bothe, Ronzitti and Rosas (eds), n 38 above, at 380–2.

[41] See Art II, (9), CWC. Activities using chemical weapons for training purposes would therefore be allowed under this provision.

the discretion attached to it.[42] However, this does not weaken the conclusion that the Convention has dramatically expanded the prohibition on chemical weapons contained in the 1925 Protocol.[43]

Most significantly, the Convention establishes a permanent Organisation on the Prohibition of Chemical Weapons[44] based in The Hague with the task, inter alia, of monitoring implementation of the treaty through various mechanisms, including routine on-site inspections, submission of periodic reports by member states, and challenge inspections of any facility or location, both private and public, when a state party suspects that the facility is not in compliance with the Convention.[45] Importantly, the Organisation is also entrusted with the task of 'consider[ing] measures to make use of advances in science and technology'.[46]

Despite the fact that the CWC was welcomed as a major achievement in the field of arms control, in particular because its verification mechanism was among the most intricate and intrusive ever designed for a disarmament regime, the implementation of the Convention has not been up to expectations, especially with respect to the compliance and sanctioning mechanism in case of breach. Also the impact of reservations and ill-drafted national measures of implementation[47] has considerably diminished the efficacy of the Convention. Moreover, an important category of chemicals, herbicides, does not fall under the Convention's scope, because the treaty does not apply to chemicals causing harm to the natural environment (as will be spelt out below, the use of herbicides is prohibited as a method of warfare if it provokes widespread, long-lasting and severe damage to the environment).

In any case, this Convention is of paramount importance as it prohibits weapons that are liable to be used in combat (certainly more than just biological weapons) because of their higher predictability and effectiveness.[48]

---

[42] If properly implemented, this criterion would serve to close the gap left open by the technological developments related to chemicals. As rightly underlined by D Feakes, as the focus of the CWC is on the purpose of the agents rather than on the properties of the actual chemicals, only those technological developments that are used for the purposes prohibited by the Convention are banned. However, the only way to make this criterion effective is by way of appropriate national implementation: General Purpose Criterion, *Open Forum on the Chemical Weapons Convention*, above in 13, 26.

[43] On this issue see extensively Gioia, n 39 above, at 383 ff.

[44] Cf A De Guttry, 'The Organisation for the Prohibition of Chemical Weapons' in Bothe, Ronzitti and Rosas (eds), n 33 above, at 119–1.

[45] Cf the very complex Art VIII of the CWC as well as the Annex on Implementation and Verification.

[46] Art VIII(A)6 CWC.

[47] Eg, according to s 307 of the US Chemical Convention Implementation Act of 1998, the President may refuse inspections on national security grounds.

[48] The effects of chemical weapons can occur within seconds of exposure or as long as several hours afterwards. However these weapons are more costly and in principle less deadly than biological weapons.

All in all, the 1925 Protocol, the BWC and the CWC present important shortcomings which strongly impact upon the efficacy of the prohibitions they contain in at least four respects: the lack of precise determination of the *ratione materiae* scope of the prohibition; the authorisation to use the weapons for peaceful purposes; lack or insufficiency of the monitoring regime; and, more generally, the limited number of states parties to them. Therefore, the question of the compatibility of offensive military applications of biotechnologies with IHL treaty law cannot be easily dismissed. For our purposes, the crucial question is whether such conventions are sufficiently comprehensive to cover the new developments which are taking place—or will occur in the future—in the biotechnological sector.

2.2. Ability of the BWC and CWC to Cover Biotechnological Developments

2.2.1 The BWC

Because none of the above analysed treaties has been specifically drafted with biotechnology in mind and explicitly prohibits weapons manufactured through recourse to this branch of life sciences, the question arises as to the ability of those treaties to cover biotechnologies which are aimed at use for offensive military activities. The question is particularly pertinent for the 1925 Geneva Protocol and the BWC because, as biotechnology started to develop *after* those treaties were concluded, it is arguably inappropriate to manipulate the language of the treaties to attempt to fit those technologies into their purview.[49] However, several factors warrant the conclusion that the 1925 Protocol and the BWC also cover offensive military applications of biotechnology.

In the first place, Article I of the BWC prohibits offensive uses of microbial or other biological agents, or toxins 'whatever their origin or method of production'. The comprehensive terms in which the provision is couched suggest its application to any technique used in the life science. This means that the creation of new pathogens and the manipulation of already known ones through the techniques used in biotechnology, such as DNA recombinant or genetic engineering, may be deemed to be covered by this provision.

Secondly, the black letter of the BWC bears witness to the fact that technological developments were not intended by the drafting parties to limit the scope of prohibition of the Convention. On the contrary, Article

---

[49] This argument is put forward by Pinson, n 2 above, at 303, in respect of nanotechnology.

XII of the BWC provides that five years after the entry into force of the BWC at the latest a review conference was to be convened where 'new scientific and technological developments relevant to the Convention'[50] were to be taken into account. It would be preposterous to argue that developments in the biotechnological field would per se not fall within the purview of the BWC, given that they do certainly constitute 'scientific and technological developments' which may be of relevance to the Convention. One may more correctly argue that the *speed* and *achievements* in this scientific and technological sector were unimaginable when the Convention was negotiated, so that a state would not have become a party to the Convention had it known what technological prospects lay ahead. However, the fact that the prohibited pathogenic agents are not listed in the BWC shows the intention of the drafters to leave open the possibility that new agents, be they newly discovered, created or mere modification of already known agents, fall under the purview of the prohibition.

Thirdly, the principle of *effet utile* requires that the BWC extend to industrial modification of microbial, biological agents and toxins which makes them deadlier or more indiscriminate than the original agent, at least as long as there is a functional identity between the two. Indeed it is difficult to deny that the use of living organisms that have been modified in order to make biological agents or toxins, for instance, faster to act, hence deadlier, are not functional equivalents to known biological agents. In fact such use of technology simply increases the effects of the agents which are prohibited under the Convention.[51]

This proposed interpretation of the inclusion of certain uses of biotechnologies under the purview of the BWC is also consonant with the purpose and object of the Convention, which is the achievement of a 'general and complete disarmament', and more specifically 'to exclude completely the possibility of bacteriological (biological) agents and toxins being used as weapons'.[52]

Finally, the comprehensiveness of the prohibition set forth in Article I BWC, read in conjunction with Article XII, is corroborated by the final

---

[50] According to Art XII: 'Five years after the entry into force of this Convention, or earlier if it is requested of Parties to the Convention by submitting a proposal to this effect to the Depositary Governments, a conference of States Parties to the Convention shall be held at Switzerland, to review the operation of the Convention, with a view to assuring of the preamble and the provisions of the Convention, including the provisions negotiations on chemical weapons, are being realized. Such review shall take into account new scientific and technological developments relevant to the Convention.'

[51] More debatable may be the question of artificially created agents which have effects previously unknown to biological agents. On these points see GH Reynolds, 'Environmental Regulations of Nanotechnology: Some Preliminary Observations' (2001) 31 *Environmental Law Reporter* 10681.

[52] See the Preamble to the BWC. According to J Goldblat, '[t]he aim of the BW Convention was not so much to remove an immediate peril, as to eliminate the possibility that scientific

declarations adopted by the states parties at the review conferences. Those declarations confirm that states parties were aware of the possible dangers that scientific and technological developments were likely to pose to the BWC since the first review conference of 1980. On that occasion, it was affirmed that 'Article I has proved sufficiently comprehensive to have covered recent scientific and technological developments relevant to the Convention',[53] but no specification concerning the type of developments of concern for the Convention was felt necessary, possibly because no main progress in the biological sciences capable of affecting weapons had yet taken place. The final declaration adopted at the successive review conference of 1986 reflects in an evident manner the rapidly increasing concern of the states parties with regard to the scope of the ban of Article I. The fast pace at which life sciences research started to advance in the 1980's urged the states parties not only to discuss the issue in the papers prepared by the depository governments,[54] but also specifically to affirm in the Final Declaration that:

> The Conference conscious of apprehensions arising from relevant scientific and technological developments, inter alia, in the fields of microbiology, genetic engineering and biotechnology, and the possibilities of their use for purposes inconsistent with the objectives and the provisions of the Convention, reaffirms that the undertaking given by the States Parties in Article I applies to all such developments.[55]

In order to leave no doubt about the fact that the BWC did extend to cover toxins of any nature and origin, the Conference further emphasised that:

> the Convention unequivocally applies to all natural or artificially created microbial or other biological agents or toxins whatever their origin or method of production. Consequently, toxins (both proteinaceous and non-protein-aceous) of a microbial, animal or vegetable nature and their synthetically pro-duced analogues are covered.[56]

and technological advances, modifying the conditions of production, storage or use of bio-logical weapons, would made these weapons militarily attractive': see n 26 above, at 261. On those grounds the author argues that the Convention is comprehensive enough to cover also progress in biotechnology.

[53] Final Declaration of the First Review Conference (1980), BWC/CONF.I/10, available a www.opbw.org/rev_cons/1rc/docs/final_dec/1RC%20Final%20Doc.pdf.

[54] See, eg, the declaration made by the representative of Canada to the Second Review Conference, BWC/CONF.II/SR.4, 17 Sept 1986, para 6, on the 'acute' problems posed by increasing advances in science, and particularly in the field of biotechnology, available at www.opbw.org/rev_cons/2rc/docs/sr_docs/BWC_CONF.II_SR.4.pdf, and Dando, n 7 above, at 132

[55] Final Declaration of the Second Review Conference (1986), BWC/CONF.II/13/II, avail-able at www.opbw.org/rev_cons/2rc/docs/final_dec/2RC%20Final%20Doc.pdf, 3.

[56] Ibid.

The awareness of the consequences that scientific and technological progress could have for the implementation of the BWC grew dramatically in the following years and became the object of major attention by the states participating in the review conferences of 1991 and 1996. The background papers prepared by the states parties unequivocally stressed the major advances in the various fields of biology, including biotechnology, but also warned against their offensive application to the army. From the discussions ensued two final declarations affirming the comprehensiveness of the ban set forth in Article I. More specifically in 1991 the ban was stated to cover 'experimentation involving open-air release of pathogens or toxins'[57] which were harmful not only to man, but also to animals or plants; whereas in 1996, the 'components' of the prohibited agents were outlawed in order to take account of the genetic sequences discoveries.[58]

The incessant attention paid to the impact of new discoveries on Article I BWC appeared to decline with the end of the 1990's. The reports presented at the review conference of 2001 show clearly that the priorities of states parties had shifted from scientific developments to verification mechanisms. The attempts to overcome the US opposition to the Protocol on the verification mechanism prepared by the ad hoc group dominated the whole conference (nor was an agreement reached on a final declaration at the conference's resumed session of 2002), and almost no attention was given to other issues. Interestingly, some states showed a changed attitude towards developments in the life sciences: as in the past, Sweden and the US in particular stressed the rapid advances in science and technology, but this time their intention was to highlight the positive impact of such developments for the states parties. In particular, the contribution of science, including biotechnology, to blocking the effects of biological weapons was emphasised thanks to new classes of drugs and vaccines.[59]

[57] More precisely, '[t]he Conference, conscious of apprehensions arising from relevant scientific and technological developments, *inter alia*, in the fields of microbiology, genetic engineering and biotechnology, and the possibilities of their use for purposes inconsistent with the objectives and provisions of the Convention, reaffirms that the undertaking given by the States parties in Article I applies to all such developments. The Conference also reaffirms that the Convention unequivocally covers all microbial agents or toxins, naturally or artificially created or altered, whatever their origin or method of production. The Conference notes that experimentation involving open-air release of pathogens or toxins harmful to man, animals or plants that has no justification for prophylactic, protective or other peaceful purposes is inconsistent with the undertakings contained in Article I': BWC/CONF.III/23, 11, available at www.opbw.org/rev_cons/3rc/docs/final_dec/3RC%20Final%20Doc.pdf.

[58] Of special interest to our study is the Background Paper on new scientific and technological developments relevant to the Convention on the Prohibition of the Development, Production and Stockpiling of Bacteriological (Biological) and Toxin Weapons and on Their Destruction, WC/CONF.IV/4, 30 Oct 1996, available at www.opbw.org/rev_cons/4rc/docs/rev_con_docs/i_docs/IV-04.pdf.

[59] Background Paper on new scientific and technological developments relevant to the Convention on the Prohibition of the Development, Production and Stockpiling of Bacteriological (Biological) and Toxin Weapons and on Their Destruction, BWC/CONF.

The new position of states towards scientific developments is borne out also by the agenda set for the three annual meetings to be convened before the next Conference of 2006. The focus of those meetings is on national measures of implementation of the Convention and on exchange of information as well as co-operation in the case of outbreaks of diseases but not on the impact of scientific developments on the BWC's scope.

All this suggests that by the end of the 1990s the states parties felt that the questions relating to the scope of Article I BWC were settled. As revealed in the final declarations of the review conferences, today there is a widespread consensus at least among the states parties[60] to the effect that the BWC covers such a wide range of pathogens as to include those created or modified by biotechnological instruments. This is further confirmed by the inclusion in the military manuals of some states parties of the express prohibitions on using biotechnological procedures for other than peaceful purposes.[61]

Now the crucial question facing the BCW is how efficaciously to monitor the non-offensive use of such pathogens—a topic that has recently been affirmed also by the UN High-level panel on Threats, Challenges and Change.[62]

2.2.2 The CWC

The complex and long debated provision dealing with the question of technological development in the CWC (Article XI) subordinates the implementation of the Convention to the economic and techno-logical developments of the parties for purposes not prohibited by the Convention. Although the language used in the text ('[t]he provisions of this Convention shall be implemented in a manner which avoids hamper-ing the economic or technological development of States Parties') is more restrictive than the corresponding text employed in the BWC (according to Article X(2) of the BWC, '[t]his Convention shall be implemented in a manner designed to avoid hampering the economic or technological development of States Parties'), it is not couched in mandatory terms.[63]

V/4, 14 Sept 2001, paras 1.3 and 5.2 (US position) and conclusion by Sweden, available at: www.opbw.org/rev_cons/5rc/docs/rev_con_docs/i_docs/V-04.pdf; see also the report by Germany, at www.opbw.org/rev_cons/5rc/docs/rev_con_docs/i_docs/V-07.pdf.

[60] Although the final declarations of review conferences are not binding on states parties, they are evidence of the *opinio* of the states regarding the question.

[61] D Fleck (ed), *Handbook on International Humanitarian Law in Armed Conflict* (Oxford, Oxford University Press, 1999), para 439, quotes the German military manual as stating that the prohibitions contained in the BWC 'apply both to biotechnological and synthetic procedures serving other than peaceful purposes. They also include genetic engineering procedures and the alteration of micro-organisms through genetic engineering.'

[62] See n 34 above.

[63] The use of the term 'avoid hampering', instead of a 'shall not hamper' formula as proposed by Egypt, leaves room for discretion: cf N Ronzitti, 'Economic and Technological Development and Trade in Chemicals' in Bothe, N. Ronzitti and Rosas (eds.), n 38 above, at 536.

However, a decision on the extent to which technological development impacts upon the Convention cannot be made unilaterally, but has to be agreed by the Council. As stated in the first review conference of 2003:

> The definitions contained in Article II, in particular of the terms 'chemical weapons' and 'chemical weapons production facility', were found to adequately cover these developments [in science and technology] and to provide for the application of the Convention's prohibitions to any toxic chemical, except where such a chemical is intended for purposes not prohibited by the Convention, and as long as the types and quantities involved are consistent with such purposes. The First Review Conference *noted*, however, that science is rapidly advancing. New chemicals may have to be assessed in relation to their relevance to the Schedules of Chemicals of the Convention. The First Review Conference *requested* the Council to consider the developments in relation to additional chemicals that may be relevant to the Convention, and assess, *inter alia*, whether these compounds should be considered in the context of the Schedules of Chemicals [emphases in the original].[64]

The caution used in the text of the CWC, as well as in the language of the first review conference relating to developments in technology, may be explained by reference to the existence of a comprehensive verification regime under the CWC which is unknown to the BWC, and hence to fear that biotechnological, chemical and pharmaceutical industries may be negatively affected by an extensive ban. On account of this, one has to value the fact that in 2003 the Council was entrusted with the task of taking account of progress in science, thus paving the way for the achievement of a consensus on the scope of Article II as has been the case for the BWC. Though this initiative is indicative of the will of states parties to resolve the question at a multilateral level, the results of the exercise are hard to foresee.[65]

In light of the above, although no conventional prescription specifically prohibits the use of offensive military applications of biotechnologies, there seems to be general agreement that they fall under the scope of prohibitions set forth in the BWC. There are reasons to believe that a similar result will also be achieved with respect to the CWC, but the question is currently occupying the attention of the Council and the outcome still uncertain.

---

[64] First Review Conference RC-1/5, 9 May 2003, Report of the first special session of the Conference of the States parties to review the operation of the Chemical Weapons Convention, available at www.opcw.org/docs/rc105.pdf, para 7.21, para 7.23. The Schedules of Chemicals are annexes to the Convention that regulate three different categories of chemicals according to their level of threat posed to CWC objectives.

[65] An explicit stand on the scope of the Convention so as to include developments in biotechnology, similar to the one adopted by the BWC review conference of 1986 and successive additions, would help clarify the scope *ratione materiae* of the CWC: see G Pearson and M Dando, 'Strengthening the Chemical Weapons Convention', First CWC Review Conference, Paper No 1, Aug 2002, available at www.bradford.ac.uk/acad/scwc/cwcrcp/cwcrcp_1.pdf.

3. IS THE USE OF BIOTECHNOLOGIES AS A METHOD OF WARFARE CONSONANT WITH IHL RULES AND PRINCIPLES?

The inclusion of offensive military applications of biotechnologies under the scope of existing treaties relating to weapons of mass destruction does not mean that the ban covers *any* possible applications of biotechnologies by *any* state. In view of the shortcomings of both the BWC and the CWC, especially with reference to the question of definition of the prohibited agents and the lack of universal adherence to those treaties, it becomes critical to investigate whether the application of biotechnologies for hostile purposes or in armed conflict is prohibited by other rules and principles of IHL.

For the purposes of this chapter three IHL principles are mainly called into question: first, the prohibition on causing superfluous injury or unnecessary suffering; secondly, the principle of distinction, in its articulation as the prohibition on using indiscriminate weapons; and thirdly, the protection of the environment.[66]

These 'cardinal principles',[67] which constitute the 'fabric of humanitarian law'[68] with respect to the legality of methods of warfare and cover all weapons, have long since been consolidated into customary law.[69] However, the specific content of the principles is subject to debate, given the vague terms in which they are couched and lack of an impartial body capable of enforcing observance.[70]

The rationale behind the applicability of those principles to new weapons is well illustrated by the well-known *Shimoda* judgment issued in 1963 by the Tokyo district court.[71] When addressing the question of the legality of new weapons, the judges observed that the fact that a new weapon, because of its novelty, cannot already be made the object of a direct and explicit prohibition by a positive rule of international

---

[66] The role played by the so-called Martens clause will not be specifically examined here. It is however acknowledged that this clause 'has proved to be an effective means of addressing the rapid evolution of military technology': ICJ, Advisory Opinion on the Legality of the Threat or Use of Nuclear Weapons [1996] ICJ Rep 257, para 78.

[67] Ibid.

[68] Ibid.

[69] Ibid. See also JM Henckaerts and L Doswald-Beck, *Customary International Humanitarian Law* (Cambridge, Cambridge University Press, 2005) (hereafter *ICRC Study on Customary IHL*), I, rules 70 and 71.

[70] See A Cassese, 'Means of Warfare: the Traditional and the New Law' in A Cassese (ed), *The New Humanitarian Law of Armed Conflict* (Naples, Scientifica, 1979), 165. Not even the ICJ in the *Legality of Nuclear Weapons* case has ventured into an in-depth analysis of IHL principles, despite recognition that they constitute 'intransgressible principles of international customary law': para 79. More illuminating in this respect is the case law of the ad hoc criminal tribunals, although IHL principles are analysed by those tribunals exclusively for criminal purposes.

[71] The judgment delivered by the Tokyo District Court in 1963 is reproduced at [1964] *Japanese Annual of International Law* 212.

law does not mean that it is lawful. On the contrary, the new weapon can be considered lawful only to the extent that it is consonant with the principles underlying the rules governing armed conflict. This principle is affirmed in Article 36 of Protocol I of 1977,[72] whereby states parties are under the obligation to assess the legality of new weapons in light of the Protocol as well as existing rules of international law.[73] The same rationale was shared by the International Court of Justice (ICJ) in the Advisory Opinion on the Legality of Nuclear Weapons, where it stated that IHL principles apply to all kinds of weapons: not just those of the past and those of the present, but also those of the future.[74]

3.1 The Prohibition on Employing Means and Methods of Warfare of a Nature to Cause Superfluous Injury or Unnecessary Suffering

The prohibition on employing 'weapons, projectiles and material and methods of warfare of a nature to cause superfluous injury or unnecessary suffering'[75] is a corollary of the general principle that belligerents do not have an unlimited right to choose means and methods of warfare,[76] even when dealing with lawful targets. It is specifically aimed at protecting combatants, but its precise content is controversial.[77] Yet, the importance of identifying the precise scope of the principle cannot be overestimated because its violation has the effect of outlawing a means of warfare by

---

[72] Protocol I Additional to the Geneva Conventions of 12 Aug 1949 and relating to the Protection of Victims of International Armed Conflict, available at www.icrc.org/ihl.use/ FULL/470? Open Document.

[73] Art 36 reads: '[i]n the study, development, acquisition or adoption of a new weapon, means or method of warfare, a High Contracting Party is under an obligation to determine whether its employment would, in some or all circumstances, be prohibited by this Protocol or by any other rule of international law applicable to the High Contracting Party'. On the implementation practice of such a rule, including practice of those states that have not ratified Protocol I, see J McClelland, 'The Review of Weapons in Accordance with Article 36 of Protocol I' (2003) 850 *International Review of the Red Cross* 397.

[74] ICJ, *Advisory Opinion on the Legality of Nuclear Weapons*, above n 66, 261, para 86.

[75] This rule is set forth in Art 35 (2) of Protocol I of 1977.

[76] See preamble to the S Petersburg Declaration of 1868, Art 22 of the Regulations annexed to Hague Convention IV Respecting the Laws and Customs of War on Land of 1907. The principle is now codified in Art 35 (1) of Protocol I of 1977.

[77] Y Sandoz, C Swinarski and B Zimmermann, *Commentary on the Additional Protocols of 8 June 1977 to the Geneva Conventions of 12 August 1949* (Geneva The Hague, ICRC/Nijhoff, 1987), at 403–410. The difficulties connected to the identification of the principle's content are also illustrated by the letter of Art 8(2)(b)(XX) of the 1998 International Criminal Court Statute, as it prohibits only weapons causing superfluous injury or unnecessary suffering on condition that they are subject to a comprehensive prohibition and are included in an annex to the Statute to be adopted following the amendment procedure; on this point see P Benvenuti, 'Armi, loro diffusione e crimini di guerra. Riflessioni in margine all'adozione dello Statuto della Corte penale internazionale' in *Divenire Sociale e Adeguamento del Diritto: Studi in Onore di Francesco Capotorti*, AA VV, (Milan, Giuffrè, 1999) 41.

reason of its nature, regardless of the existence of a treaty or customary rule prohibiting that very use of the means of warfare.[78]

The main problematic question relating to the application of this principle concerns the criterion by which the 'superfluous' and 'unnecessary' character of the injury and suffering can be assessed. In this regard two differing views have been propounded by scholars. According to some authors[79] the principle allows no room for military necessity because that inherently contradicts the very purpose of the principle. Therefore it makes no sense to subject the application of the principle to the proportionality (between damage and military advantage) test.[80] The principle has to be applied 'automatically and to the letter in every case to which it pertains',[81] regardless of the specific circumstances of the case, 'even in those cases where use of the means to which it refers would obviously not cause superfluous injury or unnecessary suffering'.[82] In short, the principle in itself constitutes the yardstick against which the lawfulness of means and methods of warfare is to be evaluated.

This approach appears to be adopted by the ICRC study on the criteria for assessing cases in which the unnecessary principle is violated, ie the 'SirUS project'.[83] On the basis of medical evidence, the ICRC proposed four criteria (specific disease or disability; mortality above a certain percentage; gravity of wounds; and effects for which there is no treatment) capable of measuring the effects of a weapon that cause superfluous injury and unnecessary suffering.

Strong objections were raised with regard to the ICRC project concerning the fact that no consideration was given to military necessity. These objections were seemingly advocated by the proponents of the second modality of interpretation of the principle of unnecessary suffering, according to which the nature of the injury or the intensity of suffering has to be weighed against the military necessity in each specific case.[84]

---

[78] This is by far the most widespread interpretation of the effects of the principle of unnecessary suffering and it has also been endorsed by the ICJ opinion on *Nuclear Weapons*, above n 66, 257. However some states maintain that a weapon that violates this principle is prohibited only to the extent that a treaty or customary rule outlaws it.

[79] H Meyrowitz, 'The Principle of Superfluous Injury or Unnecessary Suffering' (1994) 299 *International Review of the Red Cross* 105.

[80] In addition, according to this approach the proportionality principle is applicable only with respect to damages provoked to civilians.

[81] Meyrowitz, n 81 above, at 109.

[82] Ibid.

[83] R Coupland, 'The SIrUS Project: Towards a Determination of Which Weapons Cause "Superfluous Injury or Unnecessary Suffering"' in H Durham and T McCormack (eds), *The Changing Face of Conflict and the Efficacy of International Humanitarian Law*, (The Hague, Nijhoff, 1999), 99.

[84] W A Solf, Article 35, in M Bothe, KJ Partsch and WA Solf (eds), *New Rules for Victims of Armed Conflicts: Commentary on the Two 1977 Protocols Additional to the Geneva Conventions of 1949* (Nijhoff, 1982), 196. It is not clear what interpretation the ICRC Commentary on Protocol I favours, because it gives an account of the debates on the content of the principle of superfluous injuries which took place at the negotiating conference but merely concludes

While the first interpretation is probably more restrictive than the second one and appears to be more in conformity with the general purpose of IHL, in particular Protocol I, the second view has been adopted by the ICJ in the *Nuclear Weapons* Opinion. The ICJ has defined as unnecessary any suffering that causes 'a harm greater than that *unavoidable* to achieve legitimate military objectives'.[85] The *unavoidable* character of the use of certain weapons to achieve a legitimate military objective requires that a balance be struck between the military advantage expected by the use of a weapon and the harm it may cause.[86] This approach appears to have two implications. On the one hand, it outlaws the use of a means or method of warfare in the presence of an *alternative* means or method which is capable of achieving the same objective while causing less harm. On the other, the application of the principle must be tailored upon the nature of the objective that one intends to target in a concrete situation.[87] For example, the vulnerability of the objective is one of the factors that may lead to the violation of the principle despite the lawfulness of the objective and means used. In other words, the relevant elements for the implementation of the principle according to this view are both the factual capabilities or nature of the means or method of warfare, and the concrete circumstances or modality of their use.

The necessity to assess the type of harm that the *actual* use of a certain weapon in a *specific* context may cause leads us to our second point, namely the stages at which the principle applies. In order to make it effective, it is reasonable to suggest that the principle should apply both when a certain means and method are being designed and *before* their actual use in a specific combat scenario. In fact it would make no sense to allow a state to manufacture a weapon or train for a certain warfare method that would certainly not meet the legality criteria. This interpretation is warranted also by the use of the term 'of a nature to' in the principle as codified in Article 35(2) of Protocol I and has been explicitly recognised in Article 36 of Protocol I.

Turning to biochemical means of warfare that have been manufactured by means of biotechnology techniques, the evaluation of their lawfulness in light of the unnecessary suffering principle leads to differing results according to the outcome achieved through biotechnology. If the latter

---

that the relation of the principle to military necessity is 'not interpreted in a consistent and generally accepted manner': cf Sandoz, Swinarski and Zimmermann, n 77 above at 409–410. Cf also A Cassese, 'Weapons Causing Unnecessary Suffering: Are They Prohibited?' (1975) *Rivista di Diritto Internazionale* 12.

[85] ICJ, Advisory Opinion of 1996, above n 66, para 78.
[86] See Art 51(5)(b) and 52(2) of Protocol I.
[87] *Contra* L Doswald-Beck, 'International Humanitarian Law and the Advisory Opinion on the Legality of the Threat and use of Nuclear Weapons', (1997) 316 *International Review of the Red Cross* 45.

enhances the effects of biochemical weapons, there is no doubt that any weapon so manufactured violates the unnecessary suffering principle, given that biochemical weapons have been outlawed precisely because they inherently violate this principle.[88] This conclusion is valid in light of both interpretations of the principle we have referred to. In addition, the supporters of the second interpretation acutely wonder what military necessity could ever be so grave as to justify the severe suffering caused by weapons of mass destruction.[89]

On the other hand, the lawfulness of those military offensive applications of biotechnology that have effects which are apparently *below* the level of 'unnecessity' cannot be excluded solely on account of the technique used to manufacture them. It is clear that those applications have to be judged on the basis of the specific type of effects they yield—and different conclusions may be reached according to the interpretation approach we take.

Though there is no room here to expand on the possible specific effects of biochemical weapons which have 'reduced' effects, suffice it to observe that the predictability of biochemical weapons is extremely difficult to attain, and this characteristic is unlikely to change even through biotechnology. As evidenced in the recent instances of the use of gases by the Russian authorities to free the hostages being held in the Moscow theatre (2002) and Beslan school (Chechenya, 2004), chemicals are hard to control and their use in those specific circumstances gives rise to questions about its conformity with the unnecessary suffering principle.[90]

Hence, the prospects for the use of these weapons in conformity with the unnecessary suffering principle are, at best, very limited.

In this context it is opportune to refer again to the majority view of the ICJ in the *Nuclear Weapons* Opinion concerning the legality of those weapons in light of IHL principles, including the one under scrutiny. The ICJ, albeit concluding that the use of nuclear weapons would be 'scarcely reconcilable'[91] with respect for this principle, found that it did not have 'sufficient elements . . . to conclude with certainty that the use of nuclear weapons would necessarily be at variance with the principles and rules of law applicable in armed conflict in any circumstance'.[92] The Court grounded such conclusion on the fact that no conventional prescription on the prohibition of recourse to nuclear weapons as such existed,[93] and

---

[88] The *ICRC Study on Customary IHL*, above n 69, 237, refers to the numerous military manuals that take this approach.

[89] ICJ, *Advisory Opinion on the Legality of Nuclear Weapons*, above n 66, Dissenting Opinion of Judge Higgins, 587, para 18.

[90] But there is no evidence that the gas used in those two instances were modified through biotechnology.

[91] Ibid, para 95.

[92] Ibid.

[93] Ibid, paras 53 –63.

that the emergence of a customary rule was hampered by adherence to the policy of deterrence.[94]

It is difficult to take any of these two bases as legal grounds for establishing the conformity of an offensive military application of biotechnologies with the principle of unnecessary suffering. First, as we have seen, treaties specifically banning biochemical weapons exist and their scope (at least for the BWC) covers weapons manufactured by recourse to biotechnology. Secondly, the non-use of biological weapons in the last 60 years and the rare instances of recourse to chemical weapons in history cannot be ascribed to the policy of deterrence that characterised the international scene during the Cold War with respect to nuclear weapons. On the contrary, the lack or limited use of biochemical weapons in the past is directly linked to the *opinio juris* of states as to their 'abhorrent' nature.

In short, the prospects for states to use biotechnological weapons in compliance with the IHL principle prohibiting unnecessary suffering or superfluous injury are extremely limited, especially on account of the fact that biotechnology tends to enhance rather than reduce the effects of those agents whose weaponisation is banned.

3.2 The Principle of Distinction

The principle of distinction, which is mainly aimed at protecting civilians from attack,[95] is relevant to our purposes in its articulation of the rule prohibiting the use of indiscriminate weapons. While there is general agreement on the fact that attacks which are specifically aimed against civilians are prohibited,[96] other cases are more troublesome, in particular those consisting in attacks against a military objective when the action entails consequences for civilians. The terms employed in Article 51 of Protocol I provide some guidance on the troublesome cases, since the provision distinguishes between the attacks which are indiscriminate because

---

[94] Ibid, para 73.

[95] The principle is codified in Art 48 of Protocol I and spelt out in Art 51, (4) and (5) of the same Protocol. In the 1996 *Advisory Opinion on the Legality of Nuclear Weapons*, above n 66, para 78, the ICJ has equated the use of indiscriminate weapons with a deliberate attack on civilians. Although such equation is probably pertinent, it would be wrong to limit the scope of the principle to prohibition of 'attacks' against civilians, since other military operations, such as mine-laying, certainly fall under the purview of the principle.

[96] This rule is codified in Art 51(4)(a) of Protocol I. Cf also International Criminal Tribunal for the Former Yugoslavia, *Prosecutor. Blaskic*, judgment of 3 Mar 2000, para. 180; *Prosecutor v Kordic*, judgment of 21 Feb 2001, para 328, both available at www.un.org/cty/cases-e/index-e.htm. which specify that what are prohibited those attacks against civilians which are not justified by military necessity (a statement that needs clarification in light of the principle of distinction).

they 'employ a method or means of combat which *cannot* be *directed* at a specific military objective' (emphasis added) and, on the other, attacks employing 'a method or means of combat the *effects* of which cannot be limited as required by this Protocol' (emphasis added).[97]

The way in which the provision is couched suggests that the prohibition of the use of indiscriminate weapons is made up of two distinct rules. On the one hand, it prohibits weapons that are *inherently indiscriminate*,[98] thus laying emphasis on the type or nature of the weapon; on the other, it prescribes the *modality* of *use* of a certain weapon in an indiscriminate fashion (but the weapon per se would be lawful if used in a non—indiscriminate manner). In addition, Article 51 also specifies that those attacks which 'may be expected to cause incidental loss of civilian life, injury to civilians, damage to civilian objects, or a combination thereof, which would be excessive in relation to the concrete and direct military advantage anticipated'[99] are in any case to be considered as indiscriminate. This means that an attack has to be evaluated in light of the proportionality test.[100] In other words, the principle of distinction requires factual appreciation both of the specific capabilities of the weapon being used and the modality of its use.

According to Article 51, a weapon that is capable of targeting only combatants, for instance because it distinguishes the material from which their uniform is made, would probably not be inherently indiscriminate. On the contrary, an 'ethnic weapon' that is specifically aimed at annihilating a specific ethnic group within the population would always be prohibited.

It is debatable whether military applications of biotechnology would violate Article 51(4) of Protocol I because of their indiscriminate *nature* or on account of their uncontrollable *effects*.[101] Let us think of the use of high-precision missiles with heads containing a synthetically created virus: even if they are directed at a military barrack in a deserted area, the capacity of those hit to infect others, including civilians, easily becomes

---

[97] Respectively Art. 51(4)(b) and 4(c), of Protocol I. The same para, letter (a), also outlaws those attacks which 'are not directed at a specific military objective'.

[98] The ICJ stated that weapons 'incapable of distinguishing between civilian and military targets' are prohibited according to this principle, but did not expand on the criteria for assessing the discernment capability of a weapon: *Legality of Nuclear Weapons*, above n 66, para 78.

[99] Art 51(5)(b) of Protocol I.

[100] *Ibid*.

[101] Obviously, this depends also on the means of delivery used. According to the US Air Force Pamphlet (1997), para 6–4(b), it is because of the 'wholly indiscriminate and uncontrollable nature' of biological weapons that prohibition of that category of arms was achieved. On the fact that weapons of mass destruction (with the debatable exception of nuclear weapons) are banned because of their indiscriminate effects, but without expanding on the reasons see Benvenuti, n 77 above, at 37.

beyond control. Moreover, biotechnological weapons, especially if used in aerosol form, are susceptible to weather conditions, such as winds and rain (like chemical weapons, which for this reason are difficult to manufacture and require high amounts of the vital substance). As has been correctly noted, in such cases the weapon is likely 'to take 'a life of its own' and randomly hit combatants or civilians to a significant degree'.[102] Hence the high potential of such weapons contaminating not just combatants militates against the lawfulness of their use.

Finally, the reflections that were elaborated with respect to the reasoning of the ICJ in the *Nuclear Weapons* Opinion and the principle of unnecessary suffering apply also to the principle of distinction.

For all the above reasons, only in very rare instances may offensive military applications of biotechnology be used in keeping with the principle of distinction in its articulation as the prohibition of indiscriminate attacks, and in any case only in those circumstances in which the effects on the civilian population pass the proportionality test.

3.3 The Protection of the Environment

Although there are no treaty provisions relating to the environment that specifically prohibit the use of biotechnologies in warfare, 'states must take environmental considerations into account when assessing what is necessary and proportionate in the pursuit of legitimate military objectives'.[103] This principle has been codified and specified in Article 35(3) and Article 55 of Protocol I, which sets forth the prohibition on using weapons which are intended, or may be expected, to cause 'widespread, long-term and severe environmental damage'. The ICJ has affirmed that both provisions 'embody a general obligation to protect the natural environment',[104] a statement that can be read as meaning that the obligation has a customary character,[105] but it has failed to spell out the precise content of the above criteria.

Although more information on the impact that the use of biotechnologies in warfare may have on the environment, including on those structures that are indispensable for the survival of the population such as food crops is needed, it is likely that any such impact would be at least widespread, given the inevitable diffusion of biochemical agents over large areas. Yet,

---

[102] Doswald-Beck, n 87 above, at 41.
[103] *Nuclear Weapons Opinion*, above n 66, para 30.
[104] Ibid, para 31.
[105] The opposite view was held by the Office of the Prosecutor of the International Criminal Tribunal for the former Yugoslavia, Final Report to the Prosecutor by the Committee Established to Review the NATO Bombing Campaign Against the Federal Republic of Yugoslavia, 13 June 2000, para 15.

the threshold for the application of Article 35(3) of Protocol I would be met solely if all three criteria, namely the widespread, long-term and severe damage to the environment, were present —thus laying a particularly high burden of proof on the offended party.[106] In any case, the duty to protect the environment in armed conflict requires that any military offensive use of biotechnologies be assessed against the background of the principles of necessity and proportionality in the terms that have been analysed above in relation to the two previous IHL principles.[107] This may have the effect of lowering the threshold of Articles 35(3) and 55 of Protocol I because, when the collateral damage is not proportionate to the military advantage anticipated, such damage would be unlawful even if it were not widespread, long-lasting and severe.[108]

4. CONCLUSIONS

There are convincing factors justifying the conclusion that offensive military applications of biotechnologies fall within the purview of existing arms control treaties, in particular the BWC of 1972. This conclusion is warranted in light of both the letter of the treaty and the final documents adopted by the review conferences, the latter affirming in clear terms that technological developments, including biotechnology, fall under the scope of the Convention. Therefore, a generalised agreement seems to exist to the effect that such applications are banned by the treaty. Unfortunately, the lack of a verification mechanism and the recently evidenced difficulties in achieving consensus about it leave loopholes at the enforcement level of the treaty that may threaten the very efficacy of the prohibition.

---

[106] It is worth mentioning that the Convention on the Prohibition of Military or any Other Hostile Use of Environmental Modification Techniques of 1976 (ENMOD) also includes the three threshold requirements but provides that the criteria are alternative. In addition, it specifies that the term 'widespread damage' means 'encompassing an area on the scale of several hundred square kilometres', 'long-lasting' means 'lasting for a period of months, or approximately a season', and 'severe' damage means 'involving serious or significant disruption or harm to human life, natural and economic resources or other assets'. The ENMOD Convention may be relevant for offensive military applications of biotechnology because it prohibits, inter alia, environmental modification techniques that cause an 'upset in the ecological balance of a region': see the Understanding relating to Art II attached to the ENMOD Convention.

[107] The literature on IHL and the environment is abundant: see, eg, R Desgagné, 'The Prevention of Environmental Damage in Time of Armed Conflict: Proportionality and Precautionary Measures' (2000) *Yearbook of IHL* 109; Y Dinstein, 'Protection of the Environment in International Armed Conflict' (2001) *Max Planck Yearbook of United Nations Law* 523.

[108] M Bothe, 'The Protection of the Civilian Population and NATO Bombing on Yugoslavia: Comments on a Report to the Prosecutor of the ICTY' (2001) 12 *European Journal of International Law* 532.

It is more difficult to reach a similar conclusion on the comprehensiveness of the ban on biotechnology and weapons with respect to the CWC of 1993. This is due to the caution found in the text and the outcome of the review conference relating to the inclusion of any technological developments contrary to the Convention within its scope. Still, a trend in the direction of a multilateral consensus can be identified and is certainly to be encouraged.

In any case, loopholes in the law exist by reason of the inherent shortcomings of those conventions. It is therefore crucial that states review the legality of new weapons, including those which derive from biotechnology, according to IHL principles. This chapter has addressed the possible use of offensive military applications of biotechnologies in conformity with the principle of unnecessary suffering, the principle of distinction and, to a more limited extent, the protection of the environment.

The conclusion is that the possibility that states can use biotechnological weapons in keeping with the principle of unnecessary suffering are extremely limited, given that the interest in manufacturing such weapons is their enhanced—not reduced—effects compared to those of the weapons that are already prohibited in the BWC or CWC. Also the production of weapons whose effects on combatants are inferior to those prohibited seems of dubious legality considering the difficult in predicting the behaviour of the agents we are dealing with. On the contrary, the principle of distinction could be respected in some extreme instances if biotechnologies are specifically weaponised, taking into account the preservation of the rules articulated by the principle of distinction (for example by manufacturing a weapon that distinguishes the material used in a combatant's uniform).

As to the protection of the environment, more information is needed on the impact that these new weapons or methods of warfare could have on agriculture, health and nature. In general terms, if we take the standard of Protocol I, whereby only widespread, long-term and severe environmental damage is prohibited, some biotechnological application in the military field may be found to be legal because of the cumulative character of the type of harm which is required by the law. However, the need to respect the proportionality and necessity tests arises also in connection with this principle.

Index